Impacts of Environmental Change and Human Activities on Aquatic Ecosystems

Impacts of Environmental Change and Human Activities on Aquatic Ecosystems

Guest Editors

Jian Hu
Guilin Han
Qian Zhang

MDPI
Basel • Beijing • Wuhan • Barcelona • Belgrade • Novi Sad • Cluj • Manchester

Guest Editors
Jian Hu
Research Center for
Eco-Environmental Sciences
Chinese Academy of Sciences
Beijing
China

Guilin Han
Institute of Earth Sciences
China University of
Geosciences (Beijing)
Beijing
China

Qian Zhang
Institute of Geographic
Sciences and Natural
Resources Research
Chinese Academy of Sciences
Beijing
China

Editorial Office
MDPI AG
Grosspeteranlage 5
4052 Basel, Switzerland

This is a reprint of the Special Issue, published open access by the journal *Water* (ISSN 2073-4441), freely accessible at: https://www.mdpi.com/journal/water/special_issues/947274Z5MD.

For citation purposes, cite each article independently as indicated on the article page online and as indicated below:

Lastname, A.A.; Lastname, B.B. Article Title. *Journal Name* **Year**, *Volume Number*, Page Range.

ISBN 978-3-7258-5401-1 (Hbk)
ISBN 978-3-7258-5402-8 (PDF)
https://doi.org/10.3390/books978-3-7258-5402-8

Contents

About the Editors

Jian Hu

Jian Hu is an Associate Researcher at the Research Center for Eco-Environmental Sciences, Chinese Academy of Sciences. He received his Ph.D. from the Institute of Geochemistry, Chinese Academy of Sciences in 2005, and his research focuses on regional sustainable development and ecological risk assessment. He has led projects supported by the National Key Research and Development Program of China, the National Major Science and Technology Program, the National Natural Science Foundation of China, and the "Western Light" Program of the Chinese Academy of Sciences, among others, and has participated in several international collaborations and strategic programs. He serves on the editorial board of the *Chinese Journal of Ecology* and as Guest Editor for the *International Journal of Environmental Research and Public Health*. He has published over 40 peer-reviewed papers and co-authored three academic books.

Guilin Han

Guilin Han is a Professor at the Institute of Scientific Research, China University of Geosciences (Beijing). Her research focuses on watershed weathering and the carbon cycle, environmental pollutant characterization, the development of non-traditional stable isotope analytical techniques, and the links between geochemistry and human health. She has led more than twenty major projects, including the National Science Fund for Distinguished Young Scholars and key international cooperation programs of the National Natural Science Foundation of China. She has published over 200 peer-reviewed papers as first or corresponding author, and her contributions have been recognized through continuous inclusion in the "Global Top 2% Scientists" list (2021–2024) and as a Highly Cited Chinese Researcher by Elsevier (2023–2024). She was awarded the Kharaka Award by the International Association of Geochemistry, along with multiple national and provincial honors.

Qian Zhang

Qian Zhang is a researcher at the Institute of Geographic Sciences and Natural Resources Research, Chinese Academy of Sciences. Her research focuses on environmental geography and geochemistry, with particular interest in isotope fractionation of iron and copper in karst critical zones, geochemical behavior of rare earth elements, and ecological risks of potentially toxic elements in soils, waters, and atmospheric particulates. She has led and participated in several national and institutional research projects, including studies funded by the National Natural Science Foundation of China. Dr. Zhang has published more than fifty papers in leading journals. She has received the Special Award of the Green Mine Youth Science and Technology Prize and the Third Prize of the Beijing Invention and Innovation Competition. She also holds multiple national patents on isotopic purification devices and innovative analytical methods for pollutants.

Preface

With accelerating global climate change and the intensification of human activities, aquatic ecosystems are increasingly subjected to complex and overlapping environmental stressors. As critical components in sustaining water security, regulating regional climate, supporting biodiversity, and providing ecosystem services, the health and stability of aquatic ecosystems are directly linked to environmental quality and the sustainable development of human society. In recent years, natural processes—such as global warming, altered precipitation regimes, and extreme climate events—have significantly changed the hydrological patterns and ecological dynamics of water bodies. Simultaneously, anthropogenic pressures including urban expansion, agricultural intensification, hydraulic engineering, and pollutant emissions have led to persistent disturbances, posing substantial risks to aquatic system integrity.

This Special Issue Reprint, entitled "Impacts of Environmental Change and Human Activities on Aquatic Ecosystems", features ten original research articles that systematically examine the sources, processes, and ecological consequences of aquatic environmental changes across diverse spatial scales and hydro-ecological contexts. The selected studies encompass topics such as microplastic and antibiotic pollution, nitrate source apportionment, ecological compensation mechanisms, rare earth element distribution, microbial community responses, and phytoplankton dynamics under extreme drought. Methodologically, the contributions demonstrate the integration of stable isotope analysis, flow cytometry, ecological risk assessment, environmental modeling, and data-driven algorithms—reflecting the latest advancements in interdisciplinary aquatic research.

The Reprint draws upon case studies from internationally representative and environmentally sensitive regions, including the Nile Delta in Egypt, the Irtysh River in Kazakhstan, transboundary watersheds on the Korean Peninsula, and ecologically critical areas in China such as the Pearl River Basin, the Eastern Tiaoxi River, and the Qinghai–Tibet Plateau. By spanning urban wastewater systems, agricultural landscapes, reservoirs, and groundwater environments, this collection highlights the multifaceted interactions between pollution, climate, hydrology, and ecological resilience.

We hope this Reprint will serve as a valuable reference for researchers, environmental practitioners, and decision-makers, offering both theoretical foundations and actionable insights for ecological restoration, pollution control, and sustainable water governance. We express our sincere thanks to all contributing authors, reviewers, and the editorial team of *Water* for their support. It is our shared aspiration that this collection will stimulate further interdisciplinary dialogue and provide scientific guidance for building more resilient and adaptive aquatic ecosystems under global change.

Jian Hu, Guilin Han, and Qian Zhang
Guest Editors

water

Editorial

Impacts of Environmental Change and Human Activities on Aquatic Ecosystems

Jian Hu [1,2,*], Guilin Han [3,4,5] and Qian Zhang [2,6]

1 State Key Laboratory of Regional and Urban Ecology, Research Center for Eco-Environmental Sciences, Chinese Academy of Sciences, Beijing 100085, China
2 University of Chinese Academy of Sciences, Beijing 100049, China; zhangqian@igsnrr.ac.cn
3 Institute of Earth Sciences, China University of Geosciences (Beijing), Beijing 100083, China; hanguilin@cugb.edu.cn
4 State Key Laboratory of Geomicrobiology and Environmental Changes, China University of Geosciences (Beijing), Beijing 100083, China
5 Frontiers Science Center for Deep-Time Digital Earth, China University of Geosciences (Beijing), Beijing 100083, China
6 Institute of Geographic Sciences and Natural Resources Research, Chinese Academy of Sciences, Beijing 100101, China
* Correspondence: jianhu@rcees.ac.cn

1. Introduction

With the ongoing changes in the global climate system and the continuous intensification of human activities, aquatic ecosystems face unprecedented stresses from multiple sources [1,2]. As a vital foundation for maintaining water security, regulating regional climate, supporting biodiversity, and providing ecosystem services, the health of aquatic ecosystems is directly linked to overall environmental quality and the sustainable development of human society [2,3]. Recently, natural ecological changes such as global warming, altered precipitation patterns, and the increasing frequency of extreme climate events (e.g., floods, droughts, and heatwaves) have induced profound changes in the hydrological processes and ecological patterns of water bodies. For example, rising water temperatures may alter species' life cycles and metabolic rates. At the same time, extreme rainfall or drought can lead to wetland shrinkage, river flow interruption, or sudden increases in runoff, triggering additional changes in nutrient loads and degradation of ecological functions [4,5]. In addition, prolonged environmental stress has gradually reduced ecosystems' capacity for self-recovery, as evidenced by a decline in biodiversity and increased instability in ecological processes [6].

Human activities are degrading aquatic ecosystems at an increasing rate. In particular, urbanization and industrialization have led to the discharge of both point-source and non-point-source pollutants. Extensive use of chemical fertilizers and pesticides in agriculture can result in eutrophication, excessive algal blooms, and the deterioration of benthic habitats [7,8]. Furthermore, the construction of hydraulic engineering projects and the mismanagement of water resources have disrupted water bodies' natural connectivity and ecological flow regimes. These alterations affect the migration routes and habitat structures of aquatic organisms, ultimately leading to ecosystem degradation and the loss of essential ecological services [9,10]. In the face of the complex challenges posed by the combined effects of natural environmental changes and human activities, it is imperative to adopt a systems perspective to facilitate a deep understanding of the impact mechanisms of various stressors on the structure and function of aquatic ecosystems. There is an urgent need to establish integrated models for ecological risk assessment and pollution control, and to develop adaptive, practical strategies for ecological restoration and sustainable management to enhance the resilience and stability of aquatic ecosystems, ensure water quality safety, and maintain ecological integrity [11,12].

This Special Issue focuses on the environmental challenges facing aquatic ecosystems, aiming to promote interdisciplinary scientific research and provide theoretical foundations and practical guidance for understanding and addressing aquatic ecosystem crises in the context of global change. It also seeks to offer feasible scientific recommendations for policymakers and environmental managers to jointly advancing the sustainable development of aquatic ecosystems and deepening the progress of regional ecological civilization.

2. Overview of Advanced Developments

An et al. (Contribution 1) provided a comprehensive review of how external environmental factors influence the distribution and migration of microplastics in aquatic systems, focusing on meteorological conditions, ecosystem characteristics, and the physical properties of water bodies. The study revealed that rainfall introduces land-based microplastics into water bodies and facilitates vertical and horizontal transport; sunlight accelerates microplastic aging and fragmentation, altering their density, surface activity, and adsorption capacity; and wind promotes transboundary movement by disturbing the water surface or carrying particles. Additionally, aquatic plants, animal ingestion and metabolism, and microbial production of extracellular polymeric substances (EPSs) affect microplastics' morphology and transport pathways. Geomorphological features and suspended particulate matter in lakes and rivers also govern sedimentation, aggregation, and resuspension behaviors. This work systematically analyzes the drivers of microplastic transport from multiple dimensions, with particular focus on the often-overlooked role of natural environmental conditions such as climate and ecological context, providing strong support for water resource management and microplastic pollution control.

Bo et al. (Contribution 2) emphasized the unique role of rice paddies in protecting aquatic biodiversity. They examined the impacts of climate change, such as rising water temperatures and altered hydrological cycles, on aquatic ecosystems. The results revealed that climate change profoundly alters the character and dynamics of natural systems globally. Increased water temperatures and disruptions in the hydrological cycle have profound implications for biodiversity in freshwater environments, and may also be evident in artificial aquatic agro-ecosystems. Strategies such as reducing irrigation requirements, promoting the cultivation of drought-tolerant crops, and utilizing precision agriculture techniques are proposed to promote a sustainable balance between agriculture and ecological conservation.

Chen et al. (Contribution 3) classified the ecosystem services of the Xin'an River Basin into provisioning, regulating, and cultural services. They estimated the basin's total service value using ecosystem service valuation (ESV). The total ESV was estimated at CNY 70.271 billion, with provisioning services accounting for 22.7%, regulating services 24.6%, and cultural services 52.7%. Based on different compensation scopes, ecological compensation's upper and lower limits were calculated at CNY 4.085 billion and CNY 1.438 billion, respectively. By integrating theoretical modeling with empirical analysis, this study offers a practical framework and financial assessment for ecological compensation, supporting refined management and precise compensation in the basin.

Dong et al. (Contribution 4) systematically monitored the spatial and temporal distribution of nine antibiotics—including roxithromycin, oxytetracycline, sulfamethoxazole, and ofloxacin—in the influent and effluent of wastewater treatment plants (WWTPs) in urban and suburban areas of Tangshan City. They used risk quotient methods to assess the potential ecological risks these antibiotics pose to aquatic ecosystems and explored how antibiotic distribution correlates with factors such as season, temperature, pH, hydrolysis properties, usage patterns, and treatment processes. The results showed significant seasonal and regional variations in antibiotic concentrations. In urban WWTPs, roxithromycin was the main pollutant in spring, sulfamethoxazole dominated in summer and autumn, and ofloxacin peaked in winter. In suburban WWTPs, sulfamethoxazole and norfloxacin were the primary contaminants. Sulfamethoxazole had the highest per capita pollution load for most of the year, with ofloxacin peaking in winter. Ecological risk assessment indicated winter was the season with the highest antibiotic risk, while the other three seasons posed low risks. Roxithromycin, oxytetracycline, tetracycline, and chlortetracycline in urban WWTPs, and roxithromycin and oxytetracycline in suburban WWTPs, presented moderate risks. Key factors influencing antibiotic distribution included hydrolysis rate, temperature, pH, and sludge adsorption capacity. This study revealed multidimensional variations in antibiotic distribution, providing valuable data and theoretical foundations for assessing antibiotic pollution loads, implementing tiered ecological risk management, and improving wastewater treatment processes in both urban and suburban areas.

Kasem et al. (Contribution 5) focused on nitrate (NO_3^-) contamination in shallow groundwater in the arid southeastern Nile Delta, Egypt. They systematically applied stable isotope techniques ($\delta^{15}N/\delta^{18}O$–$NO_3^-$ and $\delta^2H/\delta^{18}O$–H_2O) combined with hydrochemical analysis to identify nitrate sources and transformation mechanisms. The results showed that groundwater recharge is primarily controlled by Nile River water, with irrigation water leakage in the west and mixing with deep groundwater in the east. Some samples exceeded safe limits for TDS, SO_4^{2-}, NO_3^-, and Mn^{2+}, with NO_3^- concentrations reaching up to 652 mg/L, far above WHO drinking water standards. The most severe pollution occurred in the central unconfined aquifer, spreading along groundwater flow paths to deeper and eastern zones. In the west and east, nitrate mainly originated from soil organic nitrogen (SON) and fertilizer nitrification products (CFs). At the same time, the central region was heavily impacted by domestic sewage inputs accompanied by denitrification closely linked to manganese biogeochemical cycling. This study is the first in the area to integrate multiple stable isotope methods and hydrochemical indicators to systematically identify nitrate sources and transformation pathways. It proposed a coupled mechanism of denitrification and manganese cycling, expanding our understanding of groundwater pollution–biogeochemical interactions and offering a replicable approach to pollution identification and control strategies in the Nile Delta and other arid groundwater systems.

In July 2023, Krupa et al. (Contribution 6) conducted systematic monitoring of the physicochemical parameters (transparency, temperature, pH, salinity (TDS), dissolved oxygen), pollutant concentrations, and biological indicators (zooplankton species richness, abundance, biomass, diversity index, etc.) in the upstream (Black Irtysh River) and downstream (Pavlodar region) sections of the Irtysh River in Kazakhstan. The results showed that nitrate nitrogen and phosphate were generally clean or lightly polluted, and ammonia nitrogen was below moderate pollution in most areas. In contrast, nitrite nitrogen (NO_2^-) exhibited moderate pollution across most regions and reached high pollution levels in Pavlodar and Aksu. Among the nine heavy metals, iron (Fe), copper (Cu), and manganese (Mn) were commonly exceeded, with localized exceedances of zinc (Zn) and chromium (Cr). Overall, the water quality ranged from clean to lightly polluted, with localized heavy pollution downstream. The community structure indicated high organic pollution but relatively low heavy metal toxicity. Elevated iron levels in the Black Irtysh River likely resulted from geological leaching, while water quality deterioration downstream was mainly attributed to industrial and urban wastewater discharge. High flow velocity, sufficient dissolved oxygen, adsorption of heavy metals by suspended particles and clay along riverbanks, and uptake of nutrients and heavy metals by aquatic plants contributed to the river's self-purification capacity. This study provides a comprehensive ecological assessment of the Irtysh River in Kazakhstan and offers scientific support for watershed management and transboundary water cooperation.

Li et al. (Contribution 7) investigated the impacts of four small cascade hydropower stations on microbial plankton community structure and ecological functions in tributaries of the Pearl River. Using flow cytometry, microbial populations were classified, and environmental indicators such as photosynthetic autotrophic capacity (PAC), bacterial activity index (BAI), viral regulatory capacity (VRC), and fungal metabolic capacity (FMC) were calculated. The results showed that along the cascade, dissolved oxygen (DO) and electrical conductivity (EC) increased, while the oxidation reduction potential (ORP) and total organic carbon (TOC) decreased. Abundances of viruses, low-nucleic-acid (LNA) bacteria, and fungi declined by 30.9%, 30.5%, and 34.9%, respectively. EC, TOC, and nitrate nitrogen (NO_3^--N) were key drivers influencing microbial abundance changes. Carbon and nitrogen nutrient levels significantly affected ecological indicators. The cascade hydropower stations significantly affected the PAC, BAI, and VRC, but had a relatively minor impact on FMC. The downward trend in VRC suggests that the stations had a weakening viral regulatory effect. This study applied high-throughput flow cytometry for microbial community classification, providing a novel technical approach, and comprehensively assessed the ecological impact of cascade hydropower stations by integrating environmental factors with ecological evaluation metrics.

Sim et al. (Contribution 8) analyzed water quality and phytoplankton community changes from 2013 to 2016 across a series of interconnected dams and reservoirs on the North Han River (Uiam, Cheongpyeong, Sambong-ri, and Paldang Lakes). The results showed that during drought periods, prolonged water residence time promoted nutrient accumulation and recycling within reservoirs, exacerbating eutrophication and water quality deterioration. Cyanobacteria became dominant, triggering harmful algal blooms. Notably, changes in upstream dam discharges directly impacted downstream reservoir water quality and ecosystem health. This study deeply explored the effects of climate change-induced extreme drought on water quality and phytoplankton dynamics, providing scientific insight into the impacts of climate-driven drought and informing integrated water management strategies.

Yang et al. (Contribution 9) investigated the concentration, distribution, and ecological risk of rare earth elements (REEs) in surface sediments of the Eastern Tiaoxi River (ETX) in eastern China. The study found that total REE concentrations ranged from 133.62 to 222.92 mg/kg, characterized by enrichment of middle rare earth elements (MREEs) and depletion of heavy rare earth elements (HREEs). REE concentrations and distributions were closely associated with elements such as Ca, Fe, Mg, and Mn, likely reflecting the influence of clay minerals, Fe-Mn oxides, and specific heavy minerals. The research also revealed significantly elevated REE levels near urbanized areas, while natural processes like soil transport and chemical weathering primarily drove REE variations elsewhere. Ecological risk assessment highlighted notable REE enrichment and moderate ecological risk in sediments near urban zones, with relatively minor impacts from agricultural areas. This study elucidates the combined effects of urbanization and natural processes on REE distribution and ecological risk in the ETX watershed, providing a scientific basis for environmental management and pollution control.

Zhao et al. (Contribution 10) focused on water quality pollution on the Qinghai–Tibet Plateau, investigating the spatial–temporal distribution patterns of water quality parameters, key influencing factors, and associated water quality risks. The study found significant differences in average water quality concentrations between the flood and dry seasons, with cadmium (Cd) levels meeting Class II water standards during the flood season, indicating that the water quality was relatively better at that time. Heavy metal risks exhibited distinct spatial patterns among different rivers—for example, the Sìgōu River showed higher heavy metal risks, highlighting uneven pollution levels across rivers. Pollution differences were mainly attributed to discharges from livestock farms and industrial enterprises, especially heavy metals. Meteorological factors had a noticeable impact on water quality, with seasonal variations between flood and dry periods further confirming climate regulation effects. By comprehensively analyzing spatial–temporal water quality variations across seasons and rivers, and applying Boruta and IFN-SPA algorithms to identify major pollutants, this study thoroughly revealed water quality dynamics and emphasized the significant role of meteorological factors, providing crucial insights for water quality management under climate change scenarios.

3. Conclusions

The papers in the Special Issue "Impacts of Environmental Change and Human Activities on Aquatic Ecosystems" in *Water* provide a contemporary snapshot of research trends in this field. The ten research articles in this collection cover studies on pollutant source identification and migration mechanisms (Contributions 1, 3, 4, and 10), water quality assessment and ecological risk evaluation (Contributions 2, 4, and 6), the impacts of climate change and extreme weather on aquatic ecosystems (Contributions 3, 6, and 9), ecosystem functions and ecological compensation (Contribution 5), as well as ecological responses and changes in aquatic biological communities (Contributions 2, 7, and 8). These works focus on surface waters, including reservoirs/rivers/lakes, groundwater, urban wastewater, integrated watersheds (ecosystem services), and agricultural ecosystems, covering transboundary rivers, plateau water systems, and urban water bodies. The studied pollutants are diverse, including nitrates (NO_3^-)/nitrogen and phosphorus nutrients, heavy metals, microplastics, antibiotics, and rare earth elements. The research methods and technical approaches employed are mature and include isotope tracing, risk assessment (RQ/ecological risk index), hydrochemical and statistical analyses, bioindicator methods (plankton/microorganisms), high-throughput technologies (flow cytometry), and algorithm-assisted analyses (Boruta/IFN-SPA). The study regions are typical and internationally representative, including Chinese river basins (Pearl River, Diaoxi River, Xin'anjiang River), international water bodies (Italian paddy fields, the Nile River in Egypt, the Irtysh River, and the Korean Peninsula border), as well as typical environments in plateau and arid regions.

Current research has made significant progress in identifying multi-source pollution. Combining isotope techniques and ecological indicators has effectively enhanced the quantitative analysis of pollution sources, becoming a hot topic in water environment studies. At the same time, the impact of climate factors on water quality and ecosystems has received increasing attention, reflecting new trends in water environment risks in the context of global change. At the management and decision-making level, introducing ecosystem assessment and compensation mechanisms provides a quantifiable basis for formulating scientifically sound policies. With advances in technology, biological response indicators such as microorganisms and plankton are gradually becoming important tools for evaluating ecological health. Overall, related studies commonly adopt multi-scale, multi-factor integrated analytical frameworks, showing a development trend of "cross-interdisciplinary integration".

Therefore, future research is recommended to further strengthen the synergistic analysis and integrated management of multiple pollutants, deeply explore the behavioral mechanisms of emerging pollutants (such as microplastics, antibiotics, and rare earth elements) in the environment, and promote the development of coupled ecological–water quality modeling technologies to support intelligent watershed governance. At the same time, water ecological protection and ecological compensation strategies tailored to regional ecological differences should be formulated to improve the systematization and precision of water environment management.

Author Contributions: Conceptualization, J.H. and G.H.; investigation, J.H., G.H. and Q.Z.; writing—original draft preparation, J.H., G.H. and Q.Z.; writing—review and editing, J.H., G.H. and Q.Z. All authors have read and agreed to the published version of the manuscript.

Funding: This research received no external funding.

Conflicts of Interest: The authors declare no conflicts of interest.

List of Contributions:

1. An, X.; Wang, Y.; Adnan, M.; Li, W.; Zhang, Y. Natural Factors of Microplastics Distribution and Migration in Water: A Review. *Water* **2024**, *16*, 1595.
2. Bo, T.; Marino, A.; Guareschi, S.; Laini, A.; Fenoglio, S. Rice Fields and Aquatic Insect Biodiversity in Italy: State of Knowledge and Perspectives in the Context of Global Change. *Water* **2025**, *17*, 845.
3. Chen, Y.; Wu, Q.; Guo, L. Ecological Compensation Based on the Ecosystem Service Value: A Case Study of the Xin'an River Basin in China. *Water* **2024**, *16*, 2923.
4. Dong, Z.; Hu, J.; Wang, P.; Han, G.; Jia, Z. Antibiotics in Wastewater Treatment Plants in Tangshan: Perspectives on Temporal Variation, Residents' Use and Ecological Risk Assessment. *Water* **2024**, *16*, 1627.
5. Kasem, A.M.; Xu, Z.; Jiang, H.; Liu, W.; Zhang, J.; Nosair, A. M. Nitrate Source and Transformation in Groundwater under Urban and Agricultural Arid Environment in the Southeastern Nile Delta, Egypt. *Water* **2024**, *16*, 22.
6. Krupa, E.; Romanova, S.; Serikova, A.; Shakhvorostova, L. A Comprehensive Assessment of the Ecological State of the Transboundary Irtysh River (Kazakhstan, Central Asia). *Water* **2024**, *16*, 973.
7. Li, P.; Luo, Z.; Zhu, X.; Dang, Z.; Zhang, D.; Sui, X. Picoplankton Groups and Their Responses to Environmental Factors in Small Cascade Hydropower Stations. *Water* **2025**, *17*, 903.
8. Sim, Y.B.; Im, J. K.; Park, C. H.; Byun, J. H.; Hwang, S.-J. Impact of Drought on the Aquatic Ecosystem of the Cascade Dam Reservoir in South Korea. *Water* **2025**, *17*, 1023.
9. Yang, K.; Zhang, Q.; Wang, B.; Liang, B.; Lin, Q.; Wang, W. Source, Transport, and Fractionation of Rare Earth Elements in Fluvial Sediments from a Typical Small Urban Basin (East Tiaoxi River, Eastern China). *Water* **2025**, *17*, 1279.
10. Zhao, X.; Ming, D.; Meng, Y.; Yang, Z.; Peng, Q. Analysis of the Water Quality of a Typical Industrial Park on the Qinghai–Tibet Plateau Using a Self-Organizing Map and Interval Fuzzy Number-Based Set-Pair Analysis. *Water* **2025**, *17*, 111.

References

1. Dudgeon, D.; Arthington, A.H.; Gessner, M.O.; Kawabata, Z.I.; Knowler, D.J.; Lévêque, C.; Naiman, R.J.; Prieur-Richard, A.H.; Soto, D.; Stiassny, M.L.J.; et al. Freshwater biodiversity: Importance, threats, status and conservation challenges. *Biol. Rev.* **2007**, *81*, 163–182. [CrossRef] [PubMed]
2. Reid, A.J.; Carlson, A.K.; Creed, I.F.; Eliason, E.J.; Gell, P.A.; Johnson, P.T.J.; Kidd, K.A.; MacCormack, T.J.; Olden, J.D.; Ormerod, S.J.; et al. Emerging threats and persistent conservation challenges for freshwater biodiversity. *Biol. Rev.* **2018**, *94*, 849–873. [CrossRef] [PubMed]
3. Vörösmarty, C.J.; McIntyre, P.B.; Gessner, M.O.; Dudgeon, D.; Prusevich, A.; Green, P.; Glidden, S.; Bunn, S.E.; Sullivan, C.A.; Liermann, C.R.; et al. Global threats to human water security and river biodiversity. *Nature* **2010**, *467*, 555–561. [CrossRef] [PubMed]
4. Nash, L.N.; Antiqueira, P.A.P.; Romero, G.Q.; de Omena, P.M.; Kratina, P. Warming of aquatic ecosystems disrupts aquatic–terrestrial linkages in the tropics. *J. Anim. Ecol.* **2021**, *90*, 1623–1634. [CrossRef] [PubMed]

5. Sabater, S.; Freixa, A.; Jiménez, L.; López-Doval, J.; Pace, G.; Pascoal, C.; Perujo, N.; Craven, D.; González-Trujillo, J.D. Extreme weather events threaten biodiversity and functions of river ecosystems: Evidence from a meta-analysis. *Biol. Rev.* **2022**, *98*, 450–461. [CrossRef] [PubMed]
6. Oberdorff, T. Time for decisive actions to protect freshwater ecosystems from global changes. *Knowl. Manag. Aquat. Ecosyst.* **2022**, *423*, 19. [CrossRef]
7. De Lima, R.L.P.; de Graaf-van Dinther, R.E.; Boogaard, F.C. Impacts of floating urbanization on water quality and aquatic ecosystems: A study based on in situ data and observations. *J. Water Clim. Change* **2022**, *13*, 1185–1203. [CrossRef]
8. Zhou, J.; Li, Y.; Lei, Q.; Feng, Q.; Luo, J.; Lindsey, S. Asynchrony between urban expansion and water environmental protection reshapes the spatial patterns of nitrogen and phosphorus concentrations and N:P stoichiometry in inland small water bodies in Changsha, China. *Front. Environ. Sci.* **2022**, *10*, 1018408. [CrossRef]
9. Nilsson, C.; Reidy, C.A.; Dynesius, M.; Revenga, C. Fragmentation and Flow Regulation of the World's Large River Systems. *Science* **2005**, *308*, 405–408. [CrossRef] [PubMed]
10. Grill, G.; Lehner, B.; Thieme, M.; Geenen, B.; Tickner, D.; Antonelli, F.; Babu, S.; Borrelli, P.; Cheng, L.; Crochetiere, H.; et al. Mapping the world's free-flowing rivers. *Nature* **2019**, *569*, 215–221. [CrossRef] [PubMed]
11. Palmer, M.A.; Bernhardt, E.S.; Allan, J.D.; Lake, P.S.; Alexander, G.; Brooks, S.; Carr, J.; Clayton, S.; Dahm, C.N.; Follstad Shah, J.; et al. Standards for ecologically successful river restoration. *J. Appl. Ecol.* **2005**, *42*, 208–217. [CrossRef]
12. Rockström, J.; Steffen, W.; Noone, K.; Persson, Å.; Chapin, F.S.; Lambin, E.F.; Lenton, T.M.; Scheffer, M.; Folke, C.; Schellnhuber, H.J.; et al. A safe operating space for humanity. *Nature* **2009**, *461*, 472–475. [CrossRef] [PubMed]

water

MDPI

Article

Source, Transport, and Fractionation of Rare Earth Elements in Fluvial Sediments from a Typical Small Urban Basin (East Tiaoxi River, Eastern China)

Kunhua Yang [1], Qian Zhang [2,*], Bei Wang [1], Bin Liang [3], Qiang Lin [4,5,6,*] and Weijiao Wang [1]

1 Institute for Frontiers and Interdisciplinary Sciences, Zhejiang University of Technology, Hangzhou 310014, China; ykh@zjut.edu.cn (K.Y.); bei.wang@mail.mcgill.ca (B.W.); weijiao_wang@zjut.edu.cn (W.W.)
2 Institute of Geographic Sciences and Natural Resources Research, Chinese Academy of Sciences, Beijing 100101, China
3 School of Environment and Energy, South China University of Technology, Guangzhou 510006, China; liangbin93@scut.edu.cn
4 State Key Laboratory of Ocean Sensing, Zhejiang University, Hangzhou 310027, China
5 Institute of Quantum Sensing, Zhejiang University, Hangzhou 310027, China
6 School of Physics, Zhejiang University, Hangzhou 310027, China
* Correspondence: zhangqian@igsnrr.ac.cn (Q.Z.); qlin@zju.edu.cn (Q.L.)

Abstract: As emerging contaminants, rare earth elements (REEs) have undergone significant anthropogenic enrichment in aquatic systems. This study investigates the REE concentrations, major metal elements, and grain size in surface sediments from the East Tiaoxi (ETX) River in eastern China, a small urban river subjected to substantial anthropogenic influences. Total REE concentrations of surface sediments ranged from 133.62 to 222.92 mg/kg with MREE enrichment and HREE depletion. REE concentration and fractionation were strongly correlated with Ca, Fe, Mg, and Mn, which may reflect the control of clay minerals, Fe-Mn oxides, and specific heavy minerals, and differences in REE behavior between riparian sediments and riverbed sediments highlighted the impact of hydrodynamic sorting and chemical weathering on REE distribution. Anthropogenic activities, particularly urbanization, were found to increase REE concentrations, especially at urban-adjacent sites (e.g., RBS2 and RS2), while natural processes such as soil transport and chemical weathering primarily contributed to REE variation at other sites. The enrichment factor and ecological risk assessment revealed that the enrichment and moderate risks associated with REEs occurred in river sediments adjacent to urbanized areas, though agricultural impacts were less pronounced. The findings emphasize the combined influence of urbanization and natural processes on REE distribution and ecological risks in the ETX River basin and underscore the need to prioritize urban-derived REE contamination in environmental management strategies.

Keywords: REE fractionation patterns; source identification; emerging contamination; ecological risk; river sediments

1. Introduction

The rare earth elements (REEs), including La–Lu, exhibit coherent geochemical behaviors. Over the past 60 years, the study of REEs has been central to earth sciences [1], with these elements serving as powerful proxies for investigating the genesis of rocks, sediments, ocean water, and other geological formations [2–4]. Additionally, REEs have been instrumental in unraveling geochemical reactions and weathering processes within aquatic systems [5]. Concurrently, REEs have attained significant strategic importance, being designated as a critical metal by numerous nations. In recent decades, the demand for REEs has increased substantially in high-tech industries and agricultural applications, especially in green energy and military sectors [6]. The effluent discharge associated with these was identified as a potential risk factor for environmental contamination. For example, Gd is commonly used in magnetic resonance imaging, and REEs have been widely used in electronic products worldwide and as fertilizer additives in China and Brazil [7–9]. Consequently, the accumulation of anthropogenic REEs in the surficial environment is associated with the increasing number of electronics, manufacturing, medical facilities, technology industries, renewable energy, fertilizers, livestock feeds, etc. [10–12].

Recently, REEs have been identified as emerging contaminants, emphasizing their significance in ecotoxicological effects, such as trophic bioaccumulation [11]. This phenomenon has been observed in cultivated soils [9] and aquatic ecosystems, such as river water and sediments [13,14]. However, within the expansion of urban and agricultural areas, the integrated natural and anthropogenic influence of REEs in fluvial systems remains unclear. Since the solid phases are the primary loads of REEs in aquatic systems, the abundance and characteristics of REEs in the sediments can offer insights into their provenance, weathering processes, and anthropogenic contributions, particularly in regions significantly impacted by urban, industrial, and agricultural activities.

The geochemical characteristics of REEs in local soils and sediments are determined mainly by the characteristics of the parent rocks [15]. For instance, higher concentrations of REEs have been observed in river sediments draining from metamorphic and igneous terranes over those from sedimentary or volcanic rocks [16]. Furthermore, REE fractionation occurs during water–particle interaction, hydrodynamic sorting, and river sediment transport processes, making it challenging to determine sediment provenance [17]. In general, the mobilization and fractionation of REEs undergo various processes in aquatic systems, including water–particle sorption/desorption, co-precipitation, and solution and surface complexation [11]. REE compounds and ions with positive charges can be strongly adsorbed by clay minerals (such as illite and smectite), organic colloids, and inorganic colloids (especially the Fe/Mn oxyhydroxides) due to their negative charges or large specific surface area [18]. At a pH of approximately 7.5, greater than 95% adsorption of REEs can be achieved by clay, implying its significance as an important control on the transport and enrichment of REEs in sedimentary systems [19]. Additionally, the behavior of REEs in aquatic systems is sensitive to environmental factors such as pH, redox potential, and salinity, thereby controlling their behavior in the water–particle interaction [16]. Consequently, bedrock weathering, the geochemical properties of sediment particles, and sediment transport dynamics influence the river sediments' REE cycles. However, with the anthropogenic inputs and disturbance, the alteration of REE abundance and fractionation patterns has been observed in modern riverine sediments, which raises concerns regarding emerging contaminants and provides insight into the REE geochemical knowledge [20,21].

The present study focuses on the East Tiaoxi (ETX) River, located in eastern China, a paradigm of a small river subject to multiple anthropogenic influences, including urban, industrial, and agricultural activities and hydraulic engineering construction. The pollution of urban and agrarian effluence has resulted in the variation of solutes in the West Tiaoxi River water [22], highlighting the profound impact of anthropogenic activities on regional aquatic ecosystems. The objectives of this study are (1) to determine the distribution of REE concentrations and the fractionation in different types of surface sediments; (2) to identify the control of major elements and texture on REE concentration and fractionation to discuss the influence of sources and transport; and (3) to evaluate the enrichment and ecological risk levels of REEs in surface sediments from the ETX River basin. This study provides a reference for exploring REE origins, transport, and fractionation processes in fluvial sediments under multiple anthropogenic influences, especially urbanization and agriculture.

2. Materials and Methods

2.1. Study Area

The ETX River (30.11–30.96° N, 119.19–120.48° E), a typical small urban river, is located south of the Taihu Lake basin, flowing through Hangzhou, a third megacity in eastern China, and Huzhou City (Figure 1a). The ETX River originates from Tianmushan Mountain, with a main channel length of approximately 158.36 km and a basin area of 2265.1 km^2. The average annual rainfall is about 1460 mm, with approximately 75% occurring between mid-May and mid-July (Meiyu period) and between August and September (typhoon season). To manage upstream flood during the rainy season and divert the Taihu Lake water back during the dry season, two significant hydraulic engineering projects were constructed in the mainstream: a right-bank (east-bank) river levee in the middle reaches and a diversion project in the lower reaches; these two separate the ETX catchment area from the Hang-Jia-Hu Plain. The ETX River plays a crucial role in maintaining the balance of aquatic ecosystems due to its strong hydrological connection with the Taihu Lake and the Hang-Jia-Hu Plain's waterway network, ultimately flowing into the Changjiang River and Qiantang River. For example, 56.8% of the suspended sediment output from Taihu Lake flows out via the Changdougang River and Daqiangang River, which are the terminal branches of the ETX River [23].

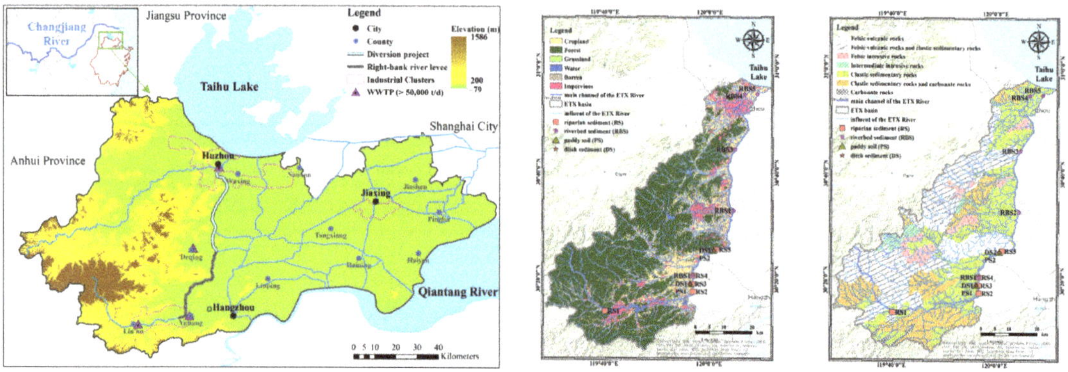

(a) (b) (c)

Figure 1. (a) The location and elevation map of the Hang-Jia-Hu area. (b) The 30 m land cover map in 2023 [24]. (c) The simplified lithology map of the ETX basin.

Over the past few decades, rapid industrialization, urbanization, and agricultural production within the ETX basin and surrounding areas have posed significant risks to water quality and aquatic ecosystems. Two critical industrial clusters are situated in the upper and lower reaches of the ETX River, and four wastewater treatment plants (WWTP; >50,000 t/d) are distributed along its main channel (Figure 1a). The dominant land-cover types in the ETX watershed are forest land, cropland, and construction land, totaling approximately 98.2% in 2023 [24]. Cropland and construction land (37.4%) are primarily concentrated along the main channel of the ETX River (Figure 1b), and paddy cultivation is the most important agricultural activity in the watershed. Additionally, the lithological map indicates that the ETX basin is dominated by felsic silicates (mainly referring to acidic lava, rhyolite, and biotite granite), clastic sedimentary rocks (primarily referring to sandstone, siltstone, and mudstone), and carbonates (referring to limestone and dolomite), with no evaporite outcrops (Figure 1c). Consequently, the ecological environment of the ETX River, particularly in its mainstream, is significantly influenced by natural conditions and anthropogenic disturbance, including hydrological connectivity, land-use types, and bedrock composition.

2.2. Sampling and Measurement

In August 2023, 14 surface fluvial sediment samples (0–5 cm) were collected along the main channel of the ETX River (Figure 1), including 10 river sediments and four paddy field sediments. Among the river sediments, five riverbed sediments (RBSs) were sampled on the bridges using a grab sampler, and five riparian sediments (RSs) were collected using a shovel at the riverbank in the upper and middle sections where RBSs were absent. Notably, RS sampling sites were located in areas where the water depth was less than 20 cm and were situated on the left bank to avoid the influence of river levee construction. The paddy field sediments, including two paddy soils and two of their adjacent ditch sediments, were collected using a shovel in the middle reaches. The outer layers of the samples in contact with a grab sampler or shovel were removed using a ceramic knife, and individual sediment samples were sealed in clean polyethylene bags. All samples were stored at approximately 4 °C and transported to the laboratory for freezing the same day.

The samples were freeze-dried in the laboratory and sieved to less than 2 mm (10 mesh). Sample pH values (sample/water ratio of 1/2.5) were determined using a pH meter (SX836, Sanxin, Shanghai, China) with a precision of ±0.01. For grain size analysis, 1.0 g of each sample was treated with 30% H_2O_2 and 2 mol/L HCl to remove organic and calcareous cementation, then dispersed in a 51 g/L $(NaPO_3)_6$ solution using an ultrasonic oscillator. Grain size was measured using a laser particle size analyzer (LAP-W2000H, EAST, Xiamen, China). Particles were classified as clay (<5 μm), silt (5–63 μm), and sand (63 μm–2 mm).

For metal element analysis, the samples were ground to 250 mesh powder using an agate mortar. Major metal elements and REEs were analyzed at the Institute of Geographic Sciences and Natural Resources Research, Chinese Academy of Sciences, China. Approximately 50 mg of each sample was digested with a mixture of HNO_3-HCl-HF in a Teflon vessel using a microwave digestion system (Multiwave PRO, Anton Paar, Graz, Austria) [14]. The digested samples were dried and redissolved in a 3% v/v HNO_3 solution. Major metal elements concentrations (Al, Ca, Fe, K, Mg, Mn, Na, and Ti) were determined using inductively coupled plasma optical emission spectroscopy (ICP-OES, Optima 5300DV, PerkinElmer, Waltham, MA, USA), and REE concentrations were measured using inductively coupled plasma mass spectrometry (ICP-MS, ELAN DRC-e, Perkin Elmer, Waltham, MA, USA). The measurement methods followed those described in a previous study for suspended sediments [25]. A certified soil reference material, GBW07449 (Institute of Geophysical and Geochemical Exploration, Langfang, China), was prepared and analyzed using the same procedure to ensure analytical accuracy. The duplicates, laboratory blanks, and standard reference materials, GBW(E)081531 (National Institute of Metrology, Beijing, China) and GBW07449, were used to ensure the data quality. The measurement accuracy was within ±5% for major metal elements and REEs compared to the standards.

2.3. Indication of REE Parameters and Enrichment Assessment

2.3.1. REE Fractionation and Anomaly Parameters

Several geochemical parameters were used to describe REE fractionation and anomalies. Total REEs (ΣREE), light REEs (ΣLREE), middle REEs (ΣMREE), and heavy REEs (ΣHREE) were defined as the sum of La–Lu, La–Nd, Sm–Dy, and Ho–Lu, respectively [26].

REE concentrations in sediments were normalized using the Post-Archean Australia Shale (PAAS) [27], reflecting the upper crust's average abundance. Normalized values (e.g., La_N) were calculated as the ratios of sample REE concentrations to PAAS values. Ratios such as $(La/Yb)_N$, $(La/Sm)_N$, and $(Sm/Yb)_N$ were used to describe fractionation among LREE, MREE, and HREE, which were defined using Equations (1)–(3) [28], respectively. Ce/Ce^* and Eu/Eu^* were used to assess Ce and Eu anomalies, which were defined using Equations (4) and (5) [29–31].

$$(La/Yb)_N = (La_{sample}/La_{PAAS})/(Yb_{sample}/Yb_{PAAS}) \quad (1)$$

$$(La/Sm)_N = (La_{sample}/La_{PAAS})/(Sm_{sample}/Sm_{PAAS}) \quad (2)$$

$$(Sm/Yb)_N = (Sm_{sample}/Sm_{PAAS})/(Yb_{sample}/Yb_{PAAS}) \quad (3)$$

$$Ce/Ce^* = Ce_N/(0.5La_N + 0.5Pr_N) \quad (4)$$

$$Eu/Eu^* = Eu_N/((Sm_N \times Gd_N)^{0.5}) \quad (5)$$

2.3.2. Assessment of Enrichment and Ecological Risk

The enrichment factor (*EF*) was used to evaluate elemental enrichment relative to background levels and identify the influence of natural and anthropogenic processes on elemental concentrations. The *EF* of a single REE was calculated as follows [32]:

$$EF_i = (C_i/C_{Al})_{sample}/(C_i/C_{Al})_{ref} \quad (6)$$

where EF_i represents the enrichment factor for REE i, C_i is the concentration of REE i, C_{Al} is the aluminum mass concentration, and ref refers to the reference material. Al was chosen as a conservative element less affected by anthropogenic sources [33]. The upper continental crust (UCC) [34] and the deep soil geochemical baseline values of China (CDS) [35] were used as reference material since the regional values of the geochemical background were unavailable. Meanwhile, this study's surface paddy soil (PS1) was used as a contrast. The concentrations of REEs and Al in these references are listed in Table 1. *EF* was categorized into <1, 1–2, 2–5, 5–20, 20–40, and >40, referring to non-enriched and slightly, moderately, significantly, strongly, and extremely enriched, respectively [28,36].

Table 1. REEs and Al concentrations of the UCC, CDS, and PS1 in this study.

	La	Ce	Pr	Nd	Sm	Eu	Gd	Tb	Dy	Ho	Er	Tm	Yb	Lu	∑REE	Al
	mg/kg	mg/kg	mg/kg	mg/kg	mg/kg	mg/kg	mg/kg	mg/kg	mg/kg	mg/kg	mg/kg	mg/kg	mg/kg	mg/kg	mg/kg	%
UCC [1]	30.00	64.00	7.10	26.00	4.50	0.88	3.80	0.64	3.50	0.80	2.30	0.33	2.20	0.32	146.37	8.04
CDS [2]	32.00	62.00	7.40	27.40	5.10	1.10	4.50	0.70	4.30	0.80	2.40	0.40	2.50	0.40	151.00	6.30
PS1 [3]	36.01	73.63	8.22	30.35	5.86	1.18	5.03	0.71	3.86	0.76	2.30	0.33	2.15	0.33	170.71	6.19

Notes: [1] UCC denotes the upper continental crust [34]. [2] CDS denotes the deep soil geochemical baseline values of China [35]. [3] PS1 denotes the paddy topsoil sample (PS1) in this study.

The potential ecological risk index (PERI) method [37] was used to assess the ecological hazards of REEs in sediments based on abundance and release effect [38,39]. The PERI for a single REE (Er_i) and the comprehensive potential ecological risk index (RI) were calculated as follows:

$$Er_i = Tr_i \times (C_i)_{sample}/(C_i)_{ref} \quad (7)$$

$$RI = \Sigma Er_i \quad (8)$$

where Tr_i is the toxicity index of the ith REE, assigned values of 1, 1, 5, 2, 5, 10, 5, 10, 5, 10, 5, 10, 5, 20 for La–Lu, respectively [38]; C_i represents the measured concentration of REE i in the samples or reference level, with the UCC [34] used as the reference.

Single-factor potential ecological risk was classified based on Er_i values: mild (<20), moderate (20–40), strength (40–80), strong (80–160), and very strong (>160) [37,38]. In this study, the measured REEs omitted Y. Therefore, the adjusted RI for the "mild ecological risk" was set to 110, calculated as RI = 150 × (94/133), and taken as the integer. The overall potential ecological risk was classified into four levels: mild (<110), moderate (110–220), strength (220–440), and strong (≥440).

2.4. Statistical Analysis and Visualization

Spearman's rank correlation coefficients were calculated to analyze the relationship between REEs and other geochemical parameters. Hierarchical clustering analysis was conducted to classify samples based on individual REE concentrations or PAAS-normalized ratios. Variables were standardized through Z-score transformation, and the average linkage method with squared Euclidean distance metric was employed.

Maps of the ETX basin were created using ArcMap 10.6. Data analysis was performed using Microsoft Office Excel 2019 and IBM SPSS Statistics 26. Graphs were generated using Origin 2021 and Python 3.11.4.

3. Results and Discussion

3.1. Characteristics of REEs in Surface Sediments

3.1.1. Distribution Characteristics of REE Concentration

The REE concentrations in various surface sediment types within the ETX basin are summarized in Table 2 and illustrated in Figure 2. The ΣREE concentrations in river sediments ranged from 133.62 to 222.92 mg/kg, averaging 186.30 mg/kg (182.10 mg/kg for RSs and 190.50 mg/kg for RBSs). These values exceed the mean concentration reported for the UCC (146.37 mg/kg) [34], Chinese catchment deep sediments (149.67 mg/kg) [40], CDS (151.00 mg/kg) [35], and the paddy field sediments in this study (151.53–170.01 mg/kg, with a mean value of 160.03 mg/kg), comparable to those of PAAS (184.77 mg/kg) [27] and Changjiang River sediments (186.59 mg/kg) [41]. These findings indicate that no significant REE pollution occurred in the surface sediments from the ETX basin. However, the ΣREE concentrations at sites RS2 (204.71 mg/kg), RBS2 (222.92 mg/kg), RBS3 (211.15 mg/kg), RBS4 (195.13 mg/kg), and RBS5 (189.68 mg/kg), flowing through the industrial clusters and high-density population areas (Figure 1), were close to those from a typical human-impacted river, the Mun River in northeast Thailand (averaged in 197.15 mg/kg) [14], and an anthropogenic-contaminated Tagus estuary in Portugal (18–210 mg/kg) [21], suggesting that these sites may be influenced by anthropogenic inputs such as industrial and urban wastewater.

Table 2. Statistics of REEs, major metal element concentrations, and other geochemical parameters of the surface sediments in the ETX basin.

Component	Units	Riparian Sediments (RSs)			Riverbed Sediments (RBSs)			Paddy Field Sediments			RSs + RBSs
		Avg [1]	Max [2]	Min [3]	Avg	Max	Min	Avg	Max	Min	Avg
La	mg/kg	39.08	44.17	36.81	40.20	44.72	30.16	33.59	36.01	31.15	39.64
Ce	mg/kg	79.13	89.69	73.40	82.31	97.02	56.31	69.80	73.63	65.22	80.72
Pr	mg/kg	8.80	9.65	8.49	9.22	10.25	6.76	7.66	8.22	7.41	9.01

Table 2. *Cont.*

Component	Units	Riparian Sediments (RSs)			Riverbed Sediments (RBSs)			Paddy Field Sediments			RSs + RBSs
		Avg [1]	Max [2]	Min [3]	Avg	Max	Min	Avg	Max	Min	Avg
Nd	mg/kg	31.67	35.12	30.25	33.25	38.73	23.03	27.96	30.35	26.38	32.46
Sm	mg/kg	5.92	6.67	5.43	6.53	7.99	4.23	5.40	5.86	5.18	6.22
Eu	mg/kg	1.15	1.47	0.92	1.25	1.46	0.74	1.15	1.19	1.07	1.20
Gd	mg/kg	5.06	5.76	4.30	5.58	6.82	3.54	4.64	5.03	4.36	5.32
Tb	mg/kg	0.74	0.86	0.63	0.80	1.04	0.54	0.66	0.71	0.62	0.77
Dy	mg/kg	4.07	4.44	3.57	4.56	6.01	3.15	3.62	3.86	3.33	4.32
Ho	mg/kg	0.81	0.87	0.72	0.88	1.18	0.60	0.71	0.76	0.67	0.85
Er	mg/kg	2.47	2.63	2.26	2.62	3.48	2.00	2.14	2.30	2.03	2.55
Tm	mg/kg	0.35	0.38	0.33	0.38	0.50	0.29	0.31	0.33	0.30	0.37
Yb	mg/kg	2.48	2.78	2.21	2.54	3.24	1.96	2.06	2.15	1.95	2.51
Lu	mg/kg	0.37	0.39	0.35	0.38	0.51	0.30	0.32	0.34	0.30	0.37
∑REE	mg/kg	182.10	204.71	169.67	190.50	222.92	133.62	160.03	170.71	151.53	186.30
∑LREE	mg/kg	158.68	178.63	148.94	164.98	190.70	116.27	139.01	148.21	130.16	161.83
∑MREE	mg/kg	16.94	19.19	14.84	18.72	23.32	12.20	15.48	16.63	14.67	17.83
∑HREE	mg/kg	6.48	6.89	5.89	6.80	8.90	5.15	5.54	5.87	5.26	6.64
$(La/Yb)_N$	[4]	1.17	1.23	1.02	1.18	1.32	1.02	1.21	1.28	1.08	1.17
$(La/Sm)_N$	[4]	0.96	1.01	0.90	0.91	1.03	0.81	0.90	0.97	0.86	0.93
$(Sm/Yb)_N$	[4]	1.22	1.31	1.02	1.30	1.45	1.10	1.33	1.39	1.26	1.26
Ce/Ce^*	[4]	0.98	1.00	0.96	0.98	1.05	0.91	1.00	1.04	0.99	0.98
Eu/Eu^*	[4]	0.98	1.12	0.89	0.97	1.08	0.89	1.08	1.18	1.01	0.98
Al	wt%	6.23	6.91	5.63	5.60	6.31	4.47	5.79	6.19	5.31	5.91
Ca	wt%	0.60	0.95	0.31	1.03	2.17	0.31	1.18	2.93	0.36	0.82
Fe	wt%	3.12	3.41	2.74	3.06	3.56	1.90	3.04	3.90	2.41	3.09
K	wt%	1.35	2.41	0.86	1.01	1.87	0.63	1.25	1.34	1.16	1.18
Mg	wt%	0.69	0.81	0.45	1.04	2.51	0.24	0.61	0.83	0.36	0.87
Mn	wt%	0.10	0.12	0.04	0.09	0.13	0.05	0.16	0.45	0.04	0.09
Na	wt%	0.77	1.13	0.67	0.88	0.98	0.77	0.64	0.73	0.48	0.83
Ti	wt%	0.33	0.37	0.25	0.34	0.42	0.16	0.40	0.43	0.33	0.33
Clay	%	52.78	60.72	35.04	39.26	53.86	30.68	43.74	67.79	32.23	46.02
Silt	%	38.22	41.81	29.36	53.79	67.38	31.47	55.90	66.75	32.21	46.01
pH	[4]	7.31	7.61	6.85	7.64	7.93	7.31	7.04	7.51	6.26	7.47

Notes: [1] Avg refers to the average of each component. [2] Max refers to the maximum of each component. [3] Min refers to the minimum of each component. [4] Dimensionless parameters.

The average concentration (mg/kg) of individual REEs in river sediments followed the order: Ce (80.72) > La (39.64) > Nd (32.46) > Pr (9.01) > Sm (6.22) > Gd (5.32) > Dy (4.32) > Er (2.55) > Yb (2.51) > Eu (1.20) > Ho (0.85) > Tb (0.77) > Tm (0.37) = Lu (0.37). A similar trend, albeit with reduced concentration, was observed in paddy field sediments: Ce (69.80) > La (33.59) > Nd (27.96) > Pr (7.66) > Sm (5.40) > Gd (4.64) > Dy (3.62) > Er (2.14) > Yb (2.06) > Eu (1.15) > Ho (0.71) > Tb (0.66) > Lu (0.32) > Tm (0.31). Notably, ΣREE concentrations increased progressively from paddy field sediments to RSs and RBSs, while the relative abundance of individual REEs remained consistent.

In river sediments, ΣLREE, ΣMREE, and ΣHREE accounted for 85.55–87.78% (mean 86.90%), 8.75–10.46% (mean 9.53%), and 3.28–3.99% (mean 3.57%), respectively. Similarly, paddy field sediments exhibited proportions of 85.90–87.52% (mean 86.85%), 9.15–10.35% (mean 9.68%), and 3.32–3.75% (mean 3.46%) for ΣLREE, ΣMREE, and ΣHREE, respectively. This indicates pronounced LREE enrichment and HREE depletion in the ETX basin sediments, a pattern consistent with global estuarine and riverine systems [42].

3.1.2. REE Fractionation Patterns and Anomalies

PAAS-normalized REE distribution patterns revealed distinct fractionation trends among RSs, RBSs, and paddy field sediments (Figure 3). For RSs, La_N, Sm_N, and Yb_N

had ranges of 0.96–1.16 (mean 1.02), 0.98–1.20 (mean 1.07), and 0.78–0.99 (mean 0.88), respectively. In RBSs, these values were 0.79–1.17 (mean 1.05), 0.76–1.44 (mean 1.18), and 0.69–1.15 (mean 0.90), respectively. Paddy field sediments exhibited lower normalized ratios: La_N 0.82–0.94 (mean 0.88), Sm_N 0.93–1.06 (mean 0.97), and Yb_N 0.69–0.76 (mean 0.73). Relative to PAAS, RSs and RBSs displayed MREE enrichment and HREE depletion, very close to the Changjiang River sediments [41], whereas paddy field sediments showed slight LREE depletion and pronounced HREE depletion. Notably, site RSB2 was significantly enriched in MREE and HREE (Figures 2 and 3b), indicating the additional anthropogenic inputs.

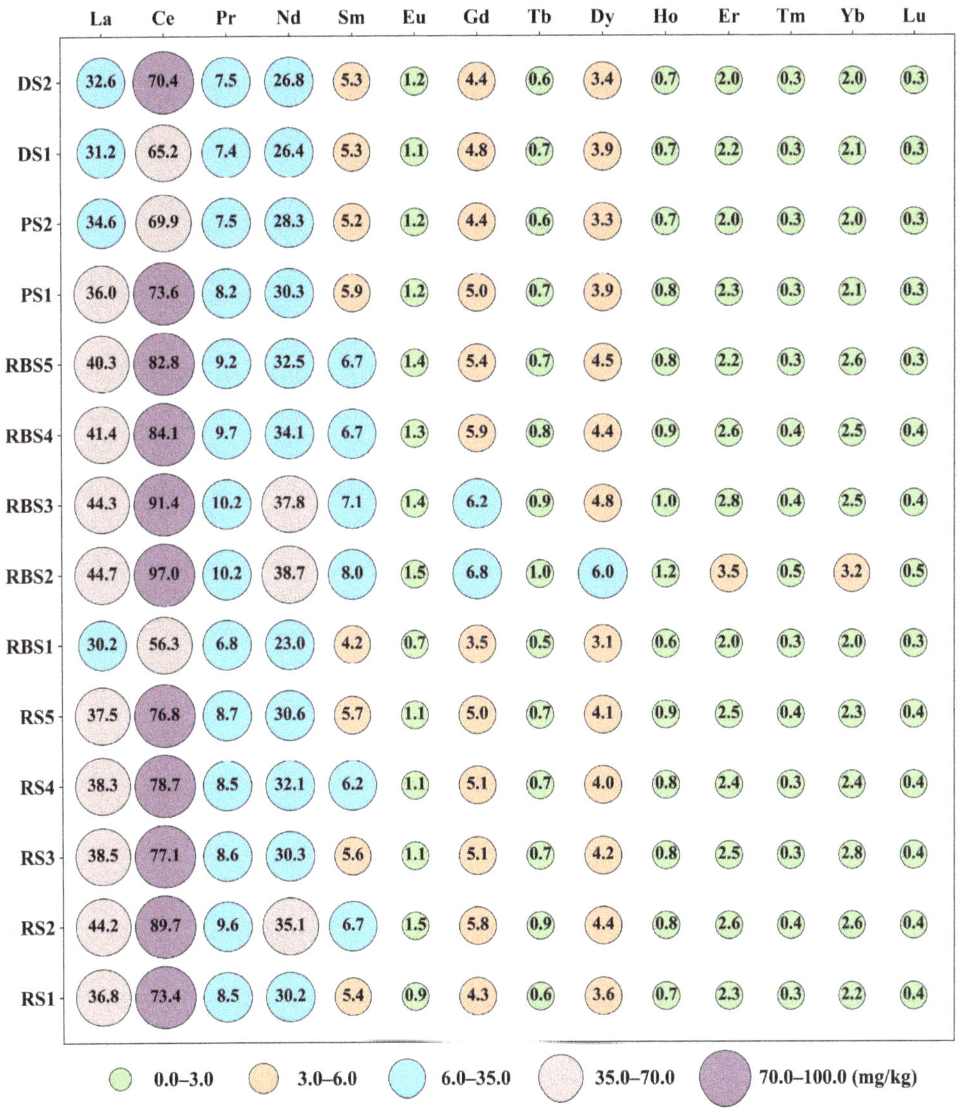

Figure 2. The REE concentrations of surface sediment samples in the ETX basin.

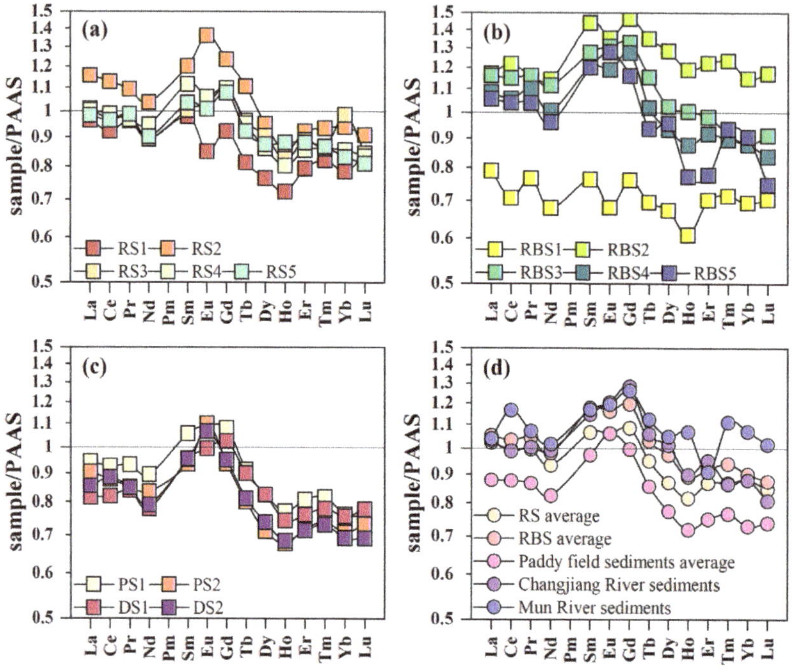

Figure 3. PAAS-normalized REE curves of surface sediment samples: (**a**) riparian sediments (RSs); (**b**) riverbed sediments (RBSs); (**c**) paddy field sediments in the ETX basin; (**d**) comparison of average PAAS-normalized REE patterns in the RSs, RBSs, paddy field sediments, Changjiang River sediments [41], and Mun River sediments [14].

The fractionation ratios $(La/Yb)_N$ and $(Sm/Yb)_N$ further quantified REE partitioning. For RSs, these ratios had ranges of 1.02–1.23 (mean 1.17) and 1.02–1.31 (mean 1.22); for RBSs, 1.02–1.32 (mean 1.18) and 1.10–1.45 (mean 1.30); and for paddy field sediments, 1.08–1.28 (mean 1.21) and 1.26–1.39 (mean 1.33). The hierarchy $(Sm/Yb)_N > (La/Yb)_N > 1$ underscores preferential HREE separated from sediments into the dissolved phase during the weathering processes due to higher dissolution of HREE over LREE under neutral and alkaline conditions [43].

Ce/Ce^* and Eu/Eu^* reflect the specific decoupling of Ce and Eu from the other REEs in the geological and environmental processes due to the special electron configuration in the outer shell and different valence states [44]. Ce/Ce^* ranged from 0.91 to 1.05 (mean 0.99), while Eu/Eu^* ranged from 0.89 to 1.18 (mean 1.01). Average Ce/Ce^* values for RSs, RBSs, and paddy field sediments were 0.98, 0.98, and 1.00, respectively; corresponding Eu/Eu^* values were 0.98, 0.97, and 1.08. The Ce/Ce^* and Eu/Eu^* values are almost equal to 1, representing no apparent Ce and Eu anomalies. However, a weak positive Eu anomaly in PS2 (1.18) and DS2 (1.12) likely reflects lithological inheritance from felsic volcanic rocks (Figure 1c), where Eu^{2+} substitution in plagioclase is common during magmatic processes [45]. In contrast, the absence of Eu anomalies in PS1 and DS1 (clastic sedimentary rock settings) may be related to Eu loss during sericite alteration.

3.2. Impacts of Geochemical Composition on REE Concentrations and Fractionation

All surface sediments in the ETX basin exhibited strong inter-element REE correlations ($r = 0.46$–0.99; $p < 0.05$, except Eu vs. Lu), consistent with natural sediment behavior [20]. The data of major elements and texture for all samples are shown in Table S1. To identify key controls on REE concentrations and fractionation in river sediments, Spearman's rank correlation analyses were performed between REE parameters (ΣREE, ΣLREE, ΣMREE, ΣHREE, Ce/Ce^*, Eu/Eu^*, $(La/Yb)_N$, and $(Sm/Yb)_N$) and geochemical composition (Figure 4a). Generally, the REE concentration and fractionation of river sediments were primarily related to Ca, Fe, Mg, Mn, and silt content but were weakly influenced by pH. The neutral to slightly alkaline condition of river sediments, with a pH between 6.8 and 7.9, is mainly related to the relatively high carbonate content. For example, carbonate (calcite/magnesium calcite and dolomite) in the floodplain sediments is about 1–4% adjacent to the middle reaches of the ETX River [46] and 6.8 wt% in the Changjiang River sediments [41]. However, REEs are depleted in carbonate minerals, thus causing the effect of dilution in sediments. In other words, the significant positive correlation between REE and Ca or Mg was induced by other Ca-rich minerals rather than carbonate. The positive correlation between REE and silt content for river sediments reflects the higher grain size and REE concentrations for RBSs, rather than the silty fractions enriching REEs, due to their weak correlation with RSs or RBSs. In the lower Changjiang River, major components of sediments are quartz, feldspar, and clay minerals, with about 3% being heavy minerals [41]. Among them, clay fractions can be a major host of REEs (>80%), followed by major and heavy minerals (<20%); in particular, REE-rich minerals such as zircon, sphene, apatite, monazite, and allanite form less than 10% of the heavy mineral fraction, highlighting their importance in hosting REEs [41]. The clay fractions consist of illite, kaolinite, chlorite, and smectite, in addition to Fe–Mn oxides/hydroxides, whose chemical components correspond to the positive correlation of REEs with Ca, Fe, Mg, and Mn [41,47]. Moreover, the PAAS-normalized REE distribution patterns in the ETX River sediments, i.e., MREE enrichment over LREE and HREE depletion, also suggest the control of clay fractions instead of silt fractions [16]. The positive $(Sm/Yb)_N$-silt% correlations suggest that silt content is essential in the fractionation between MREE and HREE [48]. Previous studies show that the enrichment of MREE may be governed by the substitution mechanism of Ca^{2+} by MREE [49] and reactive iron minerals such as goethite and hematite [50]. Therefore, the primary REE-hosting mineral phase in the ETX River sediments may refer to clay fractions, especially clay minerals (illite, kaolinite, chlorite, and smectite) and Fe–Mn oxides/hydroxides. The role of REE-rich heavy minerals such as zircon, sphene, apatite, monazite, and allanite is non-negligible.

(a) River sediments

	ΣREE	ΣLREE	ΣMREE	ΣHREE	Ce/Ce*	Eu/Eu*	(La/Yb)N	(Sm/Yb)N
Al	−0.042	−0.042	−0.042	0.103	0.037	0.154	0.491	0.224
Ca	0.912**	0.912**	0.912**	0.875**	0.712*	0.605	0.006	0.419
Fe	0.511	0.511	0.511	0.644*	0.387	0.031	−0.036	0.207
K	−0.596	−0.596	−0.596	−0.389	−0.334	−0.698*	−0.371	−0.669*
Mg	0.805**	0.805**	0.805**	0.683*	0.502	0.390	−0.073	0.476
Mn	0.557	0.557	0.557	0.783**	0.688*	0.335	−0.312	−0.128
Na	−0.176	−0.176	−0.176	−0.389	−0.442	−0.432	0.274	0.267
Ti	0.365	0.365	0.365	0.231	0.147	0.617	0.389	0.608
Clay%	−0.188	−0.188	−0.188	0.164	0.012	−0.332	−0.321	−0.430
Silt%	0.709*	0.709*	0.709*	0.455	0.489	0.566	0.261	0.745*
pH	0.339	0.339	0.339	0.326	0.109	0.103	−0.172	0.031

(b) RSs

	ΣREE	ΣLREE	ΣMREE	ΣHREE	Ce/Ce*	Eu/Eu*	(La/Yb)N	(Sm/Yb)N
Al	−0.300	−0.300	−0.300	−0.300	0.051	−0.359	−0.600	−0.200
Ca	0.700	0.700	0.700	1.000**	0.462	0.975**	0.100	−0.200
Fe	−0.400	−0.400	−0.400	−0.100	−0.205	−0.205	−0.700	−0.600
K	−0.462	−0.462	−0.462	−0.821	−0.500	−0.711	−0.205	0.205
Mg	0.616	0.616	0.616	0.872	0.289	0.895*	−0.359	−0.359
Mn	0.738	0.738	0.738	0.949**	0.433	0.973**	−0.158	−0.211
Na	−0.821	−0.821	−0.821	−0.872	−0.500	−0.947*	0.205	0.051
Ti	0.000	0.000	0.000	0.300	0.154	0.205	−0.600	−0.500
Clay%	−0.100	−0.100	−0.100	−0.100	−0.154	−0.051	−1.000**	−0.400
Silt%	0.400	0.400	0.400	0.500	0.667	0.410	0.200	0.100
pH	0.667	0.667	0.667	0.667	0.237	0.763	0.205	0.103

(c) RBSs

	ΣREE	ΣLREE	ΣMREE	ΣHREE	Ce/Ce*	Eu/Eu*	(La/Yb)N	(Sm/Yb)N
Al	0.300	0.300	0.300	0.300	0.051	0.600	0.800	0.900*
Ca	0.900*	0.900*	0.900*	0.900*	0.975**	0.300	−0.100	0.300
Fe	1.000**	1.000**	1.000**	1.000**	0.821	0.100	0.000	0.400
K	0.000	0.000	0.000	0.000	−0.205	−0.900*	−0.500	−0.600
Mg	0.700	0.700	0.700	0.700	0.667	0.100	−0.300	0.100
Mn	0.975**	0.975**	0.975**	0.975**	0.763	0.051	−0.051	0.359
Na	−0.400	−0.400	−0.400	−0.400	−0.410	0.200	0.600	0.300
Ti	0.051	0.051	0.051	0.051	0.132	0.821	0.564	0.667
Clay%	0.800	0.800	0.800	0.800	0.410	−0.400	−0.200	0.100
Silt%	0.100	0.100	0.100	0.100	0.359	1.000**	0.600	0.700
pH	−0.051	−0.051	−0.051	−0.051	−0.079	−0.359	−0.051	−0.205

1 0.8 0.6 0.4 0.2 0 −0.2 −0.4 −0.6 −0.8 −1.0

Figure 4. Spearman's rank correlation coefficients between the geochemical parameters of REEs and the geochemical composition in (**a**) river sediments, (**b**) RSs, and (**c**) RBSs. The bold shows the statistically significant correlation at $p < 0.05$ (*) or $p < 0.01$ (**).

The grouped correlation analyses revealed divergent controlling factors for RSs and RBSs (Figure 4b,c). Although RSs had slightly higher clay content (35.04–60.72%), with an average grain size of 6.1Φ (median: 7.8Φ), than that of RBSs (30.68–53.86%), with an average grain size of 6.0Φ (median: 7.2Φ), REEs in RBSs had a positive correlation (r = 0.80) with clay content, while it was weak for RSs. These indicate the disparity of the grain-size component controlling REE concentration in RSs and RBSs. For RSs, the REE concentrations (particularly ΣHREE) and Eu anomaly were mainly regulated by Ca, Mg, and Mn, in addition to a moderate positive REE-silt% correlation, suggesting a more important role of silty fractions, such as heavy minerals, in the REEs of RSs. Moreover, negative $(La/Yb)_N$ correlation with clay content likely reflects HREE preferential adsorption onto clay minerals, driven by increasing surface complexation constants with atomic numbers [51,52]. In contrast, Ca, Fe, Mg, Mn, and clay content were the main controlling factors of REE concentration in RBSs, which indicates the significant role of clay minerals, particularly Fe-containing fine minerals. For example, secondary minerals such as Fe-Mn (oxyhydr)oxides, illite, chlorite, and smectite can strongly adsorb REEs due to their negative charges or large specific surface area [18]. The shared Ca-Mg-Mn signatures and different textures that control REEs in RSs and RBSs suggest hydrodynamic sorting and chemical weathering during sediment transport from floodplains to riverbeds. These findings align with studies of Changjiang River floodplain sediments [53,54], emphasizing the dual controls of source lithology and weathering processes on REE distribution.

3.3. Distribution and Degree of REE Enrichment

3.3.1. Spatial Distribution Characteristics

Despite similar REE fractionation patterns across the ETX basin (Figure 3), marked spatial heterogeneity in ΣREE concentrations emerged (Figure 2), signaling potential ecological risks for local terrestrial and aquatic systems. Hierarchical cluster analysis of 15 REE concentrations and PAAS-normalized ratios yielded congruent groupings (Figure 5): Group 1 (RBS2, i.e., highest ΣREE), Group 2 (RBS1, i.e., lowest ΣREE), Group 3 (RS2, RBS3–RBS5, i.e., moderately high ΣREE), and Group 4 (remaining sites, i.e., soil-influenced). Cluster coherence implies shared provenance within groups [15]. Elevated ΣREE in Groups 1 and 3 likely reflect the addition of anthropogenic inputs, whereas Group 2 represents pristine bedrock signatures. Group 4's intermediate values suggest mixed soil and fluvial sources. Given the ETX River's role as a regional water resource and the basin's intensive land use, these spatial disparities necessitate comprehensive enrichment and ecological risk assessments.

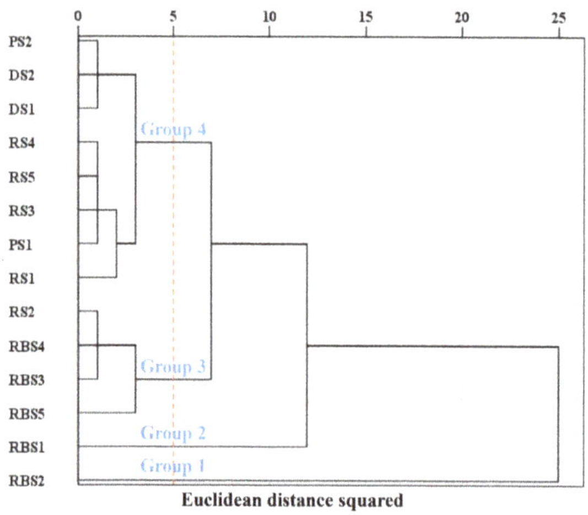

Figure 5. Cluster analysis diagram of REE concentrations (La–Lu) and the PAAS-normalized REE ratios of individual sediment samples in the ETX basin.

3.3.2. Degree of REE Enrichment and Potential Source

Although the previous study indicates that REE concentrations in sediments are often higher than maximum soil background concentrations by more than 10% [55], some river sediments in this study, including RS2 and RBS2–RBS5, seem to exceed this range, which suggests anthropogenic influences. To evaluate the degree of REE enrichment relative to the lithological background, *EF* values for all sediments were calculated using the UCC as a reference and ranged from 1.19 to 2.66 (mean 1.63), indicating slight to moderate enrichment. Notably, *EF* values showed a stronger enrichment of MREE and LREE over HREE (Figure 6a), in accord with the soils from a karst catchment in southwest China [56] and the soils from southern Konya in Turkey [57]. These areas distribute many carbonate rocks, and the enrichment of MREE may be governed by the substitution mechanism of Ca^{2+} [49]. Notably, RBS2 exhibited *EF* values comparable to multiple-anthropogenically influenced sediments from the Jiulong River estuary [42] but lower than agrarian-impacted Mun River sediments [14], which indicates potential anthropogenic inputs (perhaps urban and/or agricultural influences).

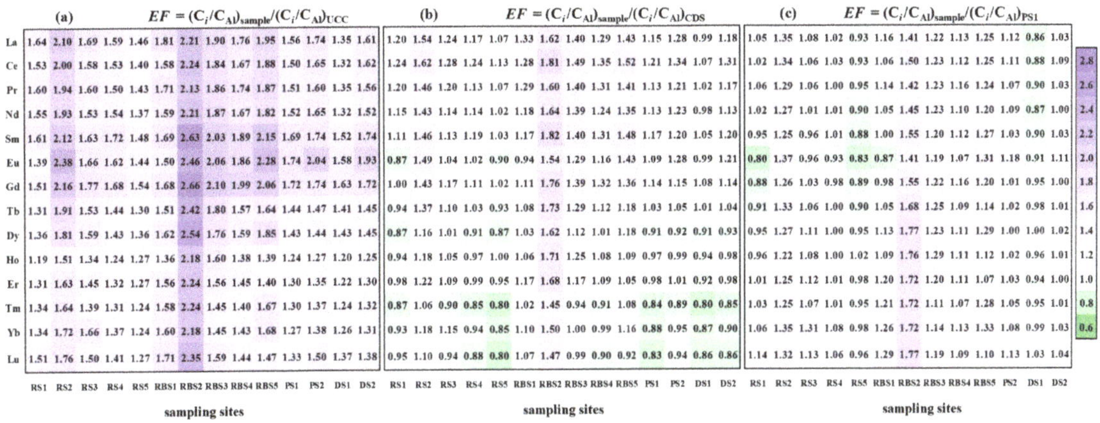

Figure 6. Heatmap of enrichment factor (*EF*) of single REE element for surface sediments in the ETX basin using three background values: (**a**) UCC [34]; (**b**) CDS [35], and (**c**) PS1 in this study.

Given the inherent variability in natural metal concentrations, the UCC and world average soils have not been used as the background values to assess anthropogenic enrichment [58]. Here, the regional baselines, i.e., CDS values, were prioritized for contamination assessment, and *EF* > 1.5 indicates pollution sources from anthropogenic activities [59]. *EF* values relative to CDS ranged from 0.80 to 1.82 (mean 1.15), reflecting slight depletion to slight enrichment (Figure 6b). RBS2 showed pronounced anthropogenic influence, while RS2 and RBS5 exhibited minor LREE contamination. Further, to exclude agricultural impacts, *EF* was recalculated against local paddy topsoil PS1 (Figure 6c). Only RBS2 displayed clear anthropogenic pollution in this case, with minor REE enrichment in RS2 and other RBS sites. Notably, it was observed that RSB2, adjacent to Deqing County, shows an obvious enrichment of MREE and HREE relative to local paddy topsoil, indicating the urban and industrial influence. Near-unity *EF* at RS3–RS5, PS2, DS1, and DS2, proximal to extensive paddy fields, suggests shared provenance between agricultural soils and adjacent river sediments. In summary, anthropogenic activities, particularly urbanization, drove marked REE enrichment at RBS2 and minor enrichment at RS2 and RBS5, while REEs at other sites were likely governed by soil and sediment transport and hydrodynamic sorting.

Single-element ecological risk indices (Er_i) for river sediments followed the order Lu (23.29) > Eu (13.64) > Tb (11.98) > Tm (11.09) > Ho (10.57) > Gd (7.00) > Sm (6.92) > Pr (6.35)

> Dy (6.17) > Yb (5.70) > Er (5.54) > Nd (2.50) > La (1.32) > Ce (1.26) (Figure 7). Paddy field sediments exhibited identical trends with reduced values: Lu (20.00) > Eu (13.04) > Tb (10.36) > Tm (9.40) > Ho (8.90) > Gd (6.11) > Sm (6.00) > Pr (5.39) > Dy (5.18) > Yb (4.68) > Er (4.65) > Nd (2.15) > La (1.12) > Ce (1.09). It was observed that Lu, Eu, and Tb largely contributed to the REE ecological risk in this study, consistent with the beach sediments from the Santa Rosalia mining region, Mexico [39], and slightly different from the order of Lu, Ho, Tb, and Eu in the Yellow River-estuary-bay sediments, China [20], and Lu, Tb, and Ho in the Mun River sediments, Thailand [14]. These regions have experienced the impacts of mining, urbanization, and industrial and agricultural activities. Thus, the ecological risks to Lu and Tb in river sediments from anthropogenic impacts are the most significant concern.

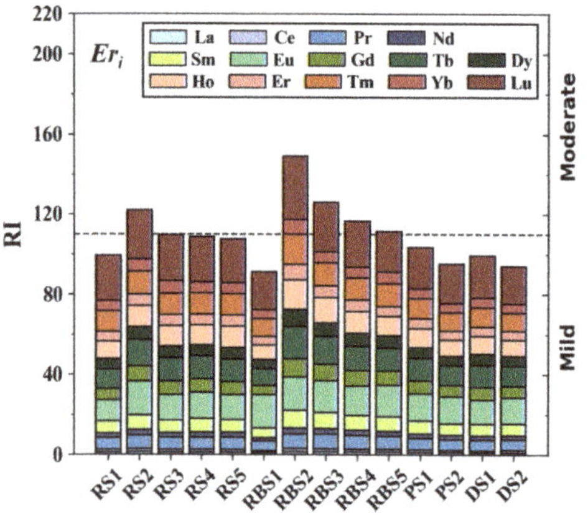

Figure 7. A stacked bar chart of *Eri* for a single REE gives the comprehensive potential ecological risk index (RI) for each sediment sample in the ETX basin.

Lu posed a moderate ecological risk at all sites except RBS1, PS2, and DS2. LREE (La–Nd), MREE (Sm–Dy), and HREE (Ho–Lu) contributed 8.85–10.75% (mean 10.08%), 36.83–43.01% (mean 40.55%), and 46.57–53.05% (mean 49.37%) to the total potential ecological risk index (RI), respectively, highlighting greater ecological risks from MREE and HREE, though their absolute concentrations (12.22–14.45%) are lower than LREE. MREE and HREE, such as Sm, Eu, Gd, Tb, Dy, and Lu, are widely used for magnets, battery alloys, electronic devices, phosphors, medical imaging, and so on [12], and are indispensable to modern society. Comprehensive RI values had ranges of 83.18–149.02 (mean 113.32) for river sediments and 94.06–103.45 (mean 103.45) for paddy field sediments, indicating elevated REE-associated risks in fluvial systems, particularly at the urban-influenced sites (RS2, RBS2–RBS5), which exhibited moderate risk. Lu, Eu, Tb, Tm, and Ho collectively contributed 52–94% of RI in river sediments and 59–64% in paddy field sediments. The accumulation of MREE and HREE in river sediments, particularly in urbanized and industrialized areas, may cause negative impacts on local aquatic ecological security through the food chain and pose a potential human health risk [60]. Considering the complex anthropogenic influence across the ETX watershed, particularly the industrial clusters and urban areas along the main channel, and reducing the impact of industrial and domestic wastewater discharges on surface water and sediments is essential for water resource management. These findings underscore urbanization, instead of agriculture, as the primary driver of REE-related ecological risks in the ETX basin.

4. Conclusions

This study analyzes and discusses the characteristics and controlling factors of the REE concentrations and fractionation of surface fluvial sediments and assesses the degree of REE enrichment in a typical urban small basin, the ETX River in eastern China. The main conclusions of the study are as follows.

The total REE concentrations (La–Lu) in surface sediments ranged from 133.62 to 222.92 mg/kg. The average ΣREE concentrations were 186.30 mg/kg for all measured sediments, 182.10 mg/kg for RSs, 190.50 mg/kg for RBSs, and 160.03 mg/kg for paddy field sediments, showing an increasing trend with a similar order among single REE concentrations. The comparison with the PAAS reveals that RSs and RBSs were enriched in MREE but depleted in HREE, while paddy field sediments were slightly depleted in LREE and markedly depleted in HREE. The relationship $(Sm/Yb)_N > (La/Yb)_N > 1$ demonstrated that the measured surface sediments of the ETX basin exhibited the evident depletion of HREE, since the preferential removal of HREE resulted from higher dissolution of HREE over LREE under neutral and alkaline conditions. The correlation analysis showed that the key factors controlling the concentrations and fractionation of REEs in river sediments include Ca, Fe, Mg, Mn, and grain size. The primary REE-hosting mineral phase in the ETX River sediments may refer to clay fractions, especially the clay minerals (illite, kaolinite, chlorite, and smectite) and Fe–Mn oxides/hydroxides, and the role of REE-rich heavy minerals, such as zircon, sphene, apatite, monazite, and allanite, is non-negligible. Differences in REE behavior between RSs and RBSs were attributed to variations in the mineral composition of the clay fraction, particularly the Fe-containing minerals. These results emphasize the impact of hydrodynamic sorting and chemical weathering on REE distribution in the ETX basin. Nevertheless, to better understand the controls and effects of mineral phases on REEs in riverine sediments, mineralogical evidence should be included in future studies.

The clustering and *EF* analyses indicated that the evident REE enrichment observed in site RBS2 and the slight REE enrichment in sites RS2 and RBS5 are attributable to anthropogenic activities, specifically urbanization. In contrast, the REEs in sites RS1 and RS3–RS5 are primarily attributed to the transport of soils. By comparison, the slight REE enrichment in RSB1, RSB3, and RBS4 may be attributed to the accumulation during the chemical weathering processes and hydrodynamic sorting. Moreover, three *EF* analyses underscore that using regional and local soils as a reference is recommended. The single-factor potential ecological risk degree was mild, except for Lu, which exhibited a moderate risk at the sites that exclude RBS1, PS2, and DS2. The river sediments of the ETX basin exhibited a higher comprehensive ecological risk of REEs compared to paddy soils, and a moderate degree of ecological risk at the sites RS2 and RBS2–RBS5 was attributed to the urban distribution. In conclusion, the potential ecological risks of REEs in surface sediments of the ETX basin were primarily attributed to urban distribution as opposed to agriculture.

Supplementary Materials: The following supporting information can be downloaded at: https://www.mdpi.com/article/10.3390/w17091279/s1, Table S1: REE concentrations, major metal element concentrations, and other geochemical parameters of the surface sediments in the ETX basin.

Author Contributions: Conceptualization, K.Y. and Q.Z.; methodology, K.Y. and Q.Z.; software, K.Y. and W.W.; validation, K.Y., Q.Z. and Q.L.; formal analysis, K.Y.; investigation, K.Y.; data curation, K.Y.; writing—original draft preparation, K.Y.; writing—review and editing, K.Y., Q.Z., B.W., B.L. and Q.L.; visualization, K.Y. and W.W.; supervision, Q.Z. and Q.L.; funding acquisition, Q.Z., B.W. and Q.L. All authors have read and agreed to the published version of the manuscript.

Funding: This research was funded by the National Natural Science Foundation of China, grant number 42203011, the Joint Funds of the National Natural Science Foundation of China, grant numbers U2341246 and U24A20319, and startup fund from the Zhejiang University of Technology grant to Bei Wang, grant number 2023044163001.

Data Availability Statement: The original contributions presented in this study are included in the article/Supplementary Materials. Further inquiries can be directed to the corresponding authors.

Acknowledgments: The authors thank the academic editor and anonymous reviewers for their constructive and thoughtful comments, which significantly improved the manuscript. They also thank Jun Li and his group members at the Zhejiang University of Technology in China for their invaluable assistance with the sample collection and pretreatment.

Conflicts of Interest: The authors declare no conflicts of interest. The funders had no role in the design of the study; in the collection, analyses, or interpretation of data; in the writing of the manuscript; or in the decision to publish the results.

References

1. Barrat, J.-A.; Bayon, G. Practical guidelines for representing and interpreting rare earth abundances in environmental and biological studies. *Chemosphere* **2024**, *352*, 141487. [CrossRef] [PubMed]
2. Saputro, S.P.; Godang, S.; Priadi, B.; Basuki, N.I.; Himawan, B. Geochemical study of Al–Fe–Ti enrichment in rock weathering: Implications for the recognizing of igneous protolith and the enrichment of REE in soil profile. *Appl. Geochem.* **2022**, *140*, 105259. [CrossRef]
3. Liu, H.; Guo, H.; Pourret, O.; Wang, Z.; Sun, Z.; Zhang, W.; Liu, M. Distribution of rare earth elements in sediments of the North China Plain: A probe of sedimentation process. *Appl. Geochem.* **2021**, *134*, 105089. [CrossRef]
4. Han, R.; Zhang, Q.; Wang, D.; Zhong, Q.; Han, G. Discrimination of brewing technologies and assessment of health risks based on rare earth elements: Evidence of fingerprint in Chinese famous vinegars. *Food Chem.* **2025**, *464*, 141539. [CrossRef]
5. Han, G.; Liu, M.; Li, X.; Zhang, Q. Sources and geochemical behaviors of rare earth elements in suspended particulate matter in a wet-dry tropical river. *Environ. Res.* **2023**, *218*, 115044. [CrossRef] [PubMed]
6. Chakhmouradian, A.R.; Wall, F. Rare Earth Elements: Minerals, Mines, Magnets (and More). *Elements* **2012**, *8*, 333–340. [CrossRef]
7. Andrade, R.L.B.; Hatje, V.; Pedreira, R.M.A.; Boening, P.; Pahnke, K. REE fractionation and human Gd footprint along the continuum between Paraguacu River to coastal South Atlantic waters. *Chem. Geol.* **2020**, *532*, 119303. [CrossRef]
8. Tommasi, F.; Thomas, P.J.; Pagano, G.; Perono, G.A.; Oral, R.; Lyons, D.M.; Toscanesi, M.; Trifuoggi, M. Review of Rare Earth Elements as Fertilizers and Feed Additives: A Knowledge Gap Analysis. *Arch. Environ. Contam. Toxicol.* **2021**, *81*, 531–540. [CrossRef]
9. Bispo, F.H.A.; de Menezes, M.D.; Fontana, A.; Sarkis, J.E.d.S.; Gonçalves, C.M.; de Carvalho, T.S.; Curi, N.; Guilherme, L.R.G. Rare earth elements (REEs): Geochemical patterns and contamination aspects in Brazilian benchmark soils. *Environ. Pollut.* **2021**, *289*, 117972. [CrossRef]
10. Costa, L.; Mirlean, N.; Johannesson, K.H. Rare earth elements as tracers of sediment contamination by fertilizer industries in Southern Brazil, Patos Lagoon Estuary. *Appl. Geochem.* **2021**, *129*, 104965. [CrossRef]
11. Gwenzi, W.; Mangori, L.; Danha, C.; Chaukura, N.; Dunjana, N.; Sanganyado, E. Sources, behaviour, and environmental and human health risks of high-technology rare earth elements as emerging contaminants. *Sci. Total Environ.* **2018**, *636*, 299–313. [CrossRef] [PubMed]
12. Dushyantha, N.; Batapola, N.; Ilankoon, I.M.S.K.; Rohitha, S.; Premasiri, R.; Abeysinghe, B.; Ratnayake, N.; Dissanayake, K. The story of rare earth elements (REEs): Occurrences, global distribution, genesis, geology, mineralogy and global production. *Ore Geol. Rev.* **2020**, *122*, 103521. [CrossRef]
13. Gao, X.; Han, G.; Liu, J.; Zhang, S. Spatial Distribution and Sources of Rare Earth Elements in Urban River Water: The Indicators of Anthropogenic Inputs. *Water* **2023**, *15*, 654. [CrossRef]
14. Ma, S.; Han, G. Rare earth elements reveal the human health and environmental concerns in the largest tributary of the Mekong river, Northeastern Thailand. *Environ. Res.* **2024**, *252*, 118968. [CrossRef]
15. Han, G.; Xu, Z.; Tang, Y.; Zhang, G. Rare Earth Element Patterns in the Karst Terrains of Guizhou Province, China: Implication for Water/Particle Interaction. *Aquat. Geochem.* **2009**, *15*, 457–484. [CrossRef]
16. Bayon, G.; Toucanne, S.; Skonieczny, C.; André, L.; Bermell, S.; Cheron, S.; Dennielou, B.; Etoubleau, J.; Freslon, N.; Gauchery, T.; et al. Rare earth elements and neodymium isotopes in world river sediments revisited. *Geochim. Cosmochim. Acta* **2015**, *170*, 17–38. [CrossRef]

17. Guo, Y.; Yang, S.; Su, N.; Yin, P.; Wang, Z. Rare earth element geochemistry of the sediments from small rivers draining southeast China. *Mar. Geol. Quat. Geol.* **2018**, *38*, 139–149.
18. Zhu, C.; Wang, Q.; Huang, X.; Yun, J.; Hu, Q.; Yang, G. Adsorption of amino acids at clay surfaces and implication for biochemical reactions: Role and impact of surface charges. *Colloids Surf. B Biointerfaces* **2019**, *183*, 110458. [CrossRef]
19. Bishop, B.A.; Alam, M.S.; Flynn, S.L.; Chen, N.; Hao, W.; Ramachandran Shivakumar, K.; Swaren, L.; Gutierrez Rueda, D.; Konhauser, K.O.; Alessi, D.S.; et al. Rare Earth Element Adsorption to Clay Minerals: Mechanistic Insights and Implications for Recovery from Secondary Sources. *Environ. Sci. Technol.* **2024**, *58*, 7217–7227. [CrossRef]
20. Liu, Z.; Gu, X.; Lian, M.; Wang, J.; Xin, M.; Wang, B.; Ouyang, W.; He, M.; Liu, X.; Lin, C. Occurrence, geochemical characteristics, enrichment, and ecological risks of rare earth elements in sediments of "the Yellow river—Estuary—bay" system. *Environ. Pollut.* **2023**, *319*, 121025. [CrossRef]
21. Brito, P.; Prego, R.; Mil-Homens, M.; Caçador, I.; Caetano, M. Sources and distribution of yttrium and rare earth elements in surface sediments from Tagus estuary, Portugal. *Sci. Total Environ.* **2018**, *621*, 317–325. [CrossRef] [PubMed]
22. Cui, Z.; Chen, C.; Chen, Q.; Huang, J. Difference in the Contribution of Driving Factors to Nitrogen Loss With Surface Runoff Between the Hill and Plain Agricultural Watersheds. *J. Geophys. Res. Biogeosci.* **2024**, *129*, e2023JG007931. [CrossRef]
23. Zhu, J.; Liu, X.; Deng, J.; Zhang, H.; Peng, J. Suspended solids transport rate of the rivers around western Lake Taihu. *Acta Sci. Circumstantiae* **2018**, *38*, 3682–3687.
24. Yang, J.; Huang, X. The 30 m annual land cover datasets and its dynamics in China from 1985 to 2023. In *Earth System Science Data*; Zenodo: Geneve, Switzerland, 2024; Volume 13.
25. Yang, K.; Han, G.; Zeng, J.; Zhou, W. Distribution, fractionation and sources of rare earth elements in suspended particulate matter in a tropical agricultural catchment, northeast Thailand. *PeerJ* **2021**, *9*, e10853. [CrossRef] [PubMed]
26. Grawunder, A.; Lonschinski, M.; Händel, M.; Wagner, S.; Merten, D.; Mirgorodsky, D.; Büchel, G. Rare earth element patterns as process indicators at the water–solid interface of a post–mining area. *Appl. Geochem.* **2018**, *96*, 138–154. [CrossRef]
27. Taylor, S.R.; McLennan, S.M. *Continental Crust: Its Composition and Evolution: An Examination of the Geochemical Record Preserved in Sedimentary Rocks*; Blackwell Scientific Publications: Oxford, UK, 1985; 312p.
28. Huang, H.; Lin, C.; Yu, R.; Yan, Y.; Hu, G.; Wang, Q. Spatial distribution and source appointment of rare earth elements in paddy soils of Jiulong River Basin, Southeast China. *J. Geochem. Explor.* **2019**, *200*, 213–220. [CrossRef]
29. Rétif, J.; Zalouk-Vergnoux, A.; Briant, N.; Poirier, L. From geochemistry to ecotoxicology of rare earth elements in aquatic environments: Diversity and uses of normalization reference materials and anomaly calculation methods. *Sci. Total Environ.* **2023**, *856*, 158890. [CrossRef]
30. Zhou, Y.; Yang, X.; Zhang, D.; Mackenzie, L.L.; Chen, B. Sedimentological and geochemical characteristics of sediments and their potential correlations to the processes of desertification along the Keriya River in the Taklamakan Desert, western China. *Geomorphology* **2021**, *375*, 107560. [CrossRef]
31. Rahlf, P.; Laukert, G.; Hathorne, E.C.; Vieira, L.H.; Frank, M. Dissolved neodymium and hafnium isotopes and rare earth elements in the Congo River Plume: Tracing and quantifying continental inputs into the southeast Atlantic. *Geochim. Cosmochim. Acta* **2021**, *294*, 192–214. [CrossRef]
32. Zeng, J.; Han, G.; Yang, K. Assessment and sources of heavy metals in suspended particulate matter in a tropical catchment, northeast Thailand. *J. Clean Prod.* **2020**, *265*, 121898. [CrossRef]
33. Galhardi, J.A.; Leles, B.P.; de Mello, J.W.V.; Wilkinson, K.J. Bioavailability of trace metals and rare earth elements (REE) from the tropical soils of a coal mining area. *Sci. Total Environ.* **2020**, *717*, 134484. [CrossRef] [PubMed]
34. Taylor, S.R.; McLennan, S.M. The geochemical evolution of the continental crust. *Rev. Geophys.* **1995**, *33*, 241–265. [CrossRef]
35. Wang, X.; Zhou, J.; Xu, S.; Chi, Q.; Nie, L.; Zhang, B.; Yao, W.; Wang, W.; Liu, H.; Liu, D.; et al. China soil geochemical baselines networks: Data characteristics. *Geol. China* **2016**, *43*, 1469–1480.
36. Sutherland, R.A. Bed sediment-associated trace metals in an urban stream, Oahu, Hawaii. *Environ. Geol.* **2000**, *39*, 611–627. [CrossRef]
37. Hakanson, L. An ecological risk index for aquatic pollution control.a sedimentological approach. *Water Res.* **1980**, *14*, 975–1001. [CrossRef]
38. Chen, H.; Chen, Z.; Chen, Z.; Ou, X.; Chen, J. Calculation of Toxicity Coefficient of Potential Ecological Risk Assessment of Rare Earth Elements. *Bull. Environ. Contam. Toxicol.* **2020**, *104*, 582–587. [CrossRef] [PubMed]
39. Godwyn-Paulson, P.; Jonathan, M.P.; Rodríguez-Espinosa, P.F.; Rodríguez-Figueroa, G.M. Rare earth element enrichments in beach sediments from Santa Rosalia mining region, Mexico: An index-based environmental approach. *Mar. Pollut. Bull.* **2022**, *174*, 113271. [CrossRef]
40. Wang, X.; Zhou, J.; Chi, Q.; Wang, W.; Zhang, B.; Nie, L.; Liu, D.; Xu, S.; Wu, H.; Gao, Y. Geochemical Background and Distribution of Rare Earth Elements in China: Implications for Potential Prospects. *Acta Geosci. Sin.* **2020**, *41*, 747–758.
41. Yang, S.Y.; Jung, H.S.; Choi, M.S.; Li, C.X. The rare earth element compositions of the Changjiang (Yangtze) and Huanghe (Yellow) river sediments. *Earth Planet. Sci. Lett.* **2002**, *201*, 407–419. [CrossRef]

42. Ma, S.; Han, G. Distribution, provenance, contamination, and probabilistic ecological risk of rare earth elements in surface sediments of Jiulong River estuary and adjacent watershed. *Ocean Coast. Manag.* **2024**, *254*, 107205. [CrossRef]
43. Han, G.; Yang, K.; Zeng, J. Spatio-Temporal Distribution and Environmental Behavior of Dissolved Rare Earth Elements (REE) in the Zhujiang River, Southwest China. *Bull. Environ. Contam. Toxicol.* **2022**, *108*, 555–562. [CrossRef] [PubMed]
44. Petrosino, P.; Sadeghi, M.; Albanese, S.; Andersson, M.; Lima, A.; De Vivo, B. REE contents in solid sample media and stream water from different geological contexts: Comparison between Italy and Sweden. *J. Geochem. Explor.* **2013**, *133*, 176–201. [CrossRef]
45. Alderton, D.H.M.; Pearce, J.A.; Potts, P.J. Rare earth element mobility during granite alteration: Evidence from southwest England. *Earth Planet. Sci. Lett.* **1980**, *49*, 149–165. [CrossRef]
46. Yan, D.; Wünnemann, B.; Gao, S.; Zhang, Y. Early Holocene tidal flat evolution in a western embayment of East China Sea, in response to sea level rise episodes. *Quat. Sci. Rev.* **2020**, *250*, 106642. [CrossRef]
47. Fu, H.; Jian, X.; Pan, H. Bias in sediment chemical weathering intensity evaluation: A numerical simulation study. *Earth-Sci. Rev.* **2023**, *246*, 104574. [CrossRef]
48. Feng, J.L.; Hu, Z.G.; Ju, J.T.; Zhu, L.P. Variations in trace element (including rare earth element) concentrations with grain sizes in loess and their implications for tracing the provenance of eolian deposits. *Quat. Int.* **2011**, *236*, 116–126. [CrossRef]
49. Yin, F.Y.; Li, W.; Xie, G.Q.; Cai, L.; Zhou, Y.Q.; Wu, J.; Meng, Y.Q. Genesis of the Wangu Au deposit in the Jiangnan orogenic belt: Constraints from texture, trace element, and in-situ Sr isotope of scheelite. *Ore Geol. Rev.* **2025**, *176*, 106375. [CrossRef]
50. Song, Y.H.; Choi, M.S. REE geochemistry of fine-grained sediments from major rivers around the Yellow Sea. *Chem. Geol.* **2009**, *266*, 328–342. [CrossRef]
51. Wood, S.A. The aqueous geochemistry of the rare-earth elements and yttrium: 1. Review of available low-temperature data for inorganic complexes and the inorganic REE speciation of natural waters. *Chem. Geol.* **1990**, *82*, 159–186. [CrossRef]
52. Liang, X.; Wu, P.; Wei, G.; Yang, Y.; Ji, S.; Ma, L.; Zhou, J.; Tan, W.; Zhu, J.; Takahashi, Y. Enrichment and fractionation of rare earth elements (REEs) in ion-adsorption-type REE deposits: Constraints of an iron (hydr)oxide-clay mineral composite. *Am. Mineral.* **2025**, *110*, 114–135. [CrossRef]
53. Yang, S.; Wang, Z. Rare Earth Element Compositions of the Sediments from the Major Tributaries and the Main Stream of the Changjiang River. *Bull. Mineral. Petrol. Geochem.* **2011**, *30*, 31–39.
54. Yang, S.; Wang, Z.; Guo, Y.; Li, C.; Cai, J. Heavy mineral compositions of the Changjiang (Yangtze River) sediments and their provenance-tracing implication. *J. Asian Earth Sci.* **2009**, *35*, 56–65. [CrossRef]
55. Tóth, A.; Sipos, P.; Jakab, G.; Szalai, Z.; Kalicz, P.; Madarász, B. Improving the reliability of using rare earth elements as soil erosion tracers. *CATENA* **2024**, *243*, 108175. [CrossRef]
56. Han, R.Y.; Xu, Z.F. Geochemical Behaviors of Rare Earth Elements (REEs) in Karst Soils under Different Land-Use Types: A Case in Yinjiang Karst Catchment, Southwest China. *Int. J. Environ. Res. Public Health* **2021**, *18*, 502. [CrossRef] [PubMed]
57. Özen, Y. Distribution and origin of rare earth elements (REEs) in topsoils and soil profiles of southern Konya (Turkey): Implication for controls on the dynamics of REEs in soils and bedrocks. *CATENA* **2024**, *246*, 108352. [CrossRef]
58. Chakraborty, B.; Bera, B.; Roy, S.h.; Adhikary, P.P.; Sengupta, D.; Shit, P.K. Assessment of non-carcinogenic health risk of heavy metal pollution: Evidences from coal mining region of eastern India. *Environ. Sci. Pollut. Res.* **2021**, *28*, 47275–47293. [CrossRef]
59. Chen, H.; Chen, R.; Teng, Y.; Wu, J. Contamination characteristics, ecological risk and source identification of trace metals in sediments of the Le'an River (China). *Ecotoxicol. Environ. Saf.* **2016**, *125*, 85–92. [CrossRef]
60. Simbanegavi, T.T.; Gwenzi, W. Chapter 11—Ecological health risks of high-technology rare earth elements. In *Emerging Contaminants in the Terrestrial-Aquatic-Atmosphere Continuum*; Gwenzi, W., Ed.; Elsevier: Amsterdam, The Netherlands, 2022; pp. 171–194. [CrossRef]

water

MDPI

Article

Impact of Drought on the Aquatic Ecosystem of the Cascade Dam Reservoir in South Korea

Youn Bo Sim [1], Jong Kwon Im [2,*], Chae Hong Park [3], Jeong Hwan Byun [4] and Soon-Jin Hwang [1,*]

1 Department of Environmental Health Science, Konkuk University, Seoul 05029, Republic of Korea; sumatra0@nate.com
2 Environmental Standards Research Division, National Institute of Environmental Research, 42 Hwangyong-ro, Incheon 22689, Republic of Korea
3 Encounter the Ecology, 248 Gwanggyojungang-ro, Suwon 16512, Republic of Korea; qkrcoghd2@gmail.com
4 Han River Environment Research Center, National Institute of Environmental Research, 42, Dumulmeori-gil 68beon-gil, Yangseo-myeon, Yangpyeong-gun, Incheon 12585, Republic of Korea; jh0130@korea.kr
* Correspondence: lim-jkjk@daum.net (J.K.I.); sjhwang@konkuk.ac.kr (S.-J.H.); Tel.: +82-32-560-8386 (J.K.I.); +82-2-450-3748 (S.-J.H.)

Abstract: Climate change has increased the frequency and intensity of extreme weather events worldwide. In South Korea, annual precipitation in 2014–2015 was only 50% of the long-term average, resulting in severe drought conditions. This drought extended water residence time in dam reservoirs, enhancing internal nutrient recycling, degrading water quality, and promoting harmful cyanophyta blooms in downstream reservoirs. Using the Standardized Precipitation Index—for drought assessment, and monthly water sampling—for environmental factors and phytoplankton analyses, this study examined the impacts of drought on water quality and phytoplankton communities in a series of interconnected dam reservoirs (Uiam, Cheongpyeong, Sambong-ri, and Paldang Lakes) within the Bukhan River system from 2013 to 2016. The prolonged residence time during drought facilitated nutrient accumulation and recycling within the reservoirs, intensifying eutrophication and water quality deterioration, alongside a pronounced cyanobacterial dominance and harmful algal blooms. These findings suggest that changes in upstream dam discharges directly influence water quality and ecosystem health in downstream reservoirs and that diverse hydrological changes associated with drought pose a significant threat to water source management. These findings may inform the development of integrated water management strategies for maintaining water quality and protecting water sources during droughts and extreme climatic events.

Keywords: cyanophyta; water residence time; drought; harmful algal bloom

1. Introduction

Extreme weather events, such as drought and heavy rainfall, have become more frequent and intense globally due to climate change. In South Korea, annual precipitation in 2014–2015 was only 50% of the long-term average, with total rainfall below 1000 mm, leading to severe drought conditions [1]. Drought significantly affects the hydrological regimes, water quality, and aquatic ecosystems of dam reservoirs by prolonging water residence time [2,3], which in turn promotes internal eutrophication [4]. Increased water residence time can enhance internal phosphorus recycling, especially in shallow lakes, thereby accelerating eutrophication [5]. Moreover, changes in inflow and outflow due to reduced precipitation not only influence the hydrological regime of downstream reservoirs and river discharge [6] but also alter nutrient concentrations [7]. Additional drought-related impacts on reservoir environments include intensified stratification [4,8], reduced dissolved oxygen levels, and increased internal nutrient cycling, thereby accelerating eutrophication processes [9]. Furthermore, drought-induced sediment resuspension and anoxic conditions can increase internal phosphorus loading, exacerbating eutrophication and cyanobacterial blooms [10,11]. Supporting this, long-term observations in reservoirs have shown that repeated drawdown and refill cycles under drought conditions can elevate phosphorus, dissolved organic carbon, and iron concentrations, reduce water clarity and oxygen levels, and shift ecosystems toward eutrophic states with increased phytoplankton biovolume [12].

The reservoirs in the Bukhan River system are artificial lakes formed by damming rivers, with expanded water surfaces that are highly susceptible to pollutant inputs from surrounding areas. Most of these reservoirs are undergoing eutrophication, which is further exacerbated by population growth and urbanization in adjacent watersheds [13,14]. Due to the monsoonal climate, where 50–60% of the annual precipitation occurs in summer, hydrological conditions (e.g., water residence time) frequently fluctuate between upstream and downstream sections of these reservoirs [15,16]. In the Bukhan River system, upstream dams (e.g., Hwacheon, Chuncheon, and Soyang dams) drain into Uiam Lake, which subsequently connects to downstream reservoirs, such as Cheongpyeong and Paldang Lakes. Consequently, variation in upstream dam discharges directly affects the water residence time and water quality of downstream reservoirs, contributing to nutrient loading and phytoplankton blooms [17,18]. A long-term study of the Andong Reservoir demonstrated that droughts, particularly reduced summer precipitation, increased water residence time, and total phosphorus concentrations, resulted in elevated chlorophyll-*a* levels and eutrophication risk [19].

In this interconnected dam reservoir ecosystem, environmental changes induced by severe drought not only degrade water quality but also alter freshwater phytoplankton community composition [20,21]. Prolonged droughts, especially when coupled with elevated water temperatures, can trigger extensive cyanobacterial blooms, including harmful species that thrive under warm, nutrient-enriched conditions with extended residence time [22–25]. A long-term study of the Daecheong Reservoir in South Korea reported that years with reduced monsoon rainfall showed increased dominance of cyanobacteria such as *Microcystis* and *Anabaena*, linked to elevated water temperatures and phosphorus levels [26]. These findings highlight the potential for drought-driven phytoplankton regime shifts under climate stress. In addition, cyanobacterial outbreaks in these reservoirs can produce odor-causing compounds, such as 2-methylisoborneol and geosmin, which adversely affect the water supply, particularly in Paldang Lake, South Korea's largest water source [27–29].

Therefore, this study investigated the effects of drought from 2013 to 2016, particularly during the low precipitation period of 2014–2015, on water quality and phytoplankton communities in the interconnected dam reservoirs (Uiam, Cheongpyeong, Sambong-ri, and Paldang Lakes) of the Bukhan River system. The objective was to elucidate how prolonged water residence times and changes in physicochemical parameters influence reservoir ecosystems, thereby informing integrated water quality management and ecosystem conservation strategies during drought conditions.

2. Materials and Methods

2.1. Study Area

The creation of Lake Uiam (St. 1) in August 1967 followed the construction of Uiam Dam, a multipurpose facility with a hydroelectric capacity of 45 MW. The dam has a height of 23 m, an embankment length of 273 m, and a total storage capacity of 80 million cubic meters, serving as a reservoir with a basin area of approximately 282.8 km^2.

Lake Cheongpyeong (St. 2) was constructed in 1943 for hydroelectric power generation with a capacity of 80 MW. This small artificial reservoir, situated in the lower Bukhan River basin, has a water surface area of 17.6 km^2 and a catchment area of 9921 km^2. The reservoir exhibits relatively stable water levels throughout the year. The corresponding dam stands 31 m tall, with an embankment length of 470 m and a total storage capacity of 185 million cubic meters.

The Namhan, Bukhan, and Gyeongancheon Rivers converge to form Lake Paldang (St 4), the man-made lake that supports recreation, river upkeep, hydropower production, and water delivery. The Namhan, Bukhan, and Gyeongancheon Rivers provide 60%, 37%, and 3% of the inflow, respectively, into the about 23,800 km^2 lake basin. The lake's high basin-to-reservoir area ratio of 652 and surface area of 36.5 km^2 make it especially susceptible to pollutant intrusions. Furthermore, Lake Paldang provides around 80% of the drinking water for the Seoul Metropolitan Area, making it a vital regional water source for the city.

Since the late 1990s, the Bukhan River system has been subject to progressively stringent water quality regulations, as it serves as a major drinking water source for the Seoul metropolitan area. In Cheongpyeong and Soyang Lakes, all cage aquaculture facilities were removed by 1999 under the Ministry of Environment's directives. In Uiam Lake, fishery and aquaculture activities were also restricted, with full implementation completed by 2016. Lake Paldang, designated as a water source protection zone in 1989, has undergone a series of comprehensive pollution control procedures. The Ministry of Environment launched a full-scale water quality improvement plan in 1998, which included removing fish farms and implementing a Total Maximum Daily Load system. By 2005, infrastructure such as sewage treatment and livestock wastewater facilities were expanded, and fishing activities were further restricted. Consequently, anthropogenic influences on the reservoir system have been substantially reduced.

The following were the study locations: the front of Uiam Dam (St. 1), the front of Cheongpyeong Dam (St. 2), the Bukhan River confluence area at Sambong-ri (St. 3), and the front of Paldang Dam (St. 4) (Figure 1).

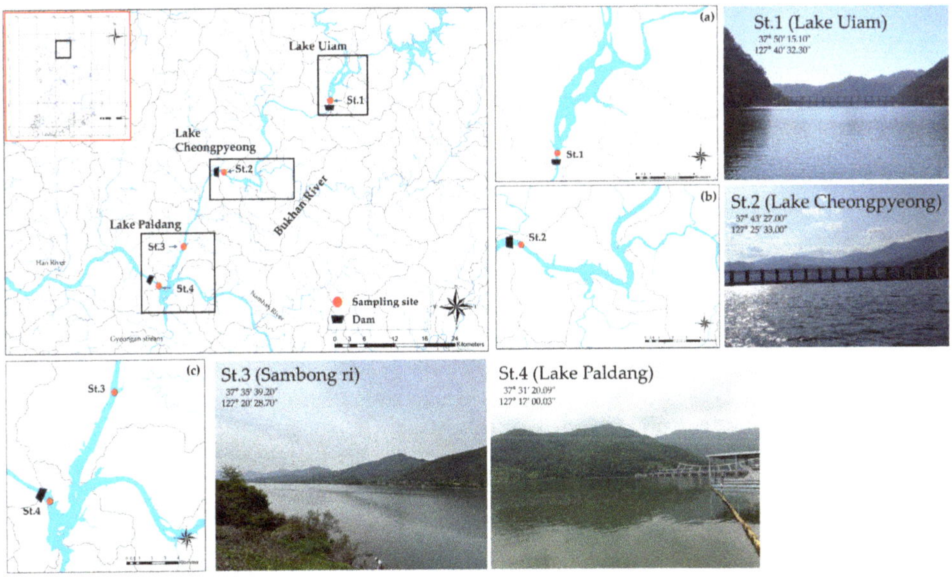

Figure 1. Locations of study sites in the Bukhan River, South Korea((**a**): Lake Uiam, (**b**): Lake Cheongpyeong, (**c**): Lake Paldang).

2.2. Data Sources for Precipitation and Hydrological Factors

Precipitation data were obtained from the Korea Meteorological Administration (KMA) for stations located within the catchment areas of Lakes Paldang and Sambong-ri (Yangpyeong Meteorological Station), Lake Cheongpyeong (Hongcheon Meteorological Station), and Lake Uiam (Chuncheon Meteorological Station) (http://www.kma.go.kr, accessed on 4 September 2023). Using cumulative or average values, daily hydrological data, including inflow, outflow, and water levels, were retrieved from the Korea Water Resources Management Information System (WAMIS, http://www.wamis.go.kr, accessed on 4 September 2023). All collected baseline data were thoroughly examined for missing values and outliers on a variable-specific basis, and only the validated data were used for subsequent analyses.

2.3. Drought Condition Assessment Using Standardized Precipitation Index (SPI)

In this study, the Standardized Precipitation Index (SPI) was used to assess drought conditions in South Korea from 2013 to 2016. SPI is a widely used index that standardizes precipitation variability over a given period to quantify drought occurrence and duration. For this analysis, SPI-6 (6-month accumulation) and SPI-12 (12-month accumulation) were applied to assess seasonal and long-term drought conditions, respectively. SPI-6 captures seasonal fluctuations and medium-term drought patterns, while SPI-12 evaluates long-term precipitation deficits and prolonged drought events.

Drought severity was classified based on the criteria proposed by McKee, et al. [30], where an SPI value of -1.0 or above indicates normal conditions, values between -1.0 and -1.5 indicate moderate drought, values between -1.5 and -2.0 indicate severe drought, and values below -2.0 denote extreme drought. The SPI analysis identified a persistent drought period from late 2014 to 2015. According to SPI-6, drought conditions intensified between May and July 2014, with SPI values decreasing from -1.05 to -2.40, marking the onset of severe drought. Furthermore, from July to October 2015, SPI-6 ranged between -1.78 and -2.78, indicating sustained severe to extreme drought conditions. The SPI-12 analysis also indicated a prolonged drought, with values reaching -1.91 in July 2014, followed by persistent values below -2.0 throughout 2015, classifying this period as an extreme long-term drought (Table S1).

Based on these analyses, the period from 2014 to 2015 was characterized by sustained drought conditions in South Korea, with 2015 exhibiting extreme drought severity according to SPI-12. This study utilized SPI-based classification to determine that 2014–2015 was a period of significant meteorological drought.

2.4. Sample Collection and Water Quality Analysis

At the study sites, water samples were taken every month from January 2013 to December 2016 for phytoplankton and environmental factor analysis. To prevent light-induced reactions in the water samples, brown glass bottles were used during the boat sampling. After collection, the samples were promptly chilled and taken to the lab for examination.

Using a digital multi-parameter sensor (YSI-EXO, YSI Inc., Yellow Springs, OH, USA), the following parameters were monitored in situ: water temperature, pH, dissolved oxygen (DO), and electrical conductivity (EC). The biological oxygen demand (BOD), chemical oxygen demand (COD), total nitrogen (TN), nitrate (NO_3-N), ammonia (NH_3-N), chlorophyll-*a* (Chl-*a*) concentration, total phosphorus (TP), phosphate (PO_4-P), suspended solids (SS), and total organic carbon (TOC) were all analyzed in the laboratory following standard procedures described in the Korea Water Quality Standard [31].

2.5. Korean Trophic State Index

The Korean Trophic State Index (TSI_{KO}), developed by the Republic of Korea's Ministry of Environment, was used to assess the trophic state of the Bukhan River system. The eutrophication evaluation index known as TSI_{KO} takes into account the water quality characteristics of South Korea's largest lakes and reservoirs. It makes use of metrics like TP and Chl-*a* as indicators of autochthonous organic matter and TOC as an indicator of allochthonous organic matter [32]. The formulas generated from water quality measurements in domestic lakes and reservoirs are represented by Equations (1) through (3).

$$TSI_{KO}(TOC) = 17.9 + 64.4\log(TOC\ mg/L) \tag{1}$$

$$TSI_{KO}(TP) = 114.6 + 43.3\log(TP\ mg/L) \tag{2}$$

$$TSI_{KO}(Chl\text{-}a) = 17.9 + 38.6\log(Chl\text{-}a\ mg/L) \tag{3}$$

Equation (4) was used to construct TSI_{KO}, with a 50% weighting factor for TOC and a 25% weighting factor for Chl-*a* and TP. Oligotrophic (≤30), mesotrophic (31–50), eutrophic (51–70), and hypertrophic (>71) were the four trophic states into which the computed TSI_{KO} values were divided.

$$TSI_{KO} = 0.5\ TSI_{KO}(TOC) + 0.25\ TSI_{KO}(TP) + 0.25\ TSI_{KO}(Chl\text{-}a) \tag{4}$$

2.6. Phytoplankton Analysis

To preserve the water samples for the analysis of dominant species and phytoplankton community composition, Lugol's solution was added to achieve a final concentration of 2%. A 1 mL aliquot was subsequently placed in a Sedgwick-Rafter counting chamber and observed under a phase-contrast microscope (Eclipse 80i; Nikon, Tokyo, Japan) at magnifications of ×100 to ×400. For species identification requiring higher resolution, microscopic examination was conducted at ×1000 magnification using a slide glass. Species identification and cell density estimation were conducted with reference to John, et al. [33], Hirose, et al. [34], Krammer [35], Krammer and Lange-Bertalot [36–38].

2.7. Statistical Analysis

A one-way analysis of variance (ANOVA) was used to analyze annual variations in physicochemical factors in the Bukhan River system, and post hoc Duncan's test was applied to identify significant differences across years. Spearman's correlation analysis was performed to examine the relationships between environmental variables and phytoplankton. Statistical analyses were conducted using SPSS 20 (IBM Corp., Armonk, NY, USA), with significance level set at $p < 0.05$. Canonical Correspondence Analysis (CCA) was conducted using PC-ORD5 (Lincoln City, OR, USA) to examine the relationship between phytoplankton communities and environmental factors. Prior to analysis, taxa present in less than 0.4% of total samples were excluded, and all environmental data were normalized using log(1 + x) transformation [39,40].

3. Results and Discussion

3.1. Precipitation and Hydrological Characteristics

At the study sites, the average annual precipitation was 1575 mm in 2013, 723 mm in 2014, 766 mm in 2015, and 1079 mm in 2016, with the 2014–2015 period receiving only 45.9% of the 2013 precipitation. Notably, summer precipitation in 2016 reached only 40.5% of the 2013 level, a critical reduction in a monsoonal climate where summer rainfall typically accounts for 50–60% of the annual total, thereby leading to severe drought conditions. In Uiam Lake, located in the mid-upper reaches of the Bukhan River system, the average annual precipitation in 2014–2015 (716 mm) was only 41.2% of that recorded in 2013 (1739 mm), with summer precipitation at merely 29.3%, indicating extreme drought conditions (Figure 2). The reduced water levels observed in the summer of 2014 were attributed to preemptive water releases in anticipation of the monsoon, and the subsequent prolonged drought prevented the recovery of water levels, reducing both inflow and outflow to approximately 30% of normal values. In contrast, Paldang Lake, which serves as a primary water source, maintained a constant full water level with minimal fluctuations due to continuous discharges from the upstream dams of the Bukhan and Namhan Rivers (Figure 2).

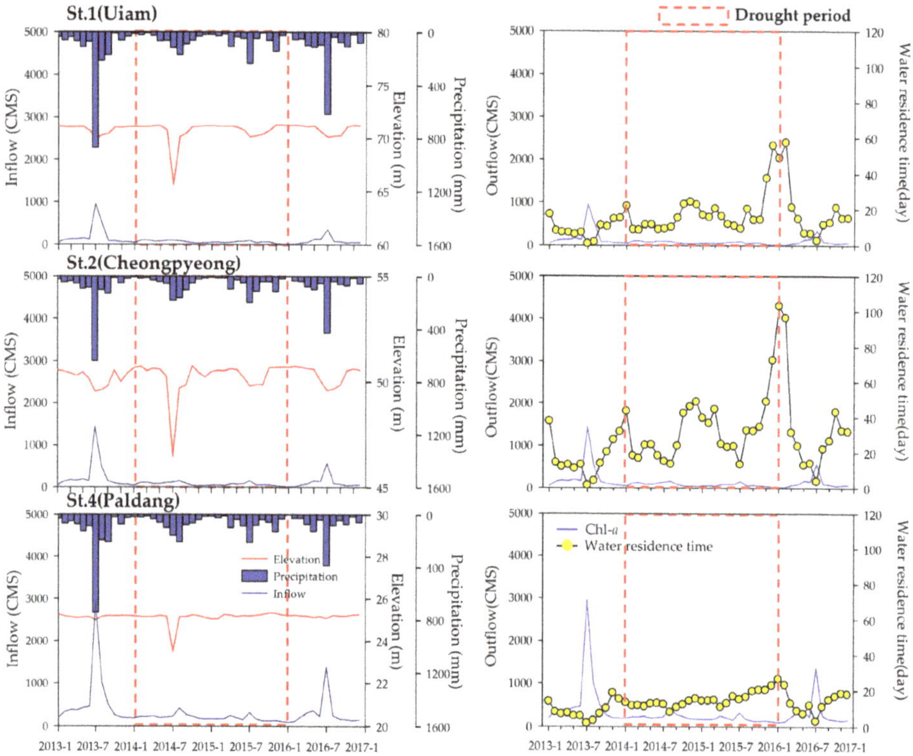

Figure 2. Annual hydrological factors in Bukhan River.

Water residence times in Uiam Lake, Cheongpyeong Lake, and Paldang Lake were approximately 2.0, 1.9, and 1.7 times higher than normal values, respectively, with differences being statistically significant ($p < 0.05$) (Figures 2 and 3). During the summer, residence times increased by more than threefold, underscoring the pronounced impact of the drought. Furthermore, as the drought continued, water residence times increased sharply during the winter (October 2015–February 2016). Although prolonged drought in dam reservoirs generally exacerbates downstream drought conditions [6,41], in the Bukhan River system, the downstream Paldang Lake, serving as a key water supply for the metropolitan area, maintained a normal high water level, suggesting that the effects of drought were more pronounced in the upstream dam reservoirs.

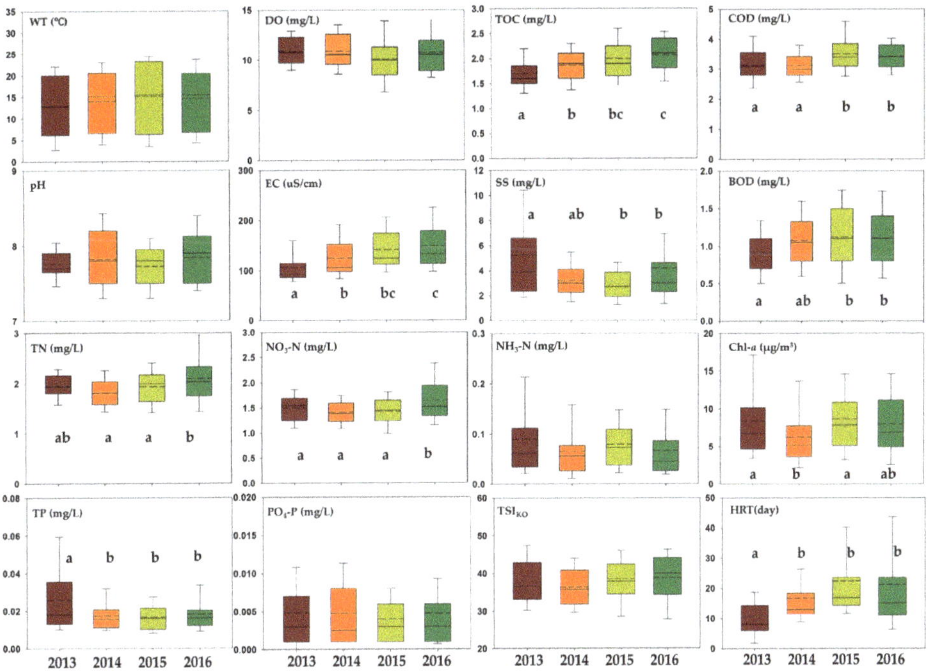

Figure 3. Changes in environmental variables over time in Bukhan River. There were significant differences among years, as determined using ANOVA. Different small letters (a, b, and c) indicate significant differences determined using Duncan's test (dashed line: average, straight line: median).

3.2. Water Quality Characteristics

Environmental parameters at the sites are presented in Figure 4 as monthly averages. In 2014, when drought conditions were severe, nutrient concentrations (e.g., TN and TP) and Chl-*a* decreased temporarily, resulting in improved water quality. However, as the drought persisted beyond 2015, indicators such as TOC, COD, BOD, and TN exhibited a deteriorating trend. TN, for example, showed significant reductions during the extreme drought period of 2014–2015 ($p < 0.05$) but increased again as precipitation recovered in 2016. This pattern suggests that TP is more influenced by pollutant inputs from rainfall than by water levels [42–44] (Figures 3 and 4). Its rapid post-drought increase may be attributed to both pollutant runoff from surrounding areas and nutrient leaching resulting from drying–reflooding cycles [45,46].

TOC and COD also displayed statistically significant differences during the drought period ($p < 0.05$) (Figures 3 and 4). Specifically, episodic rainfall following drought can lead to a sudden increase in TOC due to the flush of accumulated organic matter from surrounding soils [47], while reduced water levels during drought can result in higher concentrations of organic matter, thereby elevating TOC [48]. The TSI_{KO} (Korean Trophic State Index) also tended to deteriorate significantly between years ($p < 0.05$) (Figures 3 and 4). Notably, at downstream sites St. 3 (Sambong-ri) and St. 4 (Paldang Lake), TOC, Chl-*a*, and TP levels were high, whereas TN concentrations were low. This is attributed to the increased water residence time in downstream reservoirs during prolonged drought, which enhances internal nutrient recycling [49,50]. The increased residence time also promotes stratification, leading to enhanced algal growth and denitrification, which significantly reduces TN concentrations during drought ($p < 0.05$). Furthermore, decreased DO in the hypolimnion due to stratification can increase nutrient release from sediments, elevating TP concentrations and further exacerbating eutrophication [51]. These changes likely contribute to the development of algal blooms in downstream dam reservoirs.

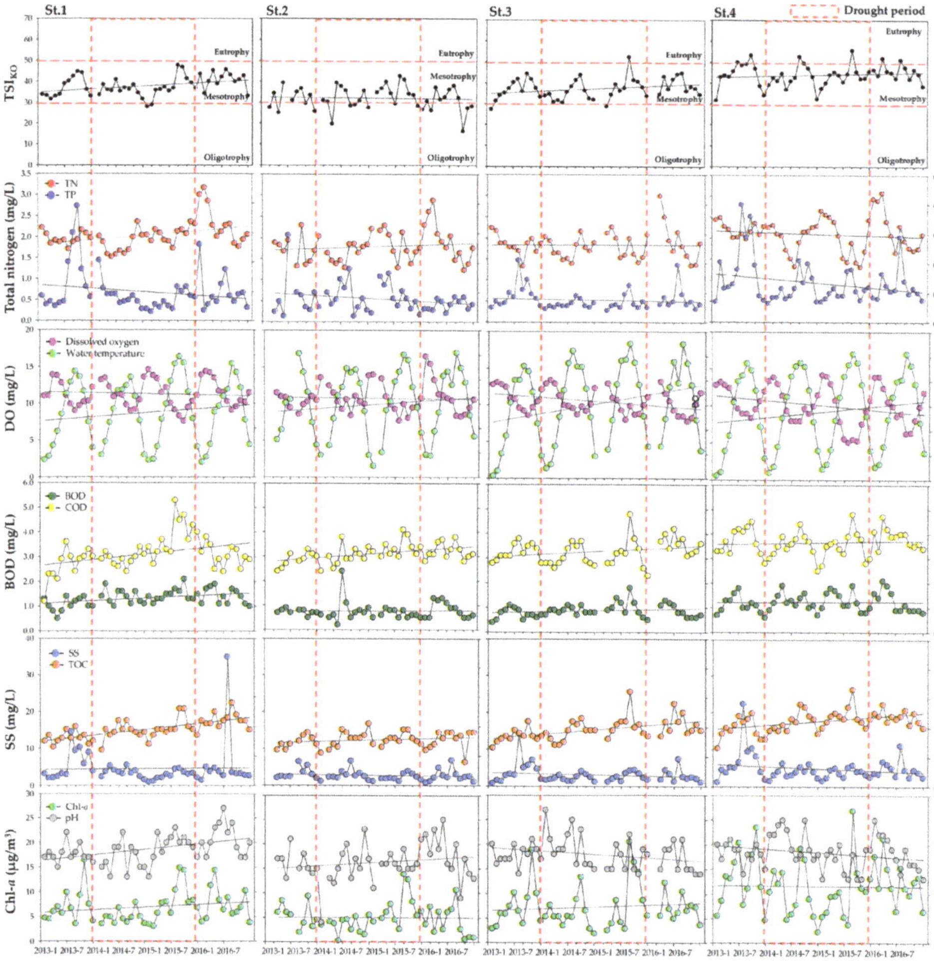

Figure 4. Annual physicochemical factors in Bukhan River.

3.3. Annual Variation in Phytoplankton Abundance and Characteristics

During the investigation period, phytoplankton cell densities at the sampling sites were as follows: St. 1, 480–5523 cells/mL (average 3502 cells/mL); St. 2, 450–7362 cells/mL (2389 cells/mL); St. 3, 520–7521 cells/mL (3086 cells/mL); and St. 4, 690–21,541 cells/mL (5260 cells/mL). The density was highest at the downstream site St. 4. The mean phytoplankton cell density for the entire Bukhan River system peaked in 2016 (3775 cells/mL), with values of 3418, 3527, and 3657 cells/mL in 2013, 2014, and 2015, respectively, with no significant differences among the years (Figure 5).

Regarding the overall relative abundance of phytoplankton groups, the Bukhan River system was dominated by Bacillariophyta, accounting for 64.7% of the total population at all sites, followed by Cryptophyta (17.8%), Cyanophyta (6.7%), Chlorophyta (5.9%), Ochrophyta (3.9%), Charophyta (0.7%), Dinophyta (0.3%), and Euglenophyta (0.02%) (Figure 5).

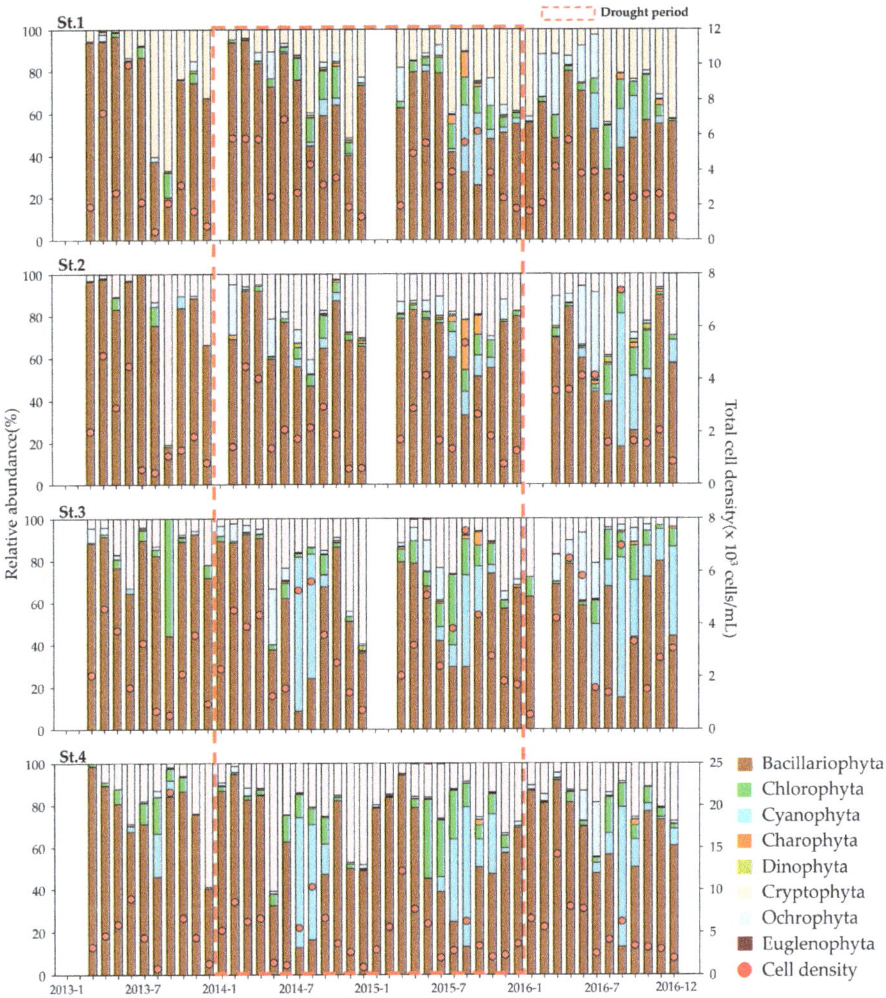

Figure 5. Annual variation in phytoplankton relative abundance and cell density in Bukhan River.

Comparisons between the reference years (2013 and 2016) and drought years (2014–2015) revealed that while the relative abundances of phytoplankton groups at the upstream sites (St. 1 and St. 2) remained relatively stable, the confluence sites (St. 3 and St. 4) exhibited an increase of over 5% in the relative abundance of Cryptophyta. *Rhodomonas*, a representative taxon within Cryptophyta, is known for its high adaptability to low water temperatures [52,53] and its insensitivity to low nutrient concentrations and light levels, likely explaining its dominance in environments with prolonged water residence times [54] (Table 1). Moreover, during the summer months (June–August), while little change was observed at upstream sites (St. 1 and St. 2), the relative dominance of Cyanophyta at the downstream sites St. 3 and St. 4 increased sharply to 14.0% and 21.2%, respectively (Figure 6). This suggests that discharge from upstream dams during drought conditions, which increases nutrient concentrations in downstream reservoirs, plays a key role in promoting cyanobacterial blooms, potentially intensifying summer cyanobacterial outbreaks [24,55].

Table 1. List of major phytoplankton species in Bukhan River.

Taxa	Species	Flood	Drought		Recovery
		2013	2014	2015	2016
Bacillariophyta	*Stephanodiscus hantzschii*	*	*	*	*
Bacillariophyta	*Asterionella formosa*	*	*	*	*
Bacillariophyta	*Fragilaria crotonensis*	*	*	*	*
Bacillariophyta	*Aulacoseira granulata* var. *angustissima*	*	*	*	*
Bacillariophyta	*Cyclotella atomus*	*	*	*	*
Bacillariophyta	*Ulnaria acus*	*	*	*	*
Bacillariophyta	*Skeletonema potamos*	*		*	*
Bacillariophyta	*Ulnaria delicatissima*	*	*	*	*
Bacillariophyta	*Aulacoseira granulata*	*	*	*	*
Bacillariophyta	*Stephanocyclus meneghinianus*	*	*	*	*
Bacillariophyta	*Urosolenia longiseta*	*	*	*	*
Bacillariophyta	*Discostella stelligera*	*	*	*	*
Bacillariophyta	*Pantocsekiella ocellata*	*	*	*	*
Bacillariophyta	*Aulacoseira ambigua*	*	*	*	*
Bacillariophyta	*Aulacoseira distans*	*	*	*	*
Charophyta	*Mougeotia* sp.	*	*	*	*
Chlorophyta	*Scenedesmus quadricauda*	*	*	*	*
Chlorophyta	*Micractinium pusillum*	*	*	*	*
Chlorophyta	*Chlamydomonas* sp.	*	*	*	*
Cryptophyta	*Rhodomonas* sp.	*	*	*	*
Cryptophyta	*Cryptomonas ovata*	*	*	*	*
Cryptophyta	*Cryptomonas* sp.	*	*	*	*
Cyanophyta	*Merismopedia tenuissima*		*	*	*
Cyanophyta	*Pseudanabaena* sp.	*	*	*	*
Cyanophyta	*Dolichospermum spiroides*			*	*
Cyanophyta	*Microcystis aeruginosa*	*	*	*	*
Cyanophyta	*Dolichospermum circinale*	*	*	*	*
Ochrophyta	*Dinobryon divergens*	*	*	*	*
Ochrophyta	*Dinobryon sertularia*	*	*	*	*

Note: * indicates the presence of the species; species that appeared at <0.4% were excluded.

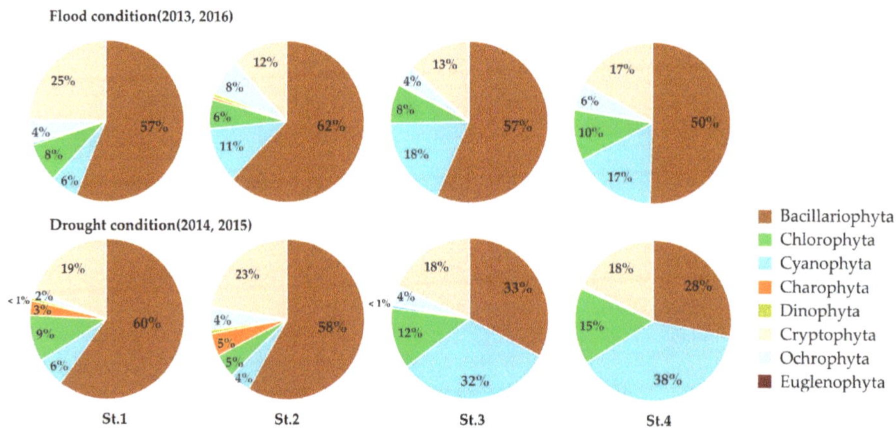

Figure 6. Annual changes in phytoplankton communities and relative abundances in the Bukhan River during summer (June–August) under flood conditions (2013, 2016) and drought conditions (2014, 2015).

3.4. CCA of the Phytoplankton Community and Environmental Factors

CCA was performed at both the group and species levels to evaluate the relationships between environmental factors and phytoplankton communities in the Bukhan River system. Axis 1 and Axis 2 explained 12.7% and 3.9% of the variation in the relationships between phytoplankton and environmental factors, respectively (Figure 7).

CCA results revealed that Bacillariophyta, including *Ulnaria* spp., *Asterionella formosa*, and *Fragilaria crotonensis*, were negatively correlated with water temperature, TP, and TOC while showing positive correlations with water residence time (Figure 7). *A. formosa* and *Ulnaria* spp., which are typically abundant in temperate lakes during the spring [56,57], demonstrate a relatively high tolerance to low water temperatures, an adaptation that favors their growth during cooler periods [58–60]. *Fragilaria crotonensis* is frequently observed under low-nutrient and silicate-deficient conditions, suggesting its competitive advantage in oligotrophic environments [61,62] (Table 1).

Cryptomonas spp. and *Rhodomonas* sp. were positively correlated with Chl-*a* and nutrient concentrations. Correlation analyses showed significant positive relationships between these taxa and TN ($r = 0.334$, $p < 0.01$), TP ($r = 0.293$, $p < 0.01$), and Chl-*a* ($r = 0.347$, $p < 0.01$), indicating their contribution to increased Chl-*a* concentrations in upstream reservoirs during spring, autumn, and drought periods (Table 2). Although *Rhodomonas* and *Cryptomonas* are abundant in the spring under low water temperatures [63,64], their ability to thrive under low temperatures, limited light, and low nutrient availability [54,65–67] suggest that they have high adaptability and can maintain year-round dominance despite environmental fluctuations due to upstream dam releases and summer monsoon events.

Cyanophyta (*Dolichospermum*, *Microcystis*, *Pseudanabaena*, and *Merismopedia*) abundances were positively correlated with water temperature, TOC, TP, precipitation, and TSI_{KO}, indicating that nutrient inputs from precipitation promote cyanobacterial dominance. A correlation analysis further revealed strong positive associations between Cyanophyta and TOC, COD, and TSI_{KO}, as well as a positive relationship with precipitation. In addition, precipitation showed a significant positive correlation with TP ($r = 0.308$, $p < 0.01$), suggesting that reduced rainfall during drought conditions enhances nutrient inputs and stimulates cyanobacterial blooms in downstream reservoirs (Table 2).

These findings imply that limited precipitation during summer droughts increases nutrient inputs in downstream dam reservoirs, favoring the proliferation of cyanobacteria. Buoyancy regulation enables cyanobacteria to thrive under favorable surface conditions, such as higher temperatures, light availability, and enhanced nutrient uptake, providing a competitive advantage over other algal groups [68–71]. Furthermore, the buoyancy control in cyanobacteria facilitates their vertical migration in the water column, allowing them to remain near the surface where optimal conditions for growth and photosynthesis are maintained while avoiding excessive light exposure, which further enhances nutrient uptake [72–74].

In particular, *Dolichospermum circinale* and *Microcystis aeruginosa*, which are managed as harmful algae in Korea, showed positive correlations with the water residence time. This suggests that prolonged drought conditions promote the proliferation of toxic cyanobacteria in downstream reservoirs, such as Paldang Lake (Table 1, Figure 7). These results are consistent with previous findings showing that extended drought enhances the growth of toxic cyanobacterial species [24,75]. Moreover, the increased abundance of *Pseudanabaena* sp. during extended drought may trigger episodes of musty odor events, adversely affecting water supply management in downstream regions [27–29].

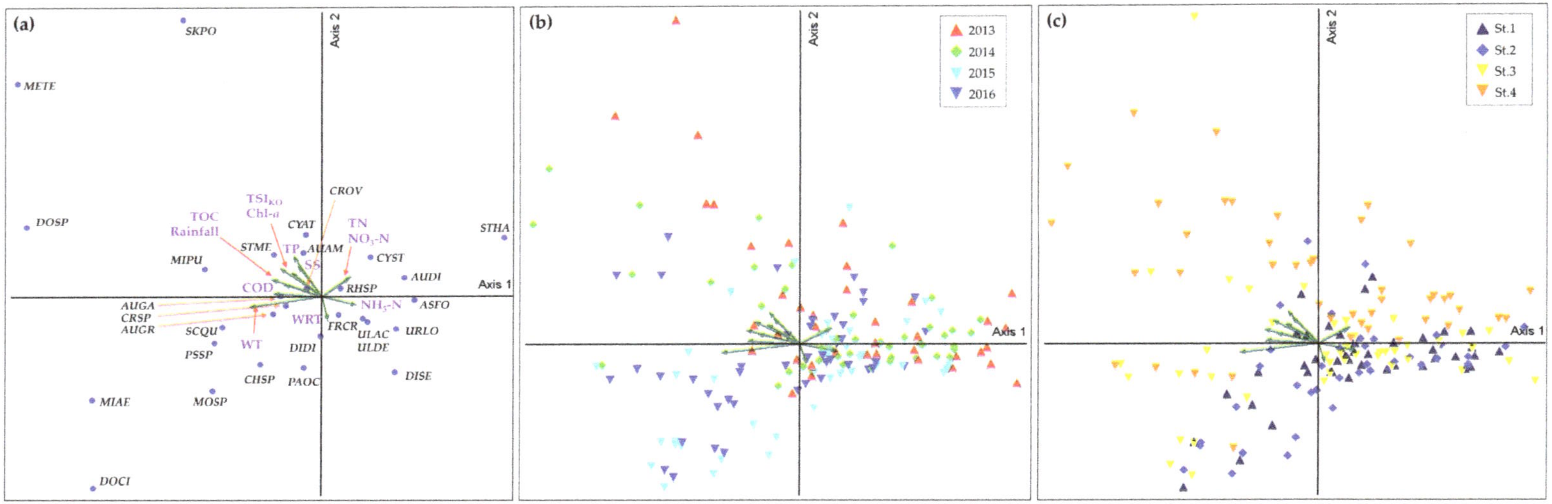

STHA	*Stephanodiscus hantzschii*	ULDE	*Ulnaria delicatissima*	URLO	*Urosolenia longiseta*	PAOC	*Pantocsekiella ocellata*
RHSP	*Rhodomonas* sp.	METE	*Merismopedia tenuissima*	DIST	*Discostella stelligera*	AUAM	*Aulacoseira ambigua*
ASFO	*Asterionella formosa*	DIDI	*Dinobryon divergens*	MIAE	*Microcystis aeruginosa*	DOCI	*Dolichospermum circinale*
FRCR	*Fragilaria crotonensis*	PSSP	*Pseudanabaena* sp.	CRSP	*Cryptomonas* sp.	AUDI	*Aulacoseira distans*
AUGA	*Aulacoseira granulata* var. *angustissima*	CROV	*Cryptomonas ovata*	SCQU	*Scenedesmus quadricauda*	CHSP	*Chlamydomonas* sp.
CYAT	*Cyclotella atomus*	DOSP	*Dolichospermum spiroides*	MIPU	*Micractinium pusillum*		
ULAC	*Ulnaria acus*	AUGR	*Aulacoseira granulata*	MOSP	*Mougeotia* sp.		
SKPO	*Skeletonema potamos*	STME	*Stephanocyclus meneghinianus*	DISE	*Dinobryon sertularia*		

Figure 7. Annual variation in canonical correspondence analysis (CCA) results for relationships between environmental parameters and phytoplankton in Bukhan River (**a**). Ordination diagrams showing differences with respect to year (**b**) and study site (**c**).

Table 2. Correlation coefficients for relationships between various taxa and environmental factors in Bukhan River.

Taxa	WT	DO	BOD	COD	SS	TOC	pH	EC	TN	NH_3-N	NO_3-N	TP	PO_4-P	Chl-*a*	TSI_{KO}	Rainfall	WRT
Bacillariophyta	−0.275 **	0.383 **	0.380 **	0.109	0.209 **	0.076	0.208 **	0.096	0.354 **	0.298 **	0.325 **	0.087	−0.120	0.455 **	0.251 **	−0.134	−0.267 **
Cyanobacteria	0.594 **	−0.555 **	0.096	0.440 **	0.090	0.433 **	−0.065	0.207 **	−0.224 **	−0.184 *	−0.220 **	0.115	0.114	0.216 **	0.318 **	0.201 **	0.090
Chlorophyta	0.601 **	−0.510 **	0.374 **	0.432 **	0.361 **	0.518 **	0.101	0.199 **	−0.013	−0.062	−0.112	0.253 **	0.124	0.398 **	0.502 **	0.396 **	−0.177 *
Charophyta	0.245 **	−0.166 *	0.122	0.228 **	−0.049	0.200 **	0.013	0.228 **	−0.021	−0.083	0.022	−0.141	−0.037	0.109	0.079	−0.049	0.148
Dinophyta	0.331 **	−0.217 **	0.327 **	0.181 *	0.171 *	0.157 *	0.150	0.167 *	0.029	−0.079	−0.055	0.035	0.134	0.165 *	0.151 *	0.150 *	0.066
Ochrophyta	−0.074	0.200 **	0.235 **	0.004	−0.091	0.060	0.256 **	0.117	0.169 *	0.173 *	0.196 *	−0.109	−0.025	0.041	0.004	−0.052	−0.023
Cryptophyta	0.166 *	−0.148	0.477 **	0.340 **	0.255 **	0.446 **	0.198 **	0.350 **	0.334 **	0.033	0.116	0.293 **	0.085	0.347 **	0.468 **	0.120	0.006
Euglenophyta	0.007	0.029	0.149	0.002	−0.049	0.058	−0.035	0.053	0.112	0.075	−0.017	−0.032	0.082	0.024	0.022	0.115	−0.031

Note(s): * $p < 0.05$, ** $p < 0.01$.

4. Conclusions

Despite being the largest watershed and water source in Korea, studies on aquatic ecosystem changes related to climate change in the Han River basin are scarce, despite its importance. This study comprehensively analyzed the impacts of severe drought conditions in the Bukhan River system on hydrological characteristics, water quality, and phytoplankton communities. During the drought, a marked reduction in precipitation led to declines in water levels as well as inflow and outflow, resulting in prolonged water residence times in upstream dam reservoirs and enhanced internal nutrient recycling. Initially, water quality improved slightly due to temporary decreases in nutrient concentrations and Chl-*a* levels; however, as the drought continued, water quality deteriorated, as evidenced by increased levels of TOC, COD, BOD, and TN. These changes could be attributed to the large-scale flush of organic matter accumulated in soils following rainfall after a drought and concentration in the water body during drought periods.

There were significant differences in the distribution and dominant phytoplankton species between upstream and downstream sites, with downstream areas exhibiting a pronounced dominance of cyanobacteria. Moreover, an increase in harmful cyanobacterial species was observed, suggesting that changes in upstream dam discharges and prolonged water residence times in downstream reservoirs contribute significantly to nutrient enrichment and subsequent harmful algal blooms.

In conclusion, this study demonstrates that severe drought conditions extend beyond a mere meteorological phenomenon to cause water quality degradation and ecosystem alterations, with substantial implications for the management of major water sources. Consequently, the development of integrated strategies that consider the multifaceted environmental changes and their interactions during drought conditions is essential for effective water resource management and conservation.

Supplementary Materials: The following supporting information can be downloaded at: https://www.mdpi.com/article/10.3390/w17071023/s1, Table S1: Annual Standardized Precipitation Index (SPI) in Bukhan River (≥ -1.0: Normal conditions, $-1.5 \leq$ SPI < -1.0: Moderate drought, $-2.0 \leq$ SPI < -1.5: Severe drought, SPI < -2.0: Extreme drought).

Author Contributions: Conceptualization, Y.B.S., J.K.I. and S.-J.H.; methodology, Y.B.S. and J.K.I.; software, Y.B.S. and J.K.I.; validation, Y.B.S. and J.K.I.; formal analysis, Y.B.S. and J.K.I.; investigation, Y.B.S., J.K.I., J.H.B., C.H.P. and S.-J.H.; resources, J.K.I., Y.B.S. and S.-J.H.; data curation, Y.B.S. and J.K.I.; writing—original draft preparation, Y.B.S. and J.K.I.; writing—review and editing, Y.B.S., J.K.I., J.H.B., C.H.P. and S.-J.H.; visualization, Y.B.S., J.K.I. and J.H.B.; supervision, J.K.I. and S.-J.H.; project administration, J.K.I. and S.-J.H.; funding acquisition, J.K.I. and S.-J.H. All authors have read and agreed to the published version of the manuscript.

Funding: This work was supported by the Korea Environment Industry & Technology Institute (KEITI) through the "Development of the polluted sediment dredging technology containing Akinete equipped with a state-of-the-art ultra-short baseline acoustic positioning ROV robot Program" (2480000243) and the National Institute of Environmental Research (NIER) [grant number NIER-2024-01-01-053] funded by the Korea Ministry of Environment (MOE).

Data Availability Statement: The data presented in this study are available on request from the corresponding authors.

Conflicts of Interest: The authors declare no conflicts of interest.

Abbreviations

The following abbreviations are used in this manuscript:

BOD	Biological oxygen demand
CCA	Canonical correspondence analysis
Chl-*a*	Chlorophyll-*a*
COD	Chemical oxygen demand
TOC	Total organic carbon
EC	Electrical conductivity
DO	Dissolved oxygen
TN	Total nitrogen
NO_3-N	Nitrate
NH_3-N	Ammonia
TP	Total phosphorus
PO_4-P	Phosphate
TSI_{KO}	Korean Trophic State Index
SS	Suspended solids
WRT	Water residence time

References

1. Mun, Y.-S.; Nam, W.-H.; Jeon, M.-G.; Kim, T.; Hong, E.-M.; Hayes, M.J.; Tsegaye, T. Application of meteorological drought index using Climate Hazards Group InfraRed Precipitation with Station (CHIRPS) based on global satellite-assisted precipitation products in Korea. *J. Korean Soc. Agric. Eng.* **2019**, *61*, 1–11.
2. Mosley, L.M. Drought impacts on the water quality of freshwater systems; review and integration. *Earth Sci. Rev.* **2015**, *140*, 203–214. [CrossRef]
3. Lake, P.S. *Drought and Aquatic Ecosystems: Effects and Responses*; John Wiley & Sons: Hoboken, NJ, USA, 2011.
4. Srifa, A.; Phlips, E.J.; Cichra, M.F.; Hendrickson, J.C. Phytoplankton dynamics in a subtropical lake dominated by cyanobacteria: Cyanobacteria 'Like it Hot' and sometimes dry. *Aquat. Ecol.* **2016**, *50*, 163–174.
5. Jeppesen, E.; Søndergaard, M.; Jensen, J.P.; Havens, K.E.; Anneville, O.; Carvalho, L.; Coveney, M.F.; Deneke, R.; Dokulil, M.T.; Foy, B. Lake responses to reduced nutrient loading–an analysis of contemporary long-term data from 35 case studies. *Freshw. Biol.* **2005**, *50*, 1747–1771.
6. Marcinkowski, P.; Grygoruk, M. Long-term downstream effects of a dam on a lowland river flow regime: Case study of the Upper Narew. *Water* **2017**, *9*, 783. [CrossRef]
7. Ingole, N.P.; An, K.-G. Modifications of nutrient regime, chlorophyll-a, and trophic state relations in Daechung Reservoir after the construction of an upper dam. *J. Ecol. Environ.* **2016**, *40*, 5.
8. Olsson, F.; Mackay, E.B.; Moore, T.; Barker, P.; Davies, S.; Hall, R.; Spears, B.; Wilkinson, J.; Jones, I.D. Annual water residence time effects on thermal structure: A potential lake restoration measure? *J. Environ. Manag.* **2022**, *314*, 115082.
9. Rocha, M.A.M.; Barros, M.U.; Costa, A.C.; de Assis de Souza Filho, F.; Lima Neto, I.E. Understanding the water quality dynamics in a large tropical reservoir under hydrological drought conditions. *Water Air Soil Pollut.* **2024**, *235*, 76.
10. Bloesch, J. Mechanisms, measurement and importance of sediment resuspension in lakes. *Mar. Freshw. Res.* **1995**, *46*, 295–304.
11. Orihel, D.M.; Baulch, H.M.; Casson, N.J.; North, R.L.; Parsons, C.T.; Seckar, D.C.; Venkiteswaran, J.J. Internal phosphorus loading in Canadian fresh waters: A critical review and data analysis. *Can. J. Fish. Aquat. Sci.* **2017**, *74*, 2005–2029. [CrossRef]
12. Luong, H.A.; Rohlfs, A.-M.; Facey, J.A.; Colville, A.; Mitrovic, S.M. Long-term study of phytoplankton dynamics in a supply reservoir reveals signs of trophic state shift linked to changes in hydrodynamics associated with flow management and extreme events. *Water Res.* **2024**, *256*, 121547. [PubMed]
13. Kim, B.C.; Park, J.H.; Hwana, G.S.; Choi, K.S. Eutrophication of large freshwater ecosystems in Korea. *Korean J. Limnol.* **1997**, *30*, 512–517.
14. Shin, J.-K.; Kang, C.-K.; Kim, H.-S.; Hwang, S.-J. Limnological characteristics of the river-type Paltang Reservoir, Korea: Hydrological and environmental factors. *Korean J. Ecol. Environ.* **2003**, *36*, 242–256.
15. An, K.-G.; Jones, J.R. Factors regulating bluegreen dominance in a reservoir directly influenced by the Asian monsoon. *Hydrobiologia* **2000**, *432*, 37–48.
16. Jung, S.; Shin, M.; Kim, J.; Eum, J.; Lee, Y.; Lee, J.; Choi, Y.; You, K.; Owen, J.; Kim, B. The effects of Asian summer monsoons on algal blooms in reservoirs. *Inland Waters* **2016**, *6*, 406–413. [CrossRef]

17. Kim, D.; Kim, Y.; Kim, B. Simulation of eutrophication in a reservoir by CE-QUAL-W2 for the evaluation of the importance of point sources and summer monsoon. *Lake Reserv. Manag.* **2019**, *35*, 64–76.
18. Jeon, H.-W.; Choi, J.-W.; An, K.-G. Spatio-temporal water quality variations at various streams of Han-River watershed and empirical models of serial impoundment reservoirs. *Korean J. Ecol. Environ.* **2012**, *45*, 378–391.
19. Mamun, M.; Atique, U.; Kim, J.Y.; An, K.-G. Seasonal water quality and algal responses to monsoon-mediated nutrient enrichment, flow regime, drought, and flood in a drinking water reservoir. *Int. J. Environ. Res. Public Health* **2021**, *18*, 10714. [CrossRef]
20. Beaver, J.R.; Casamatta, D.A.; East, T.L.; Havens, K.E.; Rodusky, A.J.; James, R.T.; Tausz, C.E.; Buccier, K.M. Extreme weather events influence the phytoplankton community structure in a large lowland subtropical lake (Lake Okeechobee, Florida, USA). *Hydrobiologia* **2013**, *709*, 213–226.
21. Williamson, C.E.; Saros, J.E.; Vincent, W.F.; Smol, J.P. Lakes and reservoirs as sentinels, integrators, and regulators of climate change. *Limnol. Oceanogr.* **2009**, *54*, 2273–2282.
22. Jang, M.T.G.; Alcântara, E.; Rodrigues, T.; Park, E.; Ogashawara, I.; Marengo, J.A. Increased chlorophyll-a concentration in Barra Bonita reservoir during extreme drought periods. *Sci. Total Environ.* **2022**, *843*, 157106.
23. Knapp, A.S.; Milewski, A.M. Spatiotemporal relationships of phytoplankton blooms, drought, and rainstorms in freshwater reservoirs. *Water* **2020**, *12*, 404. [CrossRef]
24. Grabowska, M.; Mazur-Marzec, H.; Więcko, A. How Extreme Droughts Change the Impact of Eutrophic Reservoir on Its Outflow, with Special References to Planktonic Cyanobacteria and Their Secondary Metabolites? *Water* **2025**, *17*, 86. [CrossRef]
25. Schindler, D.W. Recent advances in the understanding and management of eutrophication. *Limnol. Oceanogr.* **2006**, *51*, 356–363.
26. Kim, J.Y.; Atique, U.; Mamun, M.; An, K.-G. Long-term interannual and seasonal links between the nutrient regime, sestonic chlorophyll and dominant bluegreen algae under the varying intensity of monsoon precipitation in a drinking water reservoir. *Int. J. Environ. Res. Public Health* **2021**, *18*, 2871. [CrossRef]
27. Youn, S.J.; Im, J.K.; Byeon, M.-S.; Yu, S.J. Characteristics of cyanobacteria and odorous compounds production in lake uiam and lower gonji stream. *J. Korean Soc. Water Environ.* **2019**, *35*, 99–104.
28. Byun, J.-H.; Yu, M.; Lee, E.; Yoo, S.-J.; Kim, B.-H.; Byun, M.-S. Temporal and spatial distribution of microbial community and odor compounds in the Bukhan river system. *Korean J. Ecol. Environ.* **2018**, *51*, 299–310.
29. Byun, J.-H.; Hwang, S.J.; Kim, B.K.; Park, J.R.; Lee, J.K.; Lim, B.J. Relationship between a Dense Population of Cyanobacteria and Odorous Compounds in the North Han River System in 2014 and 2015. *Korean J. Ecol. Environ.* **2015**, *48*, 263–271.
30. McKee, T.B.; Doesken, N.J.; Kleist, J. The relationship of drought frequency and duration to time scales. In Proceedings of the 8th Conference on Applied Climatology, Anaheim, CA, USA, 17–22 January 1993; pp. 179–183.
31. MOE. *Standard Methods for the Examination Water Quality*; The Korean Ministry of Environment: Sejong, Republic of Korea, 2014.
32. Kim, B.; Kong, D. Examination of the Applicability of TOC to Korean Trophic State Index (TSI KO). *J. Korean Soc. Water Environ.* **2019**, *35*, 271–277.
33. John, D.M.; Whitton, B.A.; Brook, A.J. *The Freshwater Algal Flora of the British Isles: An Identification Guide to Freshwater and Terrestrial Algae*; Cambridge University Press: Cambridge, UK, 2002.
34. Hirose, H.; Yamagishi, T.; Akiyama, M. *Illustrations of the Japanese Freshwater Algae*; Uchidarokakuho Publ. Co., Ltd.: Tokyo, Japan, 1977.
35. Krammer, K.; Lange-Bertalot, H. *Bacillariophyceae. Naviculaceae. Süßwasserflora von Mitteleuropa*; Gustav Fischer Verlag: Stuttgart, Germany, 1986; Volume 1, p. 876. Available online: http://link.springer.com/book/9783827426154 (accessed on 4 September 2023).
36. Krammer, K.; Lange-Bertalot, H. *Süßwasserflora von Mitteleuropa, Bd 2/2. Bacillariophyceae. 2. Teil: Bacillariaceae, Epithemiaceae, Surirellaceae*; Gustav Fischer Verlag: Stuttgart, Germany, 1988.
37. Krammer, K.; Lange-Bertalot, H. *Süßwasserflora von Mitteleuropa, Bd. 02/3: Bacillariophyceae: Teil 3: Centrales, Fragilariaceae, Eunotiaceae*; Gustav Fischer Verlag: Stuttgart, Germany, 1991.
38. Krammer, K.; Lange-Bertalot, H. *Süßwasserflora von Mitteleuropa, Bd. 02/4: Bacillariophyceae: Teil 4: Achnanthaceae, Kritische Ergänzungen zu Achnanthes sl, Navicula s. Str., Gomphonema, Gesamtliteraturverzeichnis Teil 1–4, Ergänzter Nachdruck, 2004*; Spektrum Akademischer Verlag: Heidelberg, Germany, 1999. Available online: http://link.springer.com/book/9783827408389 (accessed on 4 September 2023).
39. Morabito, G.; Oggioni, A.; Caravati, E.; Panzani, P. Seasonal morphological plasticity of phytoplankton in Lago Maggiore (N. Italy). *Hydrobiologia* **2007**, *578*, 47–57.
40. Hu, R.; Han, B.; Naselli-Flores, L. Comparing biological classifications of freshwater phytoplankton: A case study from South China. *Hydrobiologia* **2013**, *701*, 219–233.
41. Schilstra, M.; Wang, W.; van Oel, P.R.; Wang, J.; Cheng, H. The effects of reservoir storage and water use on the upstream–downstream drought propagation. *J. Hydrol.* **2024**, *631*, 130668.
42. Lin, Y.-I.; Pan, S.-Y.; Chang, H.-H.; Yu, M.-S.; Lin, W.-L. Will extreme drought impact the reservoir water quality? A 30-year observational study. *Agric. Water Manag.* **2023**, *289*, 108574.

43. Moon, Y.-E.; Kim, H.-S. Inter-Annual and seasonal variations of water quality and trophic status of a reservoir with fluctuating monsoon precipitation. *Int. J. Environ. Res. Public Health* **2021**, *18*, 8499. [CrossRef]
44. HaRa, J.; Atique, U.; An, K.-G. Multiyear links between water chemistry, algal chlorophyll, drought-flood regime, and nutrient enrichment in a morphologically complex reservoir. *Int. J. Environ. Res. Public Health* **2020**, *17*, 3139. [CrossRef] [PubMed]
45. Cortez, F.; Monicelli, F.; Cavalcante, H.; Becker, V. Effects of prolonged drought on water quality after drying of a semiarid tropical reservoir, Brazil. *Limnologica* **2022**, *93*, 125959.
46. Dieter, D.; Herzog, C.; Hupfer, M. Effects of drying on phosphorus uptake in re-flooded lake sediments. *Environ. Sci. Pollut. Res.* **2015**, *22*, 17065–17081.
47. Worrall, F.; Burt, T. Trends in DOC concentration in Great Britain. *J. Hydrol.* **2007**, *346*, 81–92.
48. Lee, K.-H.; Kam, J. Spatiotemporal patterns of water volume and total organic carbon concentration of agricultural reservoirs over South Korea. *Water Res.* **2024**, *256*, 121610. [CrossRef]
49. Winton, R.S.; Calamita, E.; Wehrli, B. Reviews and syntheses: Dams, water quality and tropical reservoir stratification. *Biogeosciences* **2019**, *16*, 1657–1671.
50. Maavara, T.; Chen, Q.; Van Meter, K.; Brown, L.E.; Zhang, J.; Ni, J.; Zarfl, C. River dam impacts on biogeochemical cycling. *Nat. Rev. Earth Environ.* **2020**, *1*, 103–116.
51. Nakulopa, F.; Bärlund, I.; Borchardt, D. How a reservoir modulates downstream water quality under declining upstream loading and progressing climate change. *Sci. Total Environ.* **2024**, *912*, 169460. [CrossRef]
52. Gilabert, J. Seasonal plankton dynamics in a Mediterranean hypersaline coastal lagoon: The Mar Menor. *J. Plankton Res.* **2001**, *23*, 207–218.
53. Choi, J.-Y.; Kim, S.-K. The use of winter water temperature and food composition by the copepod *Cyclops vicinus* (Uljanin, 1875) to provide a temporal refuge from fish predation. *Biology* **2021**, *10*, 393. [CrossRef] [PubMed]
54. Lafarga-De la Cruz, F.; Valenzuela-Espinoza, E.; Millán-Núnez, R.; Trees, C.C.; Santamaría-del-Ángel, E.; Núnez-Cebrero, F. Nutrient uptake, chlorophyll a and carbon fixation by *Rhodomonas* sp.(Cryptophyceae) cultured at different irradiance and nutrient concentrations. *Aquacult. Eng.* **2006**, *35*, 51–60.
55. Nowicka-Krawczyk, P.; Żelazna-Wieczorek, J.; Skrobek, I.; Ziułkiewicz, M.; Adamski, M.; Kaminski, A.; Żmudzki, P. Persistent cyanobacteria blooms in artificial water bodies—An effect of environmental conditions or the result of anthropogenic change. *Int. J. Environ. Res. Public Health* **2022**, *19*, 6990. [CrossRef]
56. Krivtsov, V.; Bellinger, E.; Sigee, D. Changes in the elemental composition of *Asterionella formosa* during the diatom spring bloom. *J. Plankton Res.* **2000**, *22*, 169–184. [CrossRef]
57. Reynolds, C.S.; Huszar, V.; Kruk, C.; Naselli-Flores, L.; Melo, S. Towards a functional classification of the freshwater phytoplankton. *J. Plankton Res.* **2002**, *24*, 417–428. [CrossRef]
58. Gsell, A.S.; de Senerpont Domis, L.N.; Przytulska-Bartosiewicz, A.; Mooij, W.M.; van Donk, E.; Ibelings, B.W. Genotype-by-temperature interactions may help to maintain clonal diversity in *Asterionella formosa* (Bacillariophyceae). *J. Phycol.* **2012**, *48*, 1197–1208. [CrossRef]
59. Vasconcelos, V.M. Species composition and dynamics of the phytoplankton in a recently-commissioned reservoir (Azibo-Portugal). *Arch. Hydrobiol.* **1991**, *121*, 67–78. [CrossRef]
60. Bondarenko, N.; Guselnikova, N.Y. Studies on *Synedra acus* Kutz. var. radians (Kutz.) Hust.(Bacillariophyta) in culture. *Int. J. Algae* **2002**, *4*, 85–95.
61. Michel, T.J.; Saros, J.E.; Interlandi, S.J.; Wolfe, A.P. Resource requirements of four freshwater diatom taxa determined by in situ growth bioassays using natural populations from alpine lakes. *Hydrobiologia* **2006**, *568*, 235–243.
62. Bailey-Watts, A. The ecology of planktonic diatoms, especially *Fragilaria crotonensis*, associated with artificial mixing of a small Scottish loch in summer. *Diatom Res.* **1986**, *1*, 153–168.
63. Stewart, A.J.; Wetzel, R.G. Cryptophytes and other microflagellates as couplers in planktonic community dynamics. *Arch. Hydrobiol.* **1986**, *106*, 1–19.
64. Sommer, U.; Gliwicz, Z.M.; Lampert, W.; Duncan, A. The PEG-model of seasonal succession of planktonic events in fresh waters. *Arch. Hydrobiol.* **1986**, *106*, 433–471.
65. Hammer, A.; Schumann, R.; Schubert, H. Light and temperature acclimation of *Rhodomonas salina* (Cryptophyceae): Photosynthetic performance. *Aquat. Microb. Ecol.* **2002**, *29*, 287–296.
66. Lee, J.-H.; Park, J.-G.; Kim, E.-J. Trophic states and phytoplankton compositions of Dam Lakes in Korea. *Algae* **2002**, *17*, 275–281.
67. Im, J.-K.; Sim, Y.-B.; Hwang, S.-J.; Byeon, M.-S.; Kang, T.-G. Temporal and seasonal variations in a phytoplankton community structure in artificial Lake Uiam, South Korea. *Water* **2023**, *15*, 4118. [CrossRef]
68. Walsby, A. Gas vesicles. *Microbiol. Rev.* **1994**, *58*, 94–144.
69. Paerl, H.; Fulton, R., III. Ecology of harmful cyanobacteria. In *Ecology of Harmful Algae*; Springer: Berlin, Germany, 2006; pp. 95–109.

70. Agostoni, M.; Waters, C.M.; Montgomery, B.L. Regulation of biofilm formation and cellular buoyancy through modulating intracellular cyclic di-GMP levels in engineered cyanobacteria. *Biotechnol. Bioeng.* **2016**, *113*, 311–319. [PubMed]
71. Wang, Z.; Akbar, S.; Sun, Y.; Gu, L.; Zhang, L.; Lyu, K.; Huang, Y.; Yang, Z. Cyanobacterial dominance and succession: Factors, mechanisms, predictions, and managements. *J. Environ. Manag.* **2021**, *297*, 113281.
72. Kromkamp, J.; Konopka, A.; Mur, L.R. Buoyancy Regulation in a Strain of *Aphaniz. omenon flos-aquae* (Cyanophyceae): The Importance of Carbohydrate Accumulation and Gas Vesicle Collapse. *Microbiology* **1986**, *132*, 2113–2121.
73. Tashiro, Y.; Monson, R.E.; Ramsay, J.P.; Salmond, G.P. Molecular genetic and physical analysis of gas vesicles in buoyant enterobacteria. *Environ. Microbiol.* **2016**, *18*, 1264–1276. [PubMed]
74. Wei, K.; Amano, Y.; Machida, M.; Asukabe, H.; Harada, K.-i. Effects of light and potassium ion on buoyancy regulation with gas vesicle in a Cyanobacterium Microcystis aeruginosa NIES-843. *Water Air Soil Pollut.* **2018**, *229*, 1–9.
75. Gámez, T.E.; Benton, L.; Manning, S.R. Observations of two reservoirs during a drought in central Texas, USA: Strategies for detecting harmful algal blooms. *Ecol. Indic.* **2019**, *104*, 588–593.

water

MDPI

Article

Picoplankton Groups and Their Responses to Environmental Factors in Small Cascade Hydropower Stations

Peiquan Li [1,2], Zhongxin Luo [1,2], Xianfang Zhu [3], Zhengzhu Dang [3], Daxin Zhang [1,2] and Xin Sui [1,2,*]

1 China Institute of Water Resources and Hydropower Research, Beijing 100038, China; iperry@126.com (P.L.); lzhx109@126.com (Z.L.); zhangdx@iwhr.com (D.Z.)
2 National Research Center for Sustainable Hydropower Development, Beijing 100038, China
3 Key Laboratory of Water and Sediment Sciences, Ministry of Education, Department of Environmental Engineering, Peking University, Beijing 100871, China; xf_zhu@126.com (X.Z.); 1901111794@pku.edu.cn (Z.D.)
* Correspondence: suixin@iwhr.com; Tel.: +86-010-68781619

Abstract: Hydropower is a clean and renewable energy source, and cascade hydropower stations have been developed to enhance water energy utilization efficiency. While small hydropower stations have a smaller scale and environmental impact compared to large ones, the cumulative effects of cascade development on river ecosystems should not be overlooked. In this study, flow cytometry was used to classify picoplankton from water samples collected at four small cascade hydropower stations on a Pearl River tributary into six microbial groups: Virus, LNA (Low Nucleic Acid), HNA (High Nucleic Acid), Cyanobacteria, Algae, and Fungi. Four ecological assessment indices were calculated: Photosynthetic Autotrophic Capacity (PAC), Bacterial Activity Index (BAI), Virus Regulatory Capacity (VRC), and Fungal Metabolic Capacity (FMC). By analyzing trends in microbial abundance and ecological indices and their correlations with environmental factors, the results showed that along the small cascade hydropower stations, dissolved oxygen (DO) and electrical conductivity (EC) increased from 5.71 mg/L and 49.87 μS/cm upstream to 6.80 mg/L and 56.18 μS/cm downstream, respectively. In contrast, oxidation-reduction potential (ORP) and total organic carbon (TOC) concentrations decreased from 3.81 mV and 1.59 mg/L to −8.05 mV and 1.08 mg/L, respectively. Among the microbial groups, the abundance of Virus, LNA, and Fungi decreased by 30.9%, 30.5%, and 34.9%, respectively, along the cascade system. EC, TOC, and NO_3^--N were identified as key drivers of changes in the abundance of the Virus, LNA, and Fungi groups. The concentrations of carbon and nitrogen nutrients significantly influenced the ecological assessment indices. Cascade hydropower stations had a significant impact on PAC, BAI, and VRC, while their influence on FMC was relatively small. The VRC showed a decreasing trend, suggesting a weakening effect of the stations on VRC. This study offers new perspectives and methods that facilitate the rapid and quantitative assessment of the ecological impacts of cascade hydropower stations.

Keywords: small cascade hydropower station; flow cytometry; picoplankton; virus; fungi; ecological assessment index; ecological impact

1. Introduction

Small hydropower refers to hydropower stations with an installed capacity of 50,000 kW or less and is internationally recognized as a clean and renewable energy source [1]. Small hydropower stations play an important role in securing rural electricity and improving irrigation, but their operation also has a considerable impact on river ecosystems. In order to make greater use of hydropower, cascade hydropower stations have been constructed on most rivers [2]. Their development and operation have even greater effects on the river's water environment [3], which may lead to the fragmentation of river habitats, affect the diversity of species and the ecological health of rivers [4,5], and also increase the stability of the ecological functions of planktonic microorganisms [6]. Therefore, it is particularly important to objectively assess the impact of small cascade hydropower stations on river ecosystems.

Small hydropower stations are often located in mountainous areas, exhibiting typical characteristics of mountain stream ecosystems. Nanoplankton (2–20 μm) and picoplankton (0.2–2 μm) play a key role in these ecosystems [7], acting as the main drivers of material cycling, energy flow, and information transfer [8]. Furthermore, these microscopic and picoplankton respond to changes in the aquatic environment of small hydropower stations more quickly and sensitively than larger aquatic organisms such as fish and benthic organisms. Therefore, analyzing the community composition and abundance of picoplankton, as well as exploring the relationship between microbial communities and ecological environments, can provide a more effective way to assess the ecological impact of small hydropower stations.

Research on nanoplankton and picoplankton has long relied on traditional methods such as microscopic observation. Li et al. [9] classified, identified, and counted phytoplankton using optical microscopy, assessing the influence of cascade hydropower stations on the composition of phytoplankton communities. However, optical microscopy has limitations in accurately identifying and counting the tiny picoplankton, leading to an underestimation of their abundance. Scanning and transmission electron microscopes can directly observe and photograph the ultrastructure of algae, capturing the morphological features of phytoplankton smaller than 5 μm for classification and identification [10]. However, electron microscopy is only capable of species identification for single individuals and cannot facilitate quantitative analysis of phytoplankton populations, making it unsuitable for large-scale, high-throughput sample detection. Flow cytometry [11,12] and molecular biology informatics techniques [13] have been increasingly developed and applied for the monitoring and identification of microscopic organisms. In river microbiome research, amplicon-based high-throughput sequencing and quantitative Polymerase Chain Reaction (qPCR) are the most commonly used methods [14]. However, qPCR can only determine microbial abundance through relative quantification. While amplicon sequencing has made significant contributions to microbiome research, the taxonomic information derived from Operational Taxonomic Units (OTUs) or Amplicon Sequence Variants (ASVs) is still insufficiently annotated [15]. Additionally, environmental Deoxyribonucleic Acid (eDNA) technology has demonstrated significant potential for assessing river biodiversity, but the degradation of eDNA in the environment limits its applicability in research [16].

Flow cytometry (FCM), originally used in the medical field, was later introduced to the study of aquatic microorganisms. The flow cytometer can rapidly analyze large quantities of aquatic microorganisms without the need for culturing, making it suitable for counting Cyanobacteria, heterotrophic bacteria, and eukaryotic phytoplankton [17,18]. FCM can detect parameters such as the size, granularity, and concentration of cells or particles through light scattering, and by referencing standard-sized microspheres, it can classify the detected cell populations according to their size [19,20]. Additionally, based on the fluorescence emitted by fluorescent dyes or autofluorescence, aquatic microorganisms can be categorized into different microbial groups [12]. FCM enables the large-scale, rapid detection of picoplankton and nanoplankton in aquatic environments, but its application in the field of river ecology in the context of cascade hydropower stations is still relatively limited.

This research aims to assess the impact of small cascade hydropower stations on picoplankton communities and identify the key environmental factors driving their variations. FCM was used as the primary analytical technique to systematically detect and analyze picoplankton in water samples collected from four small cascade hydropower stations along a tributary of the Pearl River. Based on the cell size, granularity, and pigment autofluorescence characteristics of plankton, combined with fluorescent dye labeling, six distinct microbial groups with different ecological functions were accurately classified. The abundance trends of each microbial group across the cascade hydropower stations were analyzed, along with their responses to environmental factors. To quantitatively evaluate the ecological impact of small cascade hydropower stations on river ecosystems, microbial ecological assessment indices were developed based on flow cytometric microbial groups. This research offers new perspectives and methodologies for evaluating the ecological impacts of hydropower stations, enabling rapid and quantitative assessments of cascade hydropower stations' ecological effects. It provides valuable theoretical and practical insights for advancing ecological impact assessments.

2. Materials and Methods

2.1. Study Area and Sampling

In August 2024, water samples were collected from four run-of-river small cascade hydropower stations (A, B, C, and D) along a tributary of the Pearl River, arranged sequentially from upstream to downstream (Figure 1). The installed capacities of small hydropower stations A, B, C, and D are 3000 kW, 910 kW, 3000 kW, and 1600 kW, respectively. Four sampling sections were designated at each station: Section 1 was located 500–1000 m upstream of the dam, Section 2 was 100–200 m upstream of the dam, Section 3 was at the tailwater immediately downstream of the dam, and Section 4 was where the tailwater mixed with the original river flow. Three parallel samples were collected at each section. The water sample collection followed the methods developed by Li et al. [9] and Coggins et al. [12].

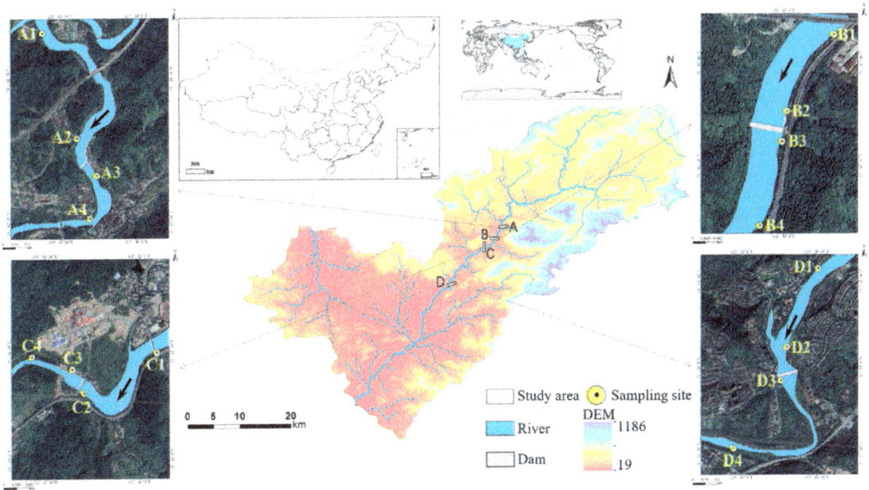

Figure 1. Geographic location of the four small cascade hydropower stations and the sampling sections. The sampling section locations are denoted by combining the hydropower station codes (A, B, C, and D) with the section numbers (1, 2, 3, and 4). The arrows in the river indicate the direction of flow.

GPS devices were used to obtain the latitude, longitude, and elevation of each sampling section. During the water sample collection, in situ measurements of pH, redox potential, water temperature, dissolved oxygen, and electrical conductivity at each sampling section were conducted using a multi-parameter water quality monitor (HACH-HQ40d, Hach Company, CO, USA). A turbidity meter and flow velocity meter were used to measure turbidity and river flow velocity, respectively. The measurements of total organic carbon (TOC), nitrate nitrogen (NO_3^--N), nitrite nitrogen (NO_2^--N), and ammonium nitrogen (NH_4^+-N) were carried out in an indoor laboratory [21,22]. For flow cytometry analysis, water samples were fixed with glutaraldehyde (final concentration 0.5%) to kill and preserve microorganisms, and then frozen and transported back to the laboratory for analysis.

2.2. Sample Preparation and Flow Cytometry Analysis

The frozen samples were thawed in the laboratory and subjected to ultrasonic treatment using an ultrasonic device to disperse biological aggregates in the natural water samples. A needle filter with a 20 μm pore size was then used to remove large particles of biological cells and impurities, preventing nozzle and pipeline blockages. Ultra-pure water was used as a blank control to eliminate instrument noise interference. Signals with VSSC channel intensity below 10^3 were considered background noise, and the fluorescence intensity threshold for the nucleic acid dye was set at 600 to remove background noise and inorganic particle interference.

In this research, the nucleic acid fluorescent dye SYBR Green I [23] was used to specifically bind with the nucleic acids of biological cells. Under excitation by a 488 nm laser on the flow cytometer (CytoFLEX SRT, Beckman Coulter, CA, USA), the biological cells bound with the SYBR Green I dye emitted green fluorescence at 520 nm. By detecting the intensity of the green fluorescence, biological cells can be distinguished from inorganic particles. The SYBR Green I stock solution used in the research had a concentration of 10,000× and was diluted 100 times with PBS to prepare the working solution. The working solution was then added to the water samples at a 1:100 volume ratio, followed by vortex mixing for 10 s and incubation in the dark for at least 10 min to allow the SYBR Green I dye to fully bind to the biological nucleic acids. After this incubation, the samples were detected using the flow cytometer. To improve the staining efficiency of SYBR Green I with viruses, the mixture was thoroughly mixed and incubated in a water bath at 80 °C for 10 min, and then allowed to cool to room temperature in the dark before being detected by the flow cytometer [24]. For fungal detection, CFW (Calcofluor White) stain solution was used to stain the chitin in the cell walls [25].

Microorganisms in water samples from different locations of four cascade hydropower stations were grouped based on their autofluorescence and staining results with fluorescent dyes. After staining with SYBR Green I, biological cells containing nucleic acids showed a positive signal in the B525 channel, with a threshold set at 1×10^3. Planktonic bacteria, which are widely distributed in aquatic ecosystems, were classified into two main functional groups: LNA (Low Nucleic Acid) and HNA (High Nucleic Acid) [26,27]. These groups were differentiated primarily by cell size and DNA content [28], with distinction based on fluorescence intensity in the B525 channel. Viruses, due to their smaller particle size and lower nucleic acid content, can be distinguished from bacteria (Figure 2a). Phytoplankton containing chlorophyll generated positive peaks in the R660 channel. Cyanobacteria, with higher phycocyanin content [29], emitted fluorescence in the Y585 yellow channel, allowing for differentiation between autotrophic eukaryotic algae and Cyanobacteria in flow cytometric analysis [30] (Figure 2b). Fungal plankton were identified by staining their cell walls with CFW, which fluoresced in the V525 channel. By combining this with the R660 channel to detect chlorophyll presence, fungi and eukaryotic algae could be distinguished (Figure 2c).

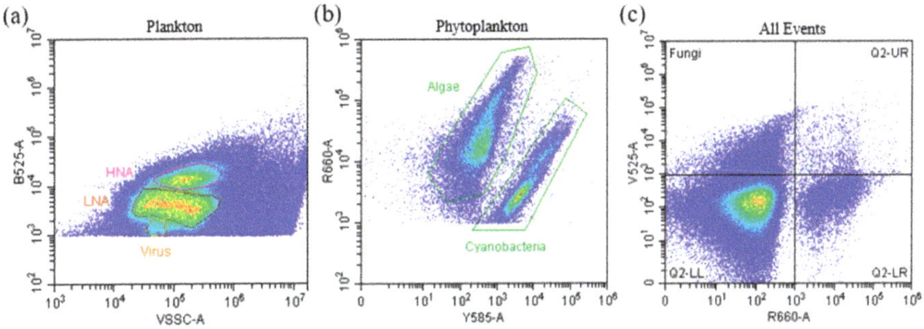

Figure 2. Microbial grouping based on flow cytometry ((**a**): Virus, LNA, and HNA groups; (**b**): Algae and Cyanobacteria groups; (**c**): Fungi group).

2.3. Absolute Abundance of Picoplankton

The quantitative analysis of planktonic microbial cells was performed using standard counting beads (Product 1426, APOGEE, Hertfordshire, UK), with a concentration of 5000 beads/μL. Flow cytometry analysis was conducted on both the standard counting beads and microbial cells under the same analysis parameters and flow rate pressure conditions. The calculation method for the quantitative analysis of microbial cells within a specific gate is as follows [31]:

$$C_p = \frac{N_p \times C_s \times t_s}{N_s \times t_p} \tag{1}$$

where C_p is the concentration of target microbial cells (cells/μL); N_p denotes the total number of target microbial cells; t_p is the analysis time for target microbial cells (s); C_s is the concentration of standard counting beads (5000/μL); N_s is the total number of standard counting beads; and t_s represents the analysis time for standard counting beads (s).

2.4. Microbial Ecological Assessment Index

Most studies on microbial structure and function rely on molecular bioinformatics approaches, using functional gene annotation and other techniques to investigate microbial functions [32]. However, in freshwater microbial community research, the scarcity of reference databases remains a challenge [33]. FCM combined with specific fluorescent dyes enables a functional analysis of microorganisms. In this study, FCM was employed to rapidly and accurately quantify different microbial groups in small cascade hydropower river systems. The quantitative distribution patterns of these microbial groups provide insights into the functional characteristics of the microecosystem. Based on the abundance data of different picoplankton in the water, a microbial assessment index can be calculated to explore the impact of the operation of small cascade hydropower stations on the river microecosystem.

The ratio of the abundance of photosynthetic autotrophic producers (Cyanobacteria and Algae) to heterotrophic consumers (LNA and HNA), decomposers (Fungi), and viruses (Virus) reflects the Photosynthetic Autotrophic Capacity (PAC) of the microecosystem.

$$PAC = \frac{C_{Cyanobacteria} + C_{Algae}}{C_{Virus} + C_{LNA} + C_{HNA} + C_{Fungi}} \tag{2}$$

HNA and LNA are two functional groups in the microecosystem that exhibit significant differences in ecology and nucleic acid content, which can currently only be detected by FCM. Hammes and Egli [26] identified and validated that the difference between HNA and LNA bacteria reflects differences in bacterial metabolic activity. Therefore, in this research, the ratio of the abundance of HNA to LNA is used to represent bacterial activity (Bacterial Activity Index, BAI) within the microecosystem.

$$BAI = \frac{C_{HNA}}{C_{LNA}} \tag{3}$$

Viruses control 20% to 40% of biological cell death and material cycling in the microecosystem, playing an important regulatory role in the structure of microbial communities [34]. The ratio of the abundance of planktonic viruses to planktonic heterotrophic bacteria represents the self-regulation capacity of the microecosystem (Virus Regulatory Capacity, VRC).

$$VRC = \frac{C_{Virus}}{C_{LNA} + C_{HNA}} \tag{4}$$

Fungi are a vital component of river ecosystems, and their key role in the degradation of organic matter is widely recognized. The ratio of the abundance of planktonic fungi to planktonic heterotrophic bacteria represents the decomposition metabolic capacity (Fungal Metabolic Capacity, FMC) of the microecosystem.

$$FMC = \frac{C_{Fungi}}{C_{LNA} + C_{HNA}} \tag{5}$$

In these calculations, C_x represents the abundance (cells/µL) of each microbial group.

2.5. Data Processing and Statistical Analysis

Basic environmental and flow cytometry data processing was performed using Microsoft Excel 2019. Boxplots comparing environmental factors and microbial group abundances at each hydropower station were generated using RStudio 2023.12.1+402. Mantel test correlation heatmaps were also generated to visualize the correlations between environmental factors, microbial group abundances, and microbial ecological assessment indices. The Kruskal–Wallis test was applied to assess differences between the various cascade hydropower stations, with $p < 0.05$ indicating significant differences between stations. In the correlation heatmaps, Spearman's correlation coefficient was used to calculate the correlation between environmental factors.

3. Results

3.1. Physicochemical Properties of Water Samples

The trends in the different environmental factors of cascade hydropower stations are shown in Figure 3. The turbidity ranged from 2.58 to 7.23, increasing from station A to B and then gradually decreasing, with station D having the lowest turbidity range of 2.58–3.42, indicating clearer water quality (Figure 3a). The pH of the water in stations A, B, and C ranged from 6.58 to 7.03, which is weakly acidic to neutral, while station D had a pH range of 6.97 to 7.14, indicating neutral to slightly alkaline water (Figure 3b). The oxidation-reduction potential (ORP) showed a decreasing trend along the cascade hydropower stations (Figure 3c). The temperature of the water (Twater), dissolved oxygen (DO), and electrical conductivity (EC) showed an increasing trend along the series of hydropower stations (Figure 3d–f). Twater ranged from 26.4 to 32.5 °C, indicating relatively high temperatures. Generally, mountainous rivers at the source tend to have low water temperatures and high DO, but as the river flows and human activities (such as thermal exchange during power generation) intervene, the water temperature rises. The increase in DO might be related to the higher abundance of phytoplankton in the water. The rising

EC reflects the increasing content of dissolved minerals in the water. TOC decreased along the cascade of stations (Figure 3h), possibly due to the degradation or sedimentation of organic matter in the water. The concentrations of NO_3^--N and NH_4^+-N decreased from stations A to B, and then increased again (Figure 3i,k).

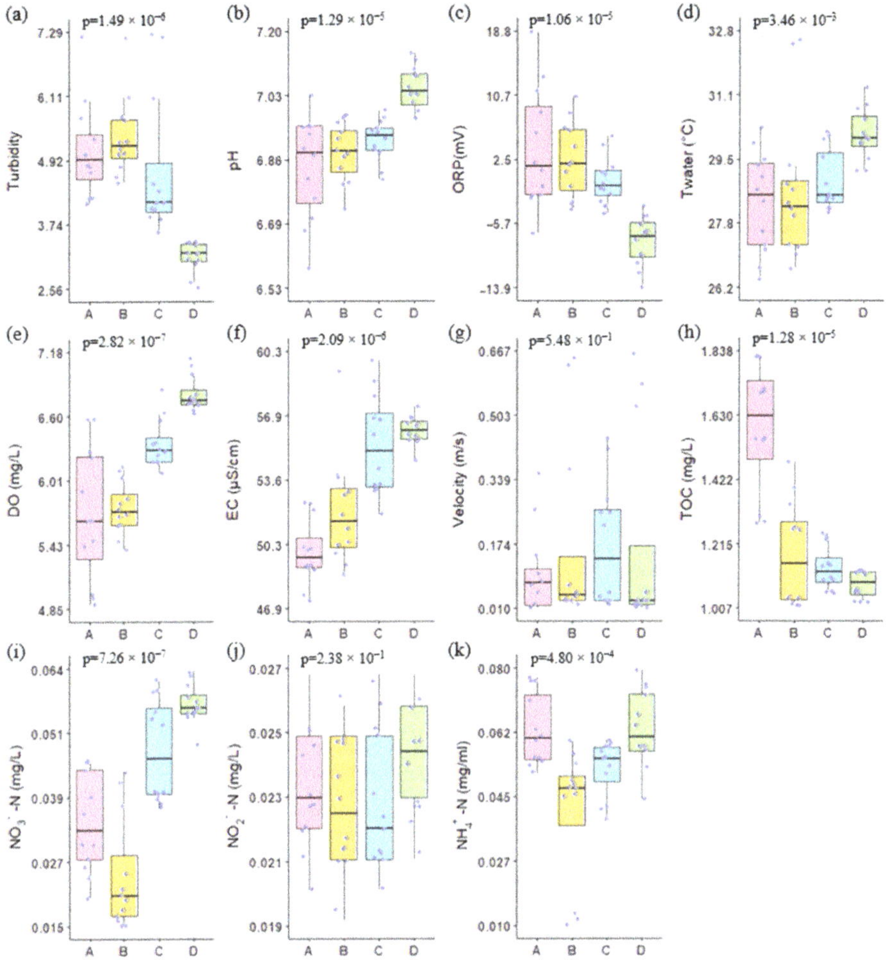

Figure 3. Trends in environmental factors of cascade hydropower stations. ((**a–k**) represent turbidity, pH, ORP, Twater, DO, EC, Velocity, TOC, NO_3^--N, NO_2^--N, and NH_4^+-N in cascade hydropower stations, respectively).

Velocity and NO_2^--N exhibited no significant differences across the four cascade hydropower stations ($p > 0.05$). As shown in Figure 3g, there were a few points with relatively high flow velocity at all four small hydropower stations. This was due to the larger flow at the tailwater discharge areas of each station, while other sampling sections had relatively slow flow velocities. The water flow near the dams upstream of each station was nearly stagnant. The distribution of NO_2^--N was more uniform at stations A, B, and C, while station D showed relatively higher values. However, the concentration did not vary significantly across the four stations, indicating that NO_2^--N remains relatively stable (Figure 3j).

3.2. Absolute Abundance of Picoplankton

Based on the cell size, granularity, and fluorescence characteristics of planktonic organisms, flow cytometry was used to classify the picoplankton in the water sample into six groups: Virus, LNA, HNA, Cyanobacteria, Algae, and Fungi (Figure 2). Among these, LNA exhibited the highest abundance, with an average of 2551 cells/µL, and the abundance of Algae and Cyanobacteria was relatively low, with averages of 47 cells/µL and 52 cells/µL, respectively. The abundance of Virus, LNA, and Fungi showed a decreasing trend along the cascade hydropower stations (Figure 4a,b,f), with LNA displaying a more pronounced trend. Virus and Fungi did not show a significant decrease between stations A, B, and C, but dropped significantly from C to D. HNA, on the other hand, increased from A to C, then decreased from C to D (Figure 4c). Cyanobacteria and Algae showed a decrease from A to B, followed by an increase from B to D (Figure 4d,e). No significant differences in the abundance of Virus and HNA were observed among the four small hydropower stations, while significant differences were found for LNA, Cyanobacteria, Algae, and Fungi. Figure 4 also reveals the differences in microbial groups across different sampling sections before and after the small hydropower stations, indicating the spatial heterogeneity of picoplankton abundance within individual small hydropower stations.

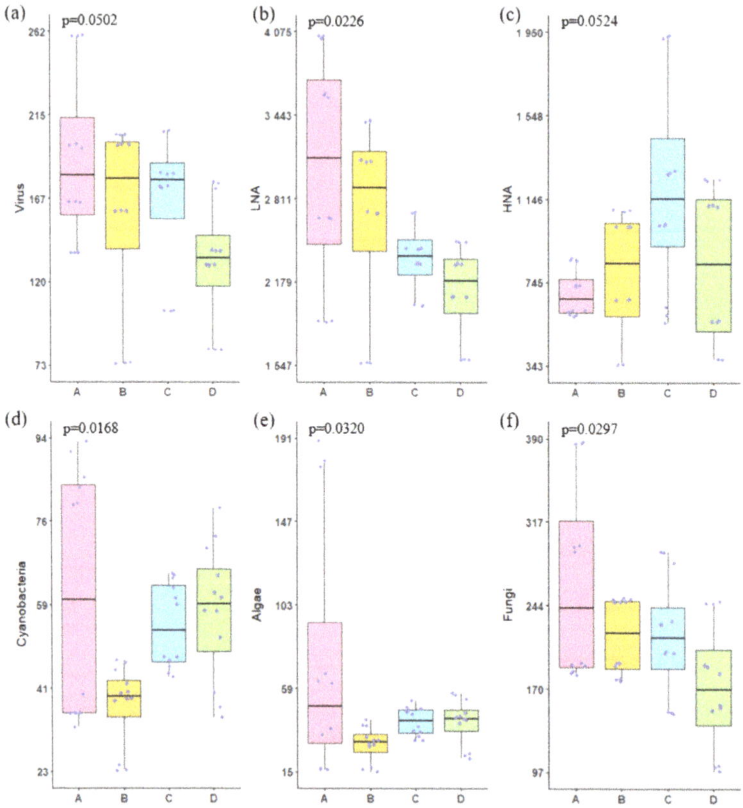

Figure 4. Trends in microbial groups abundance of cascade hydropower stations. ((**a**–**f**) represent the abundance of Virus, LNA, HNA, Cyanobacteria, Algae, and Fungi in cascade hydropower stations, respectively).

3.3. Microbial Ecological Assessment Index

The index representing the photosynthetic autotrophic capacity (PAC) of the microbial community follows a similar trend to that of Cyanobacteria and Algae, decreasing from A to B and then increasing from B to D (Figure 5a). The BAI increased from A to C and decreased from C to D (Figure 5b). The trends of VRC and FMC also mirror the abundance changes in Virus and Fungi, respectively, decreasing sequentially from A to D (Figure 5c,d). PAC, BAI, and VRC showed significant differences among the four small cascade hydropower stations, but FMC did not.

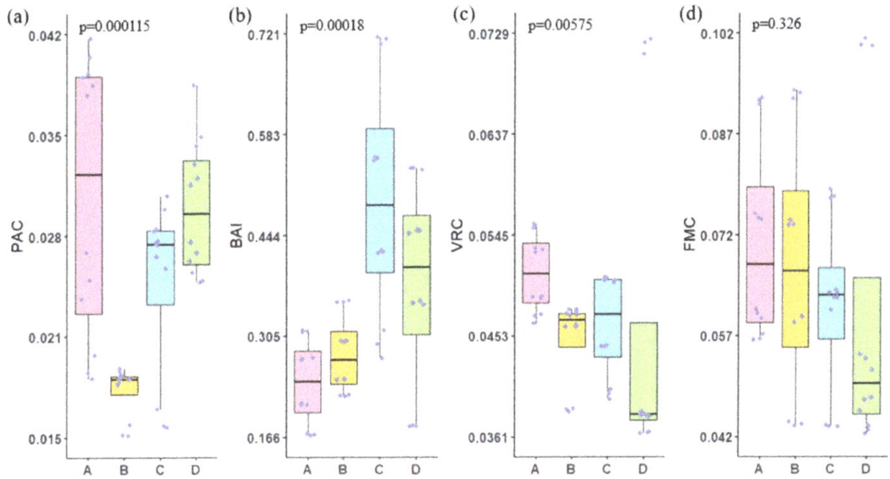

Figure 5. Trends in ecological assessment indices of cascade hydropower stations. ((**a**–**d**) represent PAC, BAI, VRC, and FMC in cascade hydropower stations, respectively).

3.4. Analysis of Key Driving Factors

The results of PCA (Principal Component Analysis) are presented in Figure 6, indicating a certain degree of similarity in environmental conditions among adjacent small hydropower stations, while gradual changes occur along the cascade hydropower system. The upstream hydropower stations, A and B, are primarily influenced by TOC, suggesting higher organic matter content in the water. In contrast, the downstream stations, C and D, are mainly affected by DO, EC, and NO_3^--N, which exhibit a strong positive correlation. Key environmental factors influencing water quality include pH, ORP, DO, EC, TOC, NO_3^--N, and NH_4^+-N.

The Mantel test is widely used in ecology and environmental sciences to explore the correlations between species and environmental factors. From Figure 7, it can be seen that virus is highly significantly correlated with EC, TOC, and NO_3^--N and significantly correlated with NO_2^--N. LNA is highly significantly correlated with EC and TOC and significantly correlated with NO_3^--N. HNA is highly significantly correlated with EC and significantly correlated with turbidity. Algae is highly significantly correlated with TOC. Cyanobacteria did not show any significant correlation with environmental factors. Fungi is highly significantly correlated with EC and significantly correlated with TOC and NO_3^--N. Overall, EC, TOC, and NO_3^--N show significant correlations with the abundance of Virus, LNA, and, Fungi, which may be key driving factors influencing the abundance changes in heterotrophic consumers and decomposers in the water.

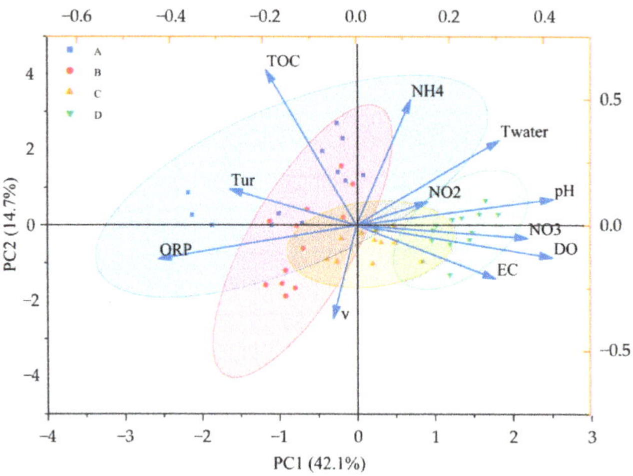

Figure 6. PCA of environmental factors in small hydropower stations.

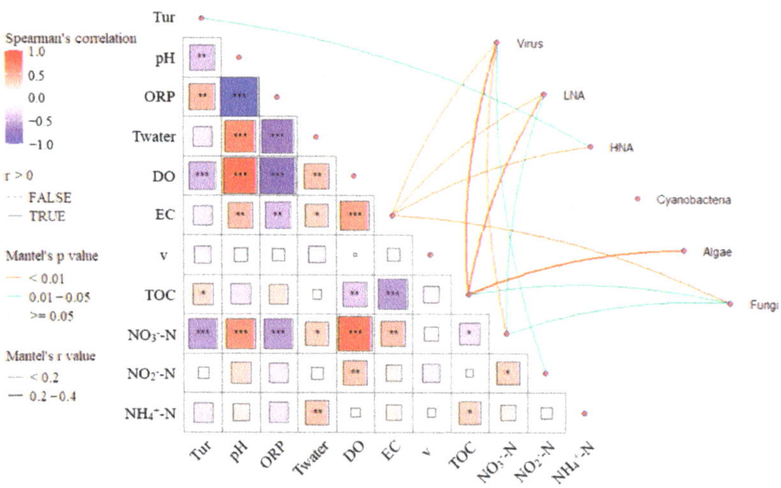

Figure 7. Correlation between microbial group abundances and environmental factors (* represents $p < 0.05$, ** represents $p < 0.01$, and *** represents $p < 0.001$, with no indication for $p > 0.05$).

It is evident from Equation (2) that PAC is significantly correlated with Cyanobacteria and Algae. Figure 7 shows that Algae is significantly correlated with TOC, whereas in Figure 8, PAC is not only highly significantly correlated with TOC but also significantly correlated with NO_2^--N. The BAI is significantly correlated with turbidity, echoing the significant correlation between HNA and turbidity in Figure 7. Additionally, VRC is significantly correlated with TOC, NO_3^--N, and NH_4^+-N, and FMC is significantly correlated with NO_3^--N. Overall, the microbial community ecological assessment index is primarily influenced by the concentrations of carbon and nitrogen nutrients in the water.

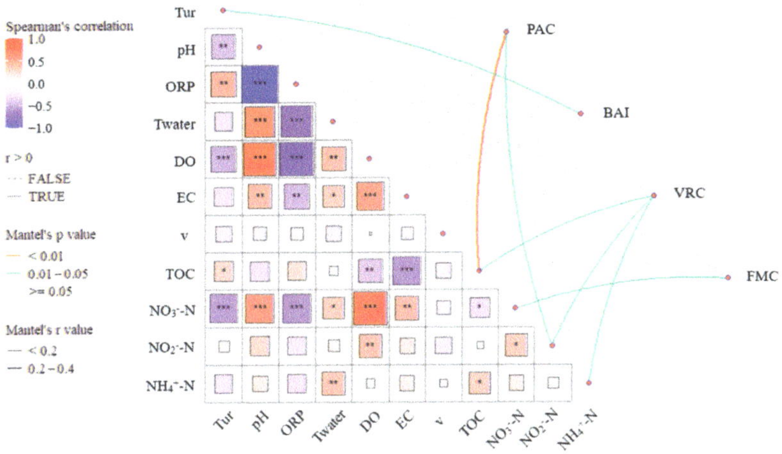

Figure 8. Correlation between ecological assessment indices and environmental factors (* represents $p < 0.05$, ** represents $p < 0.01$, and *** represents $p < 0.001$, with no indication for $p > 0.05$).

4. Discussion

4.1. Correlation Analysis Between Microbial Groups and Environmental Factors

The construction of cascade hydropower stations improves the efficiency of water energy utilization, but it also breaks the natural continuity of the river, which affects both environmental factors and microbial communities in the river ecosystem. There is a cumulative effect of the construction of cascade dams on the river's ecological factors [35], and the interception rate of suspended particulate matter increases gradually with each additional dam [36], resulting in a decrease in turbidity along the river at hydropower stations A, B, C, and D (Figure 3a), The operation of the cascade reservoirs also exerts a cumulative effect on downstream water temperature [37], causing a gradual increase in Twater (Figure 3d).

Because of the sensitivity of phytoplankton to the water environment, changes in the physicochemical factors of the water body will directly or indirectly change the community structure of phytoplankton [14], and the phytoplankton diversity can reflect the nutrient level and cleanliness of the water body to a certain extent; therefore, phytoplankton is often used as an indicator organism for water ecosystems [38,39]. The Mantel test analysis shows a highly significant correlation between Algae and TOC (Figure 7). However, since the main carbon source for phytoplankton is usually dissolved CO_2 [40], this correlation may be related to TOC serving as an indirect carbon source for phytoplankton growth [41]. Meanwhile, the abundance of Algae and Cyanobacteria shows a gradual increase (Figure 4d), and DO also rises (Figure 3e), which may be because the proliferation of Algae and Cyanobacteria contributed to the increase in DO concentrations in the river.

Bacteria play a vital role in river ecosystems, performing key functions in material cycling and energy flow. Figuring out the composition and dynamics of bacterial communities in rivers is essential for making more accurate assessments of water quality and ecosystem health. Flow cytometry can categorize planktonic bacteria into subgroups with low and high DNA content based on cell size and DNA content. Bacteria with a high DNA content have a high growth advantage and dominate in eutrophic waters, while bacteria with a low DNA content dominate in depleted waters, and the total productivity of bacterioplankton is mainly derived from high-DNA-content bacteria [28]. Wang et al. [42] conducted a study on planktonic bacteria in the cascade reservoirs of the Lancang–Mekong River Basin and pointed out that reservoir impoundment has a significant impact on planktonic bacterial communities, with dam construction leading to a reduction in the abundance of planktonic bacteria. In this study, the abundance of LNA shows a decreasing trend from A to D (Figure 4b), while the abundance of HNA increased from A to C (Figure 4c). Mantel test correlation analysis reveals that both LNA and HNA are significantly positively correlated with EC, and LNA is also significantly correlated with TOC (Figure 7), suggesting that changes in EC and TOC in the river may be detrimental to the survival of LNA bacteria [43,44]. There is a significant negative correlation between TOC and DO (Figure 7), and DO can drive bacterial community changes by altering the aerobic–anoxic transition process [45]; as DO levels increase progressively in the river, the environment becomes more favorable for the growth of aerobic bacteria.

Due to the unique nature of their life form, there is limited research on the impact and regulatory effects of environmental factors on planktonic viruses. The increase in EC (Figure 3f) could inhibit the growth of certain host bacteria, indirectly leading to a decrease in virus abundance. The correlation heatmap analysis reveals a significant negative correlation between EC and TOC (Figure 7). The decrease in TOC concentration (Figure 3h) may suggest a reduction in organic matter in the water, leading to a decrease in resources supporting the growth of virus hosts, such as bacteria, which in turn reduces virus abundance. Nitrate nitrogen (NO_3^--N) is one of the common nitrogen sources in water bodies and has an important impact on the structure of microbial communities. Changes in NO_3^--N concentration could affect the types and quantities of bacterial communities [46], indirectly influencing the abundance of viruses that rely on these bacteria as hosts.

Fungal abundance in rivers is generally lower than that of bacteria, but fungi play an indispensable role in the degradation of organic matter and in maintaining nutrient balance in the water. Mantel test analysis shows that Fungi are significantly correlated with EC, TOC and NO_3^--N (Figure 7). An increase in EC is typically associated with higher concentrations of dissolved salts and minerals in the water, and the rise in salinity may be unfavorable for fungal growth [47]. TOC is crucial for sustaining microbial growth in aquatic environments. Fungi secrete various extracellular enzymes to degrade carbon-rich substances such as lignin, which they then absorb and utilize [48]. In cascade hydropower stations, the gradual decrease in TOC concentration may reduce the organic carbon sources that fungi rely on, thereby limiting their growth. Nitrogen is a major limiting factor for primary productivity in various ecosystems and is also a primary cause of environmental issues such as water eutrophication. Fungal growth and metabolism are typically influenced by nitrogen sources. Although nitrate and nitrite are common nitrogen sources in water, fungi primarily rely on organic nitrogen as their nitrogen source [49]. The variation in NO_3^--N concentration in rivers may affect the competitive survival ability of certain fungal species.

Of course, the abundance of plankton in rivers is not only influenced by changes in environmental factors but also by the interactions between different types of microorganisms within the aquatic ecosystem. Moreover, these interactions are intricate and dynamic and can be positive (e.g., mutualism), negative (e.g., competition or parasitism), or neutral (e.g., commensalism), influenced by both abiotic and biotic factors [50,51]. Therefore, the reasons behind the changes in the abundance of each microbial community are actually very complex, and the discussion provided in this paper is far from sufficient to fully explain them.

4.2. Microbial Ecological Index for Assessing the Impact of Cascade Hydropower Stations

Changes in nutrients and habitat factors drive changes in the biological structure and function of microecosystems, and such changes may affect material cycling processes of the entire aquatic microecosystems [52–54]. The continuous damming of rivers disrupts rivers' ecological networks and reduces the connectivity and biodiversity of the microecosystem [55]. It has been shown that dam construction leads to decreased nutrient transfer efficiency in the micro-food web and the accumulation of nutrients [56], which, through trophic cascading effects, further impacts the cycling of elements such as carbon, nitrogen, and phosphorus [57]. Additionally, the reduced flow velocity induced by dams promotes a shift in the microecosystem from heterotrophy to autotrophy [58].

Photosynthetic algae are the main primary producers in the microecosystem [59]. As primary producers, they play a crucial role in aquatic ecosystems, providing a rapid response to changes in water quality and serving as effective bioindicators of water environment health. The turbidity of the four small cascade hydropower stations along the river decreases progressively (Figure 3a), which facilitates the penetration of light and thereby enhances the ecosystem's photosynthetic autotrophic capacity [60]. The PAC increases from B to D (Figure 5a), indicating a gradual enhancement of photosynthetic capacity in the river's aquatic ecosystem from B to D. Photosynthesis positively affects DO levels, and the increase in DO further promotes the growth and reproduction of other plankton in the river, which helps to maintain the biodiversity of the river's microecosystem.

The BAI increases from A to C and decreases from C to D (Figure 5b), which is consistent with the trend of HNA (Figure 4c). HNA typically represents active bacterial populations, while LNA corresponds to dormant or senescent bacterial groups. In this research, the ratio of the abundance of HNA to LNA, as represented by the BAI, reflects the bacterial activity in the microecosystem; a higher BAI indicates stronger bacterial activity, indirectly suggesting a healthy aquatic environment conducive to bacterial growth and development [61]. The BAI of hydropower stations C and D is higher than that of A and B (Figure 5b), indicating that the bacterial activity in the two downstream hydropower stations is greater than that in the two upstream stations.

The main groups of planktonic viruses are bacteriophages and algal viruses, which primarily use prokaryotes as hosts. Although they are extremely small in size, they are highly active and serve as crucial regulators of the structure and function of aquatic ecosystems. They play a significant role in modulating the size, structure, and diversity of microbial populations in water [62], acting as 'regulators' in the material cycling of the ecosystem by influencing the health and behavior of their hosts [63]. Changes in the concentrations of TOC, NO_2^--N, and NH_4^+-N in the water can affect the growth of viral host microorganisms. Viruses rely on host microorganisms for replication, and when the abundance of host microorganisms decreases, the viral abundance also declines, thereby impacting the VRC. The VRC decreases progressively from A to D (Figure 5c), indicating a decline in the self-regulation capacity of the river's microecosystem in the vicinity of the cascade hydropower stations, which is detrimental to maintaining the stability of the ecosystem. This trend suggests a possible weakening effect of the stations on the VRC, making it a key consideration in hydropower ecological impact assessments.

Fungi are major decomposers in ecosystems [64], with the potential to degrade toxic substances in rivers. They can be utilized in bioremediation to restore polluted rivers [65]. In the study area, the FMC shows a decreasing trend (Figure 5d), consistent with the changes in the abundance of Fungi (Figure 4f). This suggests a reduction in the decomposition metabolic capacity of the river ecosystem in the cascade hydropower station area. However, the differences in FMC across the four small hydropower stations are not significant, indicating that the decomposition metabolic levels of the river's microecosystem between A, B, C, and D are relatively similar. The results from the Mantel test show a significant correlation between FMC and NO_3^--N (Figure 8). The development of cascade hydropower stations alters hydrodynamic conditions and nutrient distribution, and their operation plays a role in intercepting and storing NO_3^--N [66,67]. The retention effect of

NO_3^--N varies among different reservoirs, influenced by multiple factors such as reservoir age, season, and hydraulic retention time [68]. The biogeochemical cycles within the reservoir can partially offset the nitrogen interception, alleviating the downstream nitrogen imbalance [69]. Fungi can release nitrogen from organic matter through decomposition, which may directly affect the utilization rate of nitrogen in the water and influence NO_3^--N concentrations [70]. Changes in NO_3^--N concentrations, in turn, lead to fluctuations in the FMC. The metabolism of carbon and nitrogen is an important process for maintaining the health of river ecosystems. However, our understanding of the driving factors behind river metabolism remains relatively limited. Increasing evidence suggests that the metabolic mechanisms are influenced by common environmental drivers shared with the ecosystem [71].

5. Conclusions

Small cascade hydropower stations improve water energy utilization efficiency by constructing multiple small-scale hydropower plants along rivers. With their relatively low construction and operational costs, they have become increasingly popular in mountainous regions and small basin areas. However, the potential impacts of these stations on river ecosystems must be carefully considered. This study sampled water from four small cascade hydropower stations along a tributary of the Pearl River. Using flow cytometry, picoplankton in the water samples were classified into six microbial groups: Virus, LNA, HNA, Cyanobacteria, Algae, and Fungi. The absolute abundance of each group, along with trends in four ecological assessment indices—PAC, BAI, VRC, and FMC—was analyzed. Mantel tests were used to examine the correlations between microbial abundances, ecological indices, and environmental factors, providing a comprehensive assessment of the ecological impacts of small cascade hydropower stations on the aquatic environment. The main findings of this study are as follows:

1. Along the small cascade hydropower stations, DO and EC progressively increased from 5.71 mg/L and 49.87 μS/cm upstream to 6.80 mg/L and 56.18 μS/cm downstream, respectively. Meanwhile, the ORP and TOC concentrations decreased from 3.81 mV and 1.59 mg/L to −8.05 mV and 1.08 mg/L, respectively. No significant differences were observed in velocity and NO_2^--N concentrations between the four cascade hydropower stations.
2. Among the six microbial groups classified by flow cytometry, LNA had the highest abundance, with an average of 2551 cells/μL, and the abundance of Algae and Cyanobacteria was relatively low, averaging 47 cells/μL and 52 cells/μL, respectively. The abundance of Virus, LNA, and Fungi decreased by 30.9%, 30.5%, and 34.9%, respectively, along the cascade hydropower stations.
3. EC, TOC, and NO_3^--N concentrations are significantly correlated with the abundance of Virus, LNA, and Fungi. The concentrations of carbon and nitrogen nutrients significantly influence the microbial ecological assessment indices. Cascade hydropower stations significantly impact the PAC, BAI, and VRC of the river's microbial ecosystem, while their influence on the FMC is relatively small. The VRC shows a decreasing trend along the cascade hydropower stations, suggesting a possible weakening effect of the stations on VRC.

Author Contributions: Conceptualization, P.L. and Z.L.; methodology, P.L., Z.L. and Z.D.; resources, P.L. and D.Z.; writing—original draft preparation, P.L.; writing—review and editing, Z.L., D.Z., X.Z. and Z.D.; visualization, P.L., Z.L. and D.Z.; supervision, X.Z. and X.S.; funding acquisition, X.S. All authors have read and agreed to the published version of the manuscript.

Funding: National Natural Science Foundation of China (NSFC) (No. 52379084) and the Key Projects of the Joint Fund of the National Natural Science Foundation of China (NSFC) (No. U22A20557).

Data Availability Statement: The data presented in this study are available on request from the corresponding author.

Conflicts of Interest: The authors declare no conflicts of interest.

Abbreviations

The following abbreviations are used in this manuscript:

qPCR	quantitative Polymerase Chain Reaction
OTUs	Operational Taxonomic Units
ASVs	Amplicon Sequence Variants
eDNA	environmental Deoxyribonucleic Acid
FCM	Flow cytometry
CFW	Calcofluor White
LNA	Low Nucleic Acid
HNA	High Nucleic Acid
DNA	Deoxyribonucleic Acid
PAC	Photosynthetic Autotrophic Capacity
BAI	Bacterial Activity Index
VRC	Virus Regulatory Capacity
FMC	Fungal Metabolic Capacity
ORP	Oxidation-Reduction Potential
DO	Dissolved Oxygen
EC	Electrical Conductivity
TOC	Total Organic Carbon

References

1. Sachdev, H.S.; Akella, A.K.; Kumar, N. Analysis and evaluation of small hydropower plants: A bibliographical survey. *Renew. Sustain. Energy Rev.* **2015**, *51*, 1013–1022. [CrossRef]
2. Wang, B.; Yang, X.; Li, S.L.; Liang, X.; Li, X.D.; Wang, F.; Yang, M.; Liu, C.Q. Anthropogenic regulation governs nutrient cycling and biological succession in hydropower reservoirs. *Sci. Total Environ.* **2022**, *834*, 155392. [CrossRef]
3. Moran, E.F.; Lopez, M.C.; Moore, N.; Müller, N.; Hyndman, D.W. Sustainable hydropower in the 21st century. *Proc. Natl. Acad. Sci. USA* **2018**, *115*, 11891–11898. [CrossRef] [PubMed]
4. Lai, R.; Chen, X.; Zhang, L. Evaluating the impacts of small cascade hydropower from a perspective of stream health that integrates eco-environmental and hydrological values. *J. Environ. Manag.* **2022**, *305*, 114366. [CrossRef]
5. He, F.; Zarfl, C.; Tockner, K.; Olden, J.D.; Campos, Z.; Muniz, F.; Svenning, J.C.; Jähnig, S.C. Hydropower impacts on riverine biodiversity. *Nat. Rev. Earth Environ.* **2024**, *5*, 755–772. [CrossRef]
6. Li, W.; Wang, B.; Liu, N.; Yang, M.; Liu, C.Q.; Xu, S. River damming enhances ecological functional stability of planktonic microorganisms. *Front. Microbiol.* **2022**, *13*, 1049120. [CrossRef] [PubMed]
7. Gao, F.Z.; Hu, L.X.; Liu, Y.S.; Qiao, L.K.; Chen, Z.Y.; Su, J.Q.; He, L.Y.; Bai, H.; Zhu, Y.G.; Ying, G.G. Unveiling the overlooked small-sized microbiome in river ecosystems. *Water Res.* **2024**, *265*, 122302. [CrossRef]
8. Drummond, J.D.; Davies-Colley, R.J.; Stott, R.; Sukias, J.P.; Nagels, J.W.; Sharp, A.; Packman, A.I. Retention and remobilization dynamics of fine particles and microorganisms in pastoral streams. *Water Res.* **2014**, *66*, 459–472. [CrossRef]
9. Li, J.; Dong, S.; Liu, S.; Yang, Z.; Peng, M.; Zhao, C. Effects of cascading hydropower dams on the composition, biomass and biological integrity of phytoplankton assemblages in the middle Lancang-Mekong River. *Ecol. Eng.* **2013**, *60*, 316–324. [CrossRef]
10. Vaulot, D.; Eikrem, W.; Viprey, M.; Moreau, H. The diversity of small eukaryotic phytoplankton ($\leq$3 μm) in marine ecosystems. *FEMS Microbiol. Rev.* **2008**, *32*, 795–820. [CrossRef]
11. Dashkova, V.; Malashenkov, D.; Poulton, N.; Vorobjev, I.; Barteneva, N.S. Imaging flow cytometry for phytoplankton analysis. *Methods* **2017**, *112*, 188–200. [PubMed]
12. Coggins, L.X.; Larma, I.; Hinchliffe, A.; Props, R.; Ghadouani, A. Flow cytometry for rapid characterisation of microbial community dynamics in waste stabilisation ponds. *Water Res.* **2020**, *169*, 115243.

13. Di Bella, J.M.; Bao, Y.; Gloor, G.B.; Burton, J.P.; Reid, G. High throughput sequencing methods and analysis for microbiome research. *J. Microbiol. Methods* **2013**, *95*, 401–414. [PubMed]
14. Li, K.; Hu, J.; Li, T.; Liu, F.; Tao, J.; Liu, J.; Zhang, Z.; Luo, X.; Li, L.; Deng, Y.; et al. Microbial abundance and diversity investigations along rivers: Current knowledge and future directions. *Wiley Interdiscip. Rev. Water* **2021**, *8*, e1547.
15. Roots, P.; Wang, Y.; Rosenthal, A.F.; Griffin, J.S.; Sabba, F.; Petrovich, M.; Yang, F.; Kozak, J.A.; Zhang, H.; Wells, G.F. Comammox Nitrospira are the dominant ammonia oxidizers in a mainstream low dissolved oxygen nitrification reactor. *Water Res.* **2019**, *157*, 396–405. [PubMed]
16. Perry, W.B.; Seymour, M.; Orsini, L.; Jâms, I.B.; Milner, N.; Edwards, F.; Harvey, R.; de Bruyn, M.; Bista, I.; Walsh, K.; et al. An integrated spatio-temporal view of riverine biodiversity using environmental DNA metabarcoding. *Nat. Commun.* **2024**, *15*, 4372. [PubMed]
17. Marie, D.; Rigaut-Jalabert, F.; Vaulot, D. An improved protocol for flow cytometry analysis of phytoplankton cultures and natural samples. *Cytom. Part A* **2014**, *85*, 962–968.
18. Zamorska, J.; Karwowska, E.; Przystaś, W. Assessment of microbiological quality of water using culture methods, flow cytometry and luminometry. *Water* **2023**, *15*, 4077. [CrossRef]
19. Metz, S.; Lopes Dos Santos, A.; Berman, M.C.; Bigeard, E.; Licursi, M.; Not, F.; Lara, E.; Unrein, F. Diversity of photosynthetic picoeukaryotes in eutrophic shallow lakes as assessed by combining flow cytometry cell-sorting and high throughput sequencing. *FEMS Microbiol. Ecol.* **2019**, *95*, fiz038.
20. Manohar, S.M.; Shah, P.; Nair, A. Flow cytometry: Principles, applications and recent advances. *Bioanalysis* **2021**, *13*, 181–198.
21. Zhao, M.M.; Wang, S.M.; Chen, Y.P.; Wu, J.-H.; Xue, L.-G.; Fan, T.T. Pollution status of the Yellow River tributaries in middle and lower reaches. *Sci. Total Environ.* **2020**, *722*, 137861. [PubMed]
22. Liu, Z.; Wang, X.; Jia, S.; Mao, B. Multi-methods to investigate spatiotemporal variations of nitrogen-nitrate and its risks to human health in China's largest fresh water lake (Poyang Lake). *Sci. Total Environ.* **2023**, *863*, 160975.
23. Prest, E.I.; Hammes, F.; Kötzsch, S.; van Loosdrecht, M.C.M.; Vrouwenvelder, J.S. Monitoring microbiological changes in drinking water systems using a fast and reproducible flow cytometric method. *Water Res.* **2013**, *47*, 7131–7142. [PubMed]
24. Shen, C.F.; Meghrous, J.; Kamen, A. Quantitation of baculovirus particles by flow cytometry. *J. Virol. Methods* **2002**, *105*, 321–330. [PubMed]
25. Lichius, A.; Zeilinger, S. Application of membrane and cell wall selective fluorescent dyes for live-cell imaging of filamentous fungi. *J. Vis. Exp.* **2019**, *153*, e60613.
26. Hammes, F.; Egli, T. Cytometric methods for measuring bacteria in water: Advantages, pitfalls and applications. *Anal. Bioanal. Chem.* **2010**, *397*, 1083–1095.
27. Song, Y.; Wang, Y.; Mao, G.; Gao, G.; Wang, Y. Impact of planktonic low nucleic acid-content bacteria to bacterial community structure and associated ecological functions in a shallow lake. *Sci. Total Environ.* **2019**, *658*, 868–878.
28. García, F.C.; López-Urrutia, Á.; Morán, X.A.G. Automated clustering of heterotrophic bacterioplankton in flow cytometry data. *Aquat. Microb. Ecol.* **2014**, *72*, 175–185.
29. Marie, D.; Shi, X.L.; Rigaut-Jalabert, F.; Vaulot, D. Use of flow cytometric sorting to better assess the diversity of small photosynthetic eukaryotes in the English Channel. *FEMS Microbiol. Ecol.* **2010**, *72*, 165–178.
30. Thompson, A.W.; Foster, R.A.; Krupke, A.; Carter, B.J.; Musat, N.; Vaulot, D.; Kuypers, M.M.M.; Zehr, J.P. Unicellular cyanobacterium symbiotic with a single-celled eukaryotic alga. *Science* **2012**, *337*, 1546–1550.
31. Gasol, J.M.; Morán, X.A.G. Flow Cytometric Determination of Microbial Abundances and Its Use to Obtain Indices of Community Structure and Relative Activity. In *Hydrocarbon and Lipid Microbiology Protocols: Single-Cell and Single-Molecule Methods*; Springer: Heidelberg, Germany, 2016; pp. 159–187.
32. Suparna, M.; Paul, R.; Daniel, R.; Tim, U.; Jack, G.; Folker, M.; Andreas, W.; Daniel, H. Functional analysis of metagenomes and metatranscriptomes using SEED and KEGG. *BMC Bioinform.* **2011**, *12*, S21.
33. Clark, D.R.; Ferguson, R.M.W.; Harris, D.N.; Matthews Nicholass, K.J.; Prentice, H.J.; Randall, K.C.; Randell, L.; Warren, S.L.; Dumbrell, A.J. Streams of data from drops of water: 21st century molecular microbial ecology. *WIREs Water* **2018**, *5*, e1280.
34. Gaïa, M.; Meng, L.; Pelletier, E.; Forterre, P.; Vanni, C.; Fernandez-Guerra, A.; Jaillon, O.; Wincker, P.; Ogata, H.; Krupovic, M.; et al. Mirusviruses link herpesviruses to giant viruses. *Nature* **2023**, *616*, 783–789. [CrossRef]
35. Wang, B.; Zhang, H.; Liang, X.; Li, X.; Wang, F. Cumulative effects of cascade dams on river water cycle: Evidence from hydrogen and oxygen isotopes. *J. Hydrol.* **2019**, *568*, 604–610.
36. Sun, M.; Huang, K.; Shao, J.; Wu, W.; Liang, X. Effects of mountain rivers cascade hydropower stations on water ecosystems. *Res. Ecol.* **2022**, *4*, 17–26.
37. Wang, Z.; Ma, J.; Yu, S.; Xu, Y.; Tao, Z.; Zhang, J.; Xiao, R.; Wei, H.; Liu, D. Analysis of water temperature variations in the Yangtze River's upper and middle reaches in the context of cascade hydropower development. *Water* **2024**, *16*, 1669. [CrossRef]
38. Chang, E.; Zhao, Y.; Wei, Q.; Shi, S.; Jiang, Z. Isolation of high-quality RNA from Platycladus orientalis and other Cupressaceae plants. *Electron. J. Biotechnol.* **2016**, *23*, 21–27.

39. Dembowska, E.A. The use of phytoplankton in the assessment of water quality in the lower section of Poland's largest river. *Water* **2021**, *13*, 3471. [CrossRef]
40. Ly, Q.V.; Nguyen, X.C.; Lê, N.C.; Truong, T.-D.; Hoang, T.-H.T.; Park, T.J.; Maqbool, T.; Pyo, J.; Cho, K.H.; Lee, K.-S.; et al. Application of Machine Learning for eutrophication analysis and algal bloom prediction in an urban river: A 10-year study of the Han River, South Korea. *Sci. Total Environ.* **2021**, *797*, 149040. [CrossRef]
41. Huang, Y.; Luo, L.; Xu, K.; Wang, X.C. Characteristics of external carbon uptake by microalgae growth and associated effects on algal biomass composition. *Bioresour. Technol.* **2019**, *292*, 121887. [CrossRef]
42. Wang, X.; Wang, C.; Wang, P.; Chen, J.; Miao, L.; Feng, T.; Yuan, Q.; Liu, S. How bacterioplankton community can go with cascade damming in the highly regulated Lancang-Mekong River Basin. *Mol. Ecol.* **2018**, *27*, 4444–4458. [PubMed]
43. Liu, L.; Yang, J.; Yu, X.; Chen, G.; Yu, Z. Patterns in the composition of microbial communities from a subtropical river: Effects of environmental, spatial and temporal factors. *PLoS ONE* **2013**, *8*, e81232.
44. Qi, H.; Lv, J.; Liao, J.; Jin, J.; Ren, Y.; Tao, Y.; Wang, D.; Alvarez, P.J.; Yu, P. Metagenomic insights into microalgae-bacterium-virus interactions and viral functions in phycosphere facing environmental fluctuations. *Water Res.* **2025**, *268*, 122676.
45. Liu, X.; Hu, S.; Sun, R.; Wu, Y. Dissolved oxygen disturbs nitrate transformation by modifying microbial community, co-occurrence networks, and functional genes during aerobic-anoxic transition. *Sci. Total Environ.* **2021**, *790*, 148245.
46. Wang, P.; Zhao, J.; Xiao, H.; Yang, W.; Yu, X. Bacterial community composition shaped by water chemistry and geographic distance in an anthropogenically disturbed river. *Sci. Total Environ.* **2019**, *655*, 61–69. [CrossRef]
47. Liu, L.; Wu, Y.; Yin, M.; Ma, X.; Yu, X.; Guo, X.; Du, N.; Eller, F.; Guo, W. Soil salinity, not plant genotype or geographical distance, shapes soil microbial community of a reed wetland at a fine scale in the Yellow River Delta. *Sci. Total Environ.* **2023**, *856*, 159136.
48. Bailey, V.L.; Smith, J.L.; Bolton, H., Jr. Fungal-to-bacterial ratios in soils investigated for enhanced C sequestration. *Soil Biol. Biochem.* **2002**, *34*, 997–1007.
49. Grossart, H.P.; Van den Wyngaert, S.; Kagami, M.; Wurzbacher, C.; Cunliffe, M.; Rojas-Jimenez, K. Fungi in aquatic ecosystems. *Nat. Rev. Microbiol.* **2019**, *17*, 339–354. [PubMed]
50. Faust, K.; Raes, J. Microbial interactions: From networks to models. *Nat. Rev. Microbiol.* **2012**, *10*, 538–550.
51. Yu, Z.; Gan, Z.; Tawfik, A.; Meng, F. Exploring interspecific interaction variability in microbiota: A review. *Eng. Microbiol.* **2024**, *4*, 100178.
52. Luo, Z.; Li, S.; Hou, K.; Ji, G. Spatial and seasonal bacterioplankton community dynamics in the main channel of the Middle Route of South-to-North Water Diversion Project. *Res. Microbiol.* **2019**, *170*, 24–34.
53. Pang, Y.; Ji, G. Biotic factors drive distinct DNRA potential rates and contributions in typical Chinese shallow lake sediments. *Environ. Pollut.* **2019**, *254*, 112903. [CrossRef] [PubMed]
54. Zhikharev, V.; Vodeneeva, E.; Kudrin, I.; Gavrilko, D.; Startseva, N.; Kulizin, P.; Erina, O.; Tereshina, M.; Okhapkin, A.; Shurganova, G. The species structure of plankton communities as a response to changes in the trophic gradient of the mouth areas of large tributaries to a lowland reservoir. *Water* **2022**, *15*, 74. [CrossRef]
55. Fan, H.; He, D.; Wang, H. Environmental consequences of damming the mainstream Lancang-Mekong River: A review. *Earth-Sci. Rev.* **2015**, *146*, 77–91. [CrossRef]
56. Yang, N.; Li, Y.; Zhang, W.; Lin, L.; Qian, B.; Wang, L.; Niu, L.; Zhang, H. Cascade dam impoundments restrain the trophic transfer efficiencies in benthic microbial food web. *Water Res.* **2020**, *170*, 115351.
57. Li, Z.; Lu, L.; Guo, J.; Yang, J.; Zhang, J.; He, B.; Xu, L. Responses of spatial-temporal dynamics of bacterioplankton community to large-scale reservoir operation: A case study in the Three Gorges Reservoir, China. *Sci. Rep.* **2017**, *7*, 42469.
58. Yang, N.; Li, Y.; Lin, L.; Zhang, W.; Wang, L.; Niu, L.; Zhang, H. Dam-induced flow velocity decrease leads to the transition from heterotrophic to autotrophic system through modifying microbial food web dynamics. *Environ. Res.* **2022**, *212*, 113568. [PubMed]
59. Ramanan, R.; Kim, B.H.; Cho, D.H.; Oh, H.M.; Kim, H.S. Algae–bacteria interactions: Evolution, ecology and emerging applications. *Biotechnol. Adv.* **2016**, *34*, 14–29.
60. Liu, Y.; Li, C.; Jian, S.; Miao, S.; Li, K.; Guan, H.; Mao, Y.; Wang, Z.; Li, C. Hydrodynamics regulate longitudinal plankton community structure in an alpine cascade reservoir system. *Front. Microbiol.* **2021**, *12*, 749888. [CrossRef]
61. Santos, M.; Oliveira, H.; Pereira, J.L.; Pereira, M.J.; Gonçalves, F.J.; Vidal, T. Flow cytometry analysis of low/high DNA content (LNA/HNA) bacteria as bioindicator of water quality evaluation. *Ecol. Indic.* **2019**, *103*, 774–781.
62. Johannessen, T.V.; Larsen, A.; Bratbak, G.; Pagarete, A.; Edvardsen, B.; Egge, E.D.; Sandaa, R.-A. Seasonal dynamics of haptophytes and dsDNA algal viruses suggest complex virus-host relationship. *Viruses* **2017**, *9*, 84. [CrossRef] [PubMed]
63. Gao, Y.; Lu, Y.; Dungait, J.A.J.; Liu, J.; Lin, S.; Jia, J.; Yu, G. The "regulator" function of viruses on ecosystem carbon cycling in the anthropocene. *Front. Public Health* **2022**, *10*, 858615.
64. Martínez, A.; Larrañaga, A.; Pérez, J.; Descals, E.; Pozo, J. Temperature affects leaf litter decomposition in low-order forest streams: Field and microcosm approaches. *FEMS Microbiol. Ecol.* **2014**, *87*, 257–267. [CrossRef]
65. Shishir, T.; Mahbub, N.; Kamal, N. Review on bioremediation: A tool to resurrect the polluted rivers. *Pollution* **2019**, *5*, 555–568.

66. Hua, R.; Zhang, Y. Assessment of water quality improvements using the hydrodynamic simulation approach in regulated cascade reservoirs: A case study of drinking water sources of Shenzhen, China. *Water* **2017**, *9*, 825. [CrossRef]
67. Wang, X.; Wang, P.; Wang, C.; Chen, J.; Hu, B.; Yuan, Q.; Du, C.; Xing, X. Cascade damming impacts on microbial mediated nitrogen cycling in rivers. *Sci. Total Environ.* **2023**, *903*, 166533.
68. Bao, Y.; Wang, Y.; Hu, M.; Hu, P.; Wu, N.; Qu, X.; Liu, X.; Huang, W.; Wen, J.; Li, S. Deciphering the impact of cascade reservoirs on nitrogen transport and nitrate transformation: Insights from multiple isotope analysis and machine learning. *Water Res.* **2024**, *268*, 122638. [PubMed]
69. Chen, Q.; Chen, Y.; Lin, Y.; Zhang, J.; Ni, J.; Xia, J.; Xiao, L.; Feng, T.; Ma, H. Does a hydropower reservoir cascade really harm downstream nutrient regimes. *Sci. Bull.* **2024**, *69*, 661–670.
70. Jia, Y.; Hu, X.; Kang, W.; Dong, X. Unveiling microbial nitrogen metabolism in rivers using a machine learning approach. *Environ. Sci. Technol.* **2024**, *58*, 6605–6615. [CrossRef]
71. Battin, T.J.; Lauerwald, R.; Bernhardt, E.S.; Bertuzzo, E.; Gener, L.G.; Hall, R.O., Jr.; Hotchkiss, E.R.; Maavara, T.; Pavelsky, T.M.; Ran, L. River ecosystem metabolism and carbon biogeochemistry in a changing world. *Nature* **2023**, *613*, 449–459.

water

MDPI

Article

Analysis of the Water Quality of a Typical Industrial Park on the Qinghai–Tibet Plateau Using a Self-Organizing Map and Interval Fuzzy Number-Based Set-Pair Analysis

Xiaoyuan Zhao [1,2,†], Di Ming [1,2,†], Yingyi Meng [1,2], Zhiping Yang [3,*] and Qin Peng [1,2]

1 Key Laboratory of Land Surface Pattern and Simulation, Institute of Geographic Sciences and Natural Resources Research, Chinese Academy of Sciences, Beijing 100101, China
2 College of Resources and Environment, University of Chinese Academy of Sciences, Beijing 100049, China
3 Jiangxi Research Academy of Ecological Civilization, Nanchang 330036, China
* Correspondence: yangzhiping8998@163.com
† These authors contributed equally to this work and should be considered co-first authors.

Abstract: The Qinghai–Tibet Plateau (QTP) serves as the origin for several major rivers in Asia and acts as a crucial ecological barrier in China, characterized by its regional conservation significance. Production activities in the industrial park in this special geographical environment may exacerbate its environmental vulnerability. We examined the spatial and temporal patterns of water quality parameters, identified the factors influencing water quality, and evaluated the associated risks using various analytical methods, including the Boruta algorithm and interval fuzzy number-based set-pair analysis (IFN-SPA). The results showed that the average concentrations in the flood season and dry season were significantly different. The average value of Cd in the flood season belonged to the water quality standard of Class II. Different heavy metals show different spatial distribution characteristics, and the reason for the difference comes from livestock farms and industrial enterprises. The results for the flood season and dry season were different, which further proves that meteorological factors can influence water quality. The risk of heavy metals in different rivers presents different spatial distribution characteristics; for example, the risk of heavy metals in the Sigou River is higher. The water quality assessment results indicate the need to develop a well-structured evaluation framework for managing and controlling river water pollution in the future.

Keywords: water quality; driving factor; risk assessment; Qinghai–Tibet Plateau

1. Introduction

Water plays a crucial role in the geographical environment, linking and interacting with the atmosphere, biosphere, pedosphere, and lithosphere in a dynamic process through the water cycle. Water is also essential to human survival and economic and social development. However, with the development of society, water quality and quantity may change due to the surrounding environment such as habituated areas and industrial park [1]. Previous research has shown that certain activities can lead to a decline in water quality, such as the discharge of domestic sewage from urban areas, livestock waste from rural areas, the excessive use of pesticides and fertilizers, and wastewater from intensive industrial activities [2]. The bad water environment caused the death of fish and shrimp. Moreover, fishing, agriculture, and the economy can be restricted as well [3].

A self-organizing map (SOM) can uncover hidden information due to its powerful classification ability compared to other traditional mathematical analysis methods [4]. It is an unsupervised learning technique with strong clustering and visualization abilities, allowing it to detect the distribution patterns of pollutants in water. A previous study revealed vegetation cover's influence on water quality indexes in the Liaohe River Basin in China using an SOM [5]. SOMs have excellent performance in dealing with nonlinear problems [4]. As classical competitive neural networks, SOMs produce the best matching units and obtain good classification and visualization results for input samples. They have been applied to the pattern recognition of heavy metals and pollutants in water or soil. The Analytical Hierarchy Process (AHP), fuzzy AHP, and numerical models have been extensively used in the risk assessment of water quality and geohazards. These methods are beneficial because they enable the quantification of water resources and risks. However, they face challenges and uncertainties in determining model parameters (such as variable weights) under real-world conditions, which can lead to inaccuracies in water quality risk assessments. The Set-Pair Analysis (SPA) method has been widely utilized for assessing the risks related to water quality and water resources, as demonstrated in the water pollution risk assessment conducted in Shiyan City using SPA combined with the K-means clustering technique [6]. Given the challenges in determining the relevant factors, Lyu et al. [7,8] enhanced SPA by integrating Interval Fuzzy Numbers (IFN). This modification addressed the limitations of the original SPA and has been successfully applied in water quality evaluation [7] and risk assessment during shield tunneling [8]. The results from these studies confirm the reliability and applicability of modified SPA. The combination of SOM and IFN-SPA can effectively analyze the spatial clustering of water quality and water quality risk.

Water quality risks are influenced by interaction between factors such as geographical location and human management. These locations include climate-related locations (i.e., temperature, precipitation, potential evapotranspiration, leaching, and loss of elements from the crust) [9,10]. Other factors include land use and the type and amount of pollutants discharged (i.e., sewage and plastics). In theory, the climate determines the state of water quality [11]. For instance, heavy rainfall leads to agricultural runoff and the release of nutrients into water bodies. Additionally, higher temperatures accelerate metabolic processes, resulting in an increase in bacteria and phytoplankton [12]. Water quality can be affected by changes in land use, for example, from farmland to cities. Agricultural runoff is a significant source of TN and TP in the water bodies in the study area, while domestic sewage discharge also contributes to water pollution. Recognizing these key drivers is essential for enhancing water quality.

The Qinghai–Tibet Plateau is known as the "Headwater of Asia", because many rivers in southeast area originate from here [13]. In recent years, people have recognized that it can protect and construct the national ecological security barrier. Freshwater resources are critical resources following energy resources. The level of water resource development on the Qinghai–Tibet Plateau is below the average level in China due to the constraints of the environment, transportation, and social–historical factors. The Qinghai–Tibet Plateau is less influenced by human activities and has fewer pollutants. It is known as "the last pure land on earth". However, the ecological environment has become more sensitive and fragile due to unique environmental conditions and human activities such as geothermal and mineral exploitation, urban sewage, and wastewater discharge. Production activities in the industrial park in this special geographical environment may exacerbate this environmental vulnerability. Moreover, the agricultural soil is usually irrigated by the surrounding water. Toxic metals may accumulate in croplands and enter the human body through the food chain. Thus, it is necessary to monitor river water quality around industrial parks and explore the industrial activities' impact on water quality. Previous studies have shown that the water quality indexes around industrial parks are significantly higher than other sites, with water quality decreasing to "Undrinkable". Thus, exploring the connection between industrial parks and water quality is an innovative approach. Our objectives are to (1) examine the spatial distribution of heavy metal concentrations in water around a typical industrial park; (2) categorize water quality indicators using SOM; (3) identify the relationship between environmental auxiliary variables and water quality indexes using the Boruta algorithm; and (4) assess the water quality risk using IFN-SPA. Our findings may provide enlightenment for making policies to improve environmental conditions around the industrial park on the Qinghai–Tibet Plateau.

2. Material and Methods

2.1. Study Area

The Datong Beichuan Industrial Park is situated in the Sanjiangyuan region of the Tibetan Plateau, a critical ecological zone that serves as the headwater region for the Yangtze, Yellow, and Lancang rivers, often referred to as the "Water Tower of China". This area is not only a key hydrological source for millions of people across China but also a globally significant biodiversity hotspot, designated as a priority region for conservation. The Sanjiangyuan region is particularly sensitive to the impacts of climate change, with rapid shifts in temperature and precipitation patterns influencing both local ecosystems and broader climatic systems.

The study area lies in the western sector of the Sanjiangyuan region (31°36′–39°19′ N, 89°35′–103°04′ E), as shown in Figure 1. The region's climate is characterized by a typical plateau continental climate, with marked seasonal contrasts: a cold, dry winter season and a warm, wet summer season. Notably, the annual temperature variation is minimal, while diurnal temperature fluctuations are significant. This climate regime, coupled with the rugged and high-altitude topography, creates a unique environmental context. The region's elevation ranges from 2178 m to over 2835 m, which further intensifies its ecological sensitivity and the challenges faced in balancing development with environmental protection.

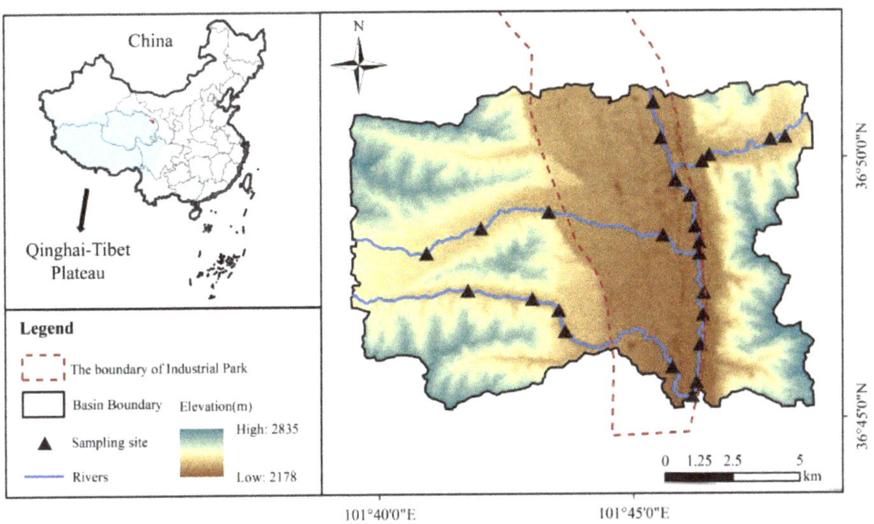

Figure 1. The location of the study area and sampling sites.

The Datong Beichuan Industrial Park, as a key industrial hub in Qinghai Province, serves as a prominent example of industrial development within this ecologically sensitive area. As a significant driver of regional economic growth, the park's expansion brings with it the need for rigorous water quality research. Conducting such studies in typical industrial parks in the Qinghai–Tibet Plateau is crucial for assessing the impact of industrial activities on local water systems. This research is essential for understanding pollution dynamics, nutrient loading, and other water quality indicators that directly influence the region's water security. By focusing on water quality, the park's development can help safeguard this vital resource, ensuring its availability for both local ecosystems and human populations and contributing to long-term ecological sustainability.

2.2. Sample Collection

River samples were collected twice, once during the flood season in September 2023 and once during the dry season in January 2024, with 25 samples taken in each season, resulting in a total of 50 samples. These water quality parameters were measured using the methods published in the National Environmental Quality Standards for Surface Water in China (GB3838-2002) [14]. The relative standard deviation (RSD) of 8 elements was less than 5.0%, the recovery rate of each element was 80%~120%, and the measured values were all within the range of standard values.

Environmental indicators were measured for both September 2023 and January 2024, including precipitation, evapotranspiration, and air temperature. Additional geographic and environmental factors were also considered: elevation, proximity to the nearest residential area, distance to the nearest road, and distance to the nearest industrial facility. Remote sensing data, including the Normalized Difference Vegetation Index (NDVI) derived from the MOD13A3 product with a 250 m resolution and 16-day composite, were incorporated into the analysis. Furthermore, heavy metal concentrations in road dust and soil within the study area were analyzed. Road dust samples (196 points) were collected, interpolated, and extracted to sampling points. Similarly, soil heavy metal concentrations were measured at 177 locations, with interpolation followed by extraction to sampling points. After small stones as well as organic debris were removed and air-dried, the soil samples were broken up and sifted through a fine sieve of 1 mm. Soil samples and road dust samples were digested in an acid mixture of 65–68% HNO_3, 36–38% HCl, and 40% HF (6 mL, 2 mL, and 3 mL) with microwave digestion.

In this study, pH was measured using a portable pH meter (PHB-4). The concentrations of Cr and Ni were determined via UV-visible spectrophotometry, while As and Hg were detected using atomic fluorescence spectroscopy. Cu, Zn, Cd, and Pb concentrations were analyzed via atomic absorption spectrophotometry.

2.3. Statistical Analyses and Models

Data analysis was performed using IBM SPSS Statistics version 27.0. The normality of each parameter was assessed with the Shapiro–Wilk test, which revealed a departure from normality, necessitating the use of non-parametric statistical methods. Visualization and graphical representations were generated using R version 4.3.1. Spatial distribution clustering and further visualization were performed with MATLAB R2023b for SOM and ArcMap 10.6 (ESRI, Redlands, CA, USA).

2.3.1. Self-Organizing Map Analysis (SOM)

Self-organizing map (SOM) analysis, a type of unsupervised learning algorithm introduced by Teuvo Kohonen, is employed to map high-dimensional data into a lower-dimensional space while preserving their topological structure [15]. This enables the retention of data point similarity within the reduced space [16]. Through a competitive learning strategy, SOM maps similar input data points to adjacent neurons, creating a topological structure on the map that reflects the relationships among the data. This capability makes SOM useful for tasks such as dimensionality reduction, visualization, and clustering.

The SOM network consists of an input layer and an output layer (competitive layer), where the output layer typically has a lower-dimensional structure, such as two dimensions. Each node is represented by a feature vector and can be regarded as a "cluster center". The learning process involves mapping each input data point to a node in the competitive layer, ensuring that data points in the input space that are geographically close are also mapped to nearby positions in the output layer. This design enhances the model's accuracy, efficiency, and visualization and minimizes human interference [17].

Applications of SOM include, but are not limited to, identifying water quality pollution patterns, assessing the spatial heterogeneity of water quality, and serving as part of water quality prediction and early warning systems. Its visualization capabilities allow researchers to gain intuitive insights into the complex structure of water quality data, which is crucial for the development of effective water management and conservation strategies.

2.3.2. Major Factor Selection

Spearman rank correlation analysis was used in this study to assess the monotonic relationships between variables in non-normally distributed datasets [18]. The method evaluates correlations by computing the ranks (i.e., the positions after sorting the data) of variables. The correlation coefficient (r_s) is determined by calculating the rank differences (d_i) and sample size (n), as follows:

$$r_s = 1 - \frac{6\sum_{i=1}^{n} d_i^2}{n(n^2-1)}$$

This analysis reveals potential nonlinear relationships between variables, providing a foundation for subsequent predictive model development. To construct an accurate and stable model, we applied the Boruta algorithm for feature selection [19,20]. Based on random forests, Boruta identifies significant features by comparing them with random attributes. Unlike traditional methods that focus on minimizing model error, Boruta emphasizes a comprehensive understanding of the factors influencing the dependent variable. The application of Boruta in this study aids in identifying key variables, thereby enhancing the predictive performance of the model. For further details on these methods, refer to the literature [19–21].

2.3.3. Interval Fuzzy Number-Based Set-Pair Analysis

This study employs interval fuzzy number-based set-pair analysis (IFN-SPA) for water quality risk assessment, with factors selected through Spearman rank correlation analysis and the Boruta algorithm [7]. IFN-SPA simplifies the membership function of set-pair analysis (SPA), using relative distance to represent the degree of importance between the samples and criteria [8]. The weights of evaluation factors are adjusted using the analytic hierarchy process (AHP), revealing the relationships between evaluation samples. IFN-SPA further calibrates the correlation between evaluation factors and water quality risk. Detailed calculations can be found in a previously published study.

3. Results and Discussion

3.1. *Statistical Overview of Water Quality Parameters*

The statistical data on water quality during the flood and dry seasons are presented in Table 1 and Figure 2. Significant differences were observed between the average concentrations of pollutants in these two seasons. The average concentration of Cr in the flood season was 0.00116 ± 0.00026 mg/L, which was above the average value of the dry season, but still fell within the Class I standard for surface water quality, as defined by GB3838-2002 [22]. Cr with the CV% (22.65% and 32.73%) indicated a certain degree of variability. The average Cu of the flood season was 0.00119 ± 0.0004 mg/L, with most monthly Cu levels exceeding this mean. And sampling sites in the dry season also presented the same pattern (Figure 2). This difference may reflect the fact that heavy metals through wastewater discharge or runoff can enter into the river. For example, pollution in the Diaojiang River is serious because there are mining and smelting activities in this area and its karst landforms [23,24]. The mean Cd concentration in the flood season was 0.00115 ± 0.00038 mg/L, which corresponded to the Class II water quality standard. Such water quality is generally suitable for drinking, aquaculture, and recreational uses like swimming, though continuous monitoring is required to prevent deterioration. However, the maximum Cd concentration reached 0.00190 mg/L, indicating severe contamination near specific sampling sites, which warrants closer attention. The CV% of Ni displayed the smallest variation among all water quality metrics, suggesting consistently low Ni levels in surface water. In terms of Pb, the concentrations were low and seemingly unthreatened, but CV% was higher than other heavy metals which reflected their greater dispersion in this area and the need to be aware of Cd contamination in other places. Heavy metal contamination in water bodies can devastate aquatic ecosystems by killing fish and shrimp, causing water to stink and eutrophication, and finally collapsing aquatic ecosystems [25]. Previous studies have indicated that many types of industries are concentrated near river systems, where industrial operations release heavy metals into the environment via effluents and emissions [26,27]. Elevated Cd concentrations in rivers may result from the presence of Cd in raw materials used during production processes in certain enterprises. In the upper reaches of the Yellow River, industries such as electroplating, tanning, chemical manufacturing, metallurgy, and refractory material production contribute to Pb, Cd, Cr, and As contamination through wastewater discharge [28–30]. Additionally, As and Pb concentrations were notably high in wastewater generated by thermal power plants [31,32]. On the other hand, agriculture can influence the river water quality through planting and breeding [28]. Natural factors (i.e., climate, geography, terrain, etc.) and human factors (i.e., pollution discharge) caused by regional differences lead to different management needs and water quality classifications [33]. Different seasonal climate factors have different impacts on water quality. The implementation of differentiated management of regional water quality objectives can better adapt to the actual situation in different seasons.

Table 1. Summary of the water quality parameters in the flood season and dry season.

	Index	pH	Cr (mg/L)	Ni (mg/L)	Cu (mg/L)	Zn (mg/L)	Cd (mg/L)	Pb (mg/L)	Hg (mg/L)	As (mg/L)
Flood Season	Max	8.1	0.00158	0.00082	0.00181	0.00781	0.00190	0.00083	0.00028	0.0021
	Min	7.15	0.00058	0.00046	0.00042	0.00223	0.00047	0.00449	0.0001	0.0011
	Mean	7.63	0.00116	0.00064	0.00119	0.00484	0.00115	0.00283	0.000196	0.0016
	SD	0.23	0.00026	0.00011	0.00040	0.00176	0.00038	0.00082	0.00007	0.0016
	CV%	2.98	22.65	17.06	33.64	36.33	32.89	29.00	35.98	23.63
Dry Season	Max	7.75	0.0018	0.0076	0.0058	0.00091	0.00091	0.00454	0.0009	0.0038
	Min	7.00	0.00052	0.00099	0.00095	0.00046	0.00046	0.00100	0.00006	0.00045
	Mean	7.45	0.00109	0.00304	0.00323	0.00067	0.00067	0.00266	0.00018	0.0018
	SD	0.23	0.00036	0.00160	0.00138	0.00013	0.00013	0.00091	0.00027	0.00085
	CV%	3.06	32.73	52.64	42.55	18.61	18.61	34.34	147.99	46.23

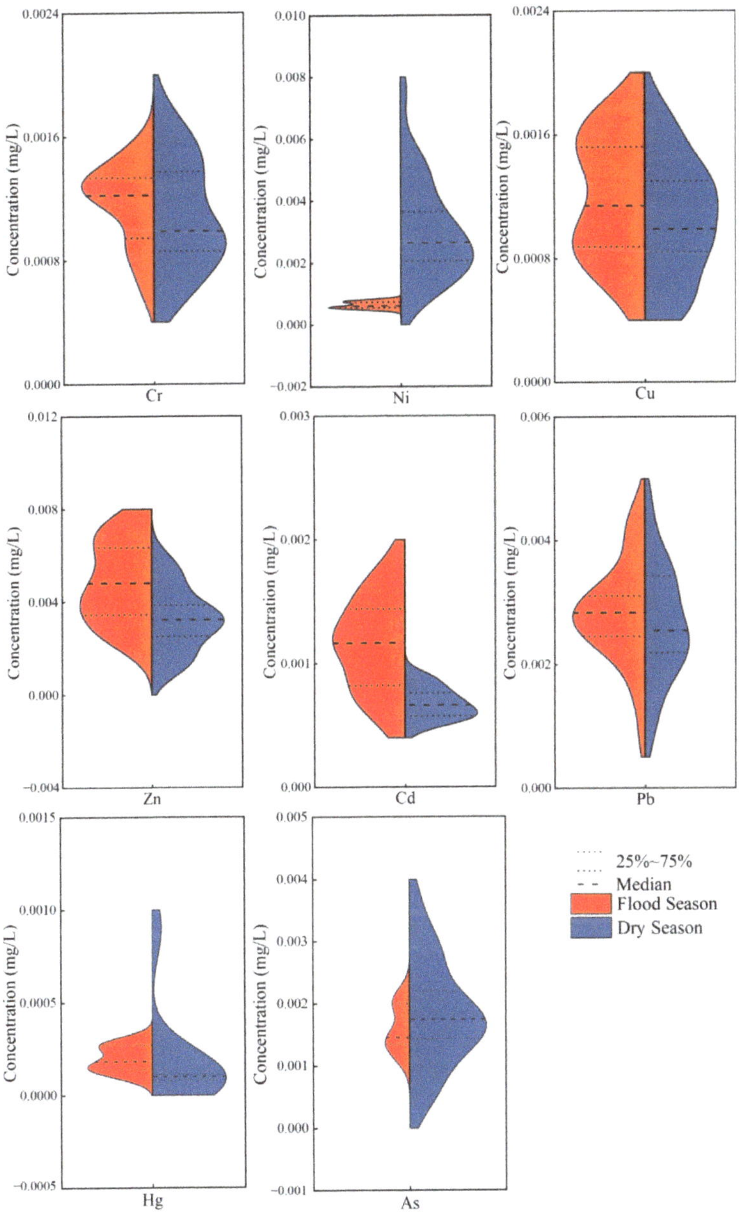

Figure 2. Comparison of heavy metal content in rivers in the flood season and dry season.

3.2. *Spatial Patterns of Water Quality Parameters*

The spatial variation in water quality indicators at each sampling site was categorized into four groups using the natural breakpoint method (Figures 3 and 4). This method can help predict the potential risks posed by heavy metals in surface water and offer recommendations for land use planning and water resource management [34]. Data were grouped into the most suitable categories to maximize the distinction between different classes [35]. The distribution of heavy metals varied significantly across different rivers, likely due to differences in the average concentrations, which could result from diverse pollution sources such as industrial activities, agriculture, and natural processes [36]. Sampling sites with high heavy metal concentrations were represented by darker colors, while lighter colors indicated lower contamination levels. It was observed that the northern rivers exhibited higher Cr concentrations compared to those in the southern region. Conversely, the southern part of the study area showed elevated Ni concentrations, potentially linked to the uneven spatial distribution of pollutants [37,38]. Higher Cu concentrations tended to occur across the entire river basin. Areas with increased heavy metal levels reflected industrial wastewater discharge and pesticide residues from agricultural activities, which entered water bodies via surface runoff [39,40]. Moreover, severe heavy metal pollution poses risks to biological health, with toxic metals accumulating in the human body and potentially leading to chronic poisoning [41]. The spatial distribution of Ni concentration has the same pattern as Cu. Notably, Zn and Cd concentrations in the southwest were higher than in other regions. A plausible explanation is the presence of livestock farms near rivers, where animal waste and decomposed organic matter enter surface waters through precipitation and seep into groundwater via irrigation drainage [42]. Furthermore, transportation activities were identified as a primary source of Cd in the environment. Cadmium is used in manufacturing brake coatings and corrosion-resistant parts [43,44]. Friction from braking leads to the corrosion of Cd layers, releasing Cd particles into the environment during operation [45]. Due to the plateau's unique and changing environmental conditions and its ecological vulnerability, human activities along the river, such as mineral exploitation, urban sewage, sewage discharge, etc., have had a significant negative impact on water quality in recent years. Pollution is therefore an inevitable consequence. The water quality of the source area will affect the middle and lower reaches of the Yellow River, so the water quality of the study area has an important impact on the safety of humans, animals, and crops.

3.3. Categorization of Water Quality Parameters

In this study, a self-organizing map (SOM) was utilized to analyze the distributional variations of water quality parameters, offering a scientific basis for effective water pollution management. The results, derived by integrating the SOM with the k-means clustering algorithm, were calculated [46,47]. The QE = 0.424 and TE = 0.1, indicating good model fitting. The matrix, which represents the distances between weight vectors of neurons and their adjacent neurons, is termed the uniform distance matrix (U-matrix). Different colors correspond to normalized variable values, with each hexagon representing a neuron on the component plane. Smaller distances between hexagons reflect greater similarity in sample features [48]. The neurons on the SOM maps showing higher values corresponded to elevated concentrations of Cu, Zn, and Pb, while Ni and As demonstrated the lowest concentrations at the same locations, as illustrated in Figure 5. Additionally, Hg and Cd predominantly exhibited lower concentrations.

The classification of water samples was conducted using the k-means clustering method. A total of fifty water samples were classified into three distinct groups, as shown in Figure 6. Samples collected during the flood season were assigned to Cluster I and Cluster III, while those from the dry season were grouped into Cluster II and Cluster III. Monitoring sites in Cluster III, characterized by higher levels of Cr and Cd, showed relatively lower concentrations of Hg. The results for the flood season and dry season were different, which further proved that meteorological factors can influence water quality. These positions with Cluster III were distributed near factories. The production activities of these factories may have an impact on the water quality of these monitoring sites. Additionally, both natural and artificial wetlands along riversides played a significant role in reducing the levels of pollutants in groundwater and surface water before they entered aquatic ecosystems. Riparian vegetation could prevent the agricultural non-point source pollution as a natural ecological barrier [49,50]. However, in lower vegetation, the function of removing organic pollutants through fixation and adsorption was weak. Most sampling positions with high NDVI had relatively lower heavy metal concentrations, which further proved the impact of vegetation.

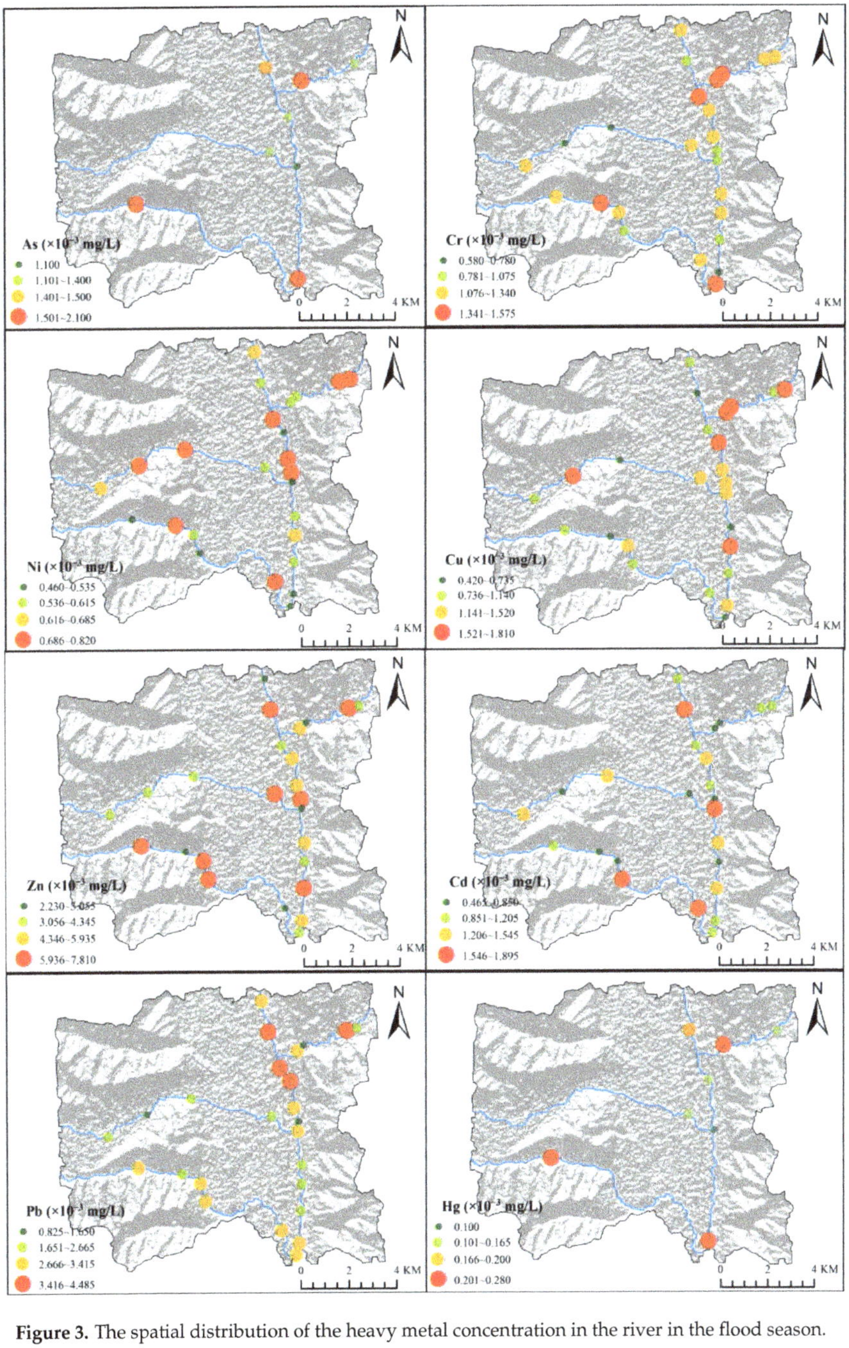

Figure 3. The spatial distribution of the heavy metal concentration in the river in the flood season.

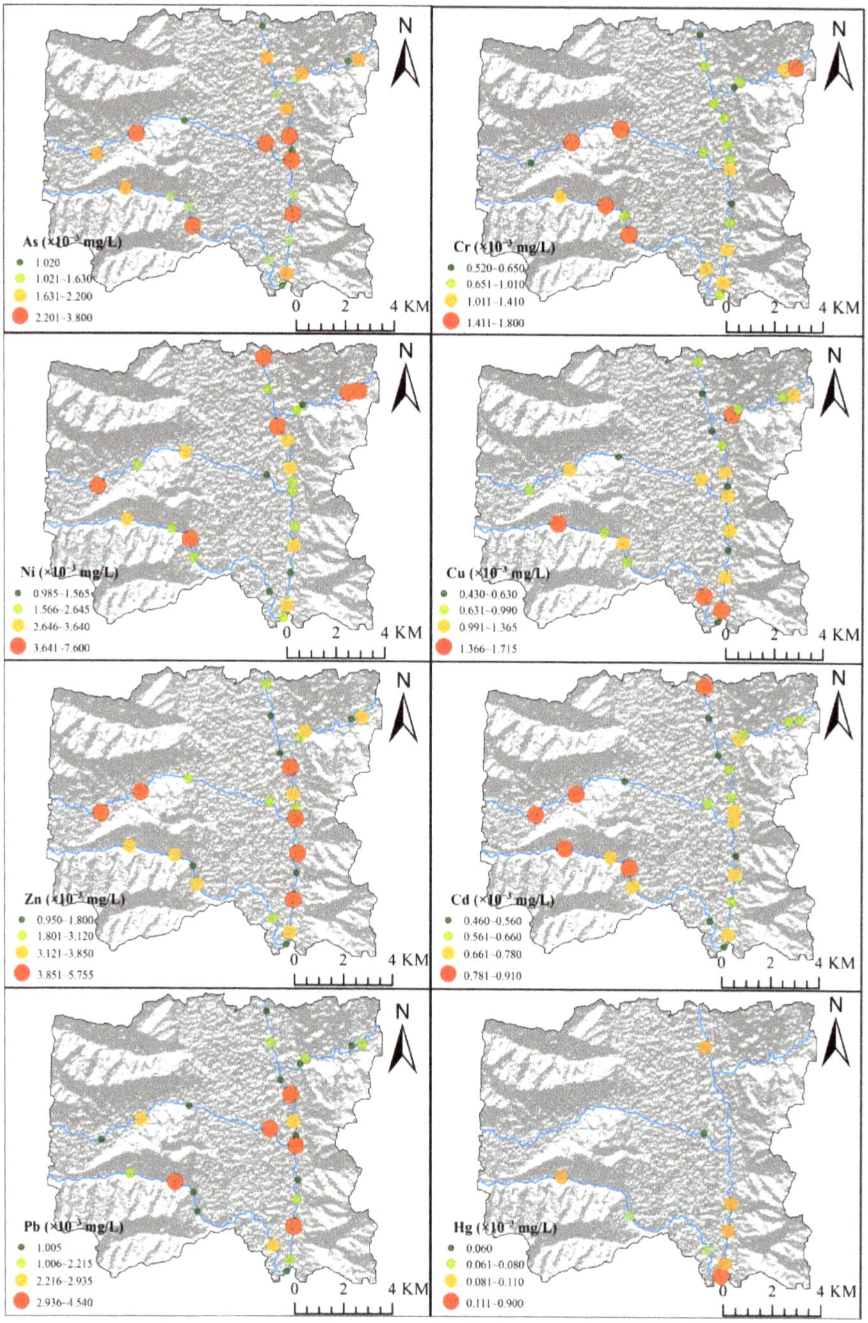

Figure 4. The spatial distribution of the heavy metal concentration in the river in the dry season.

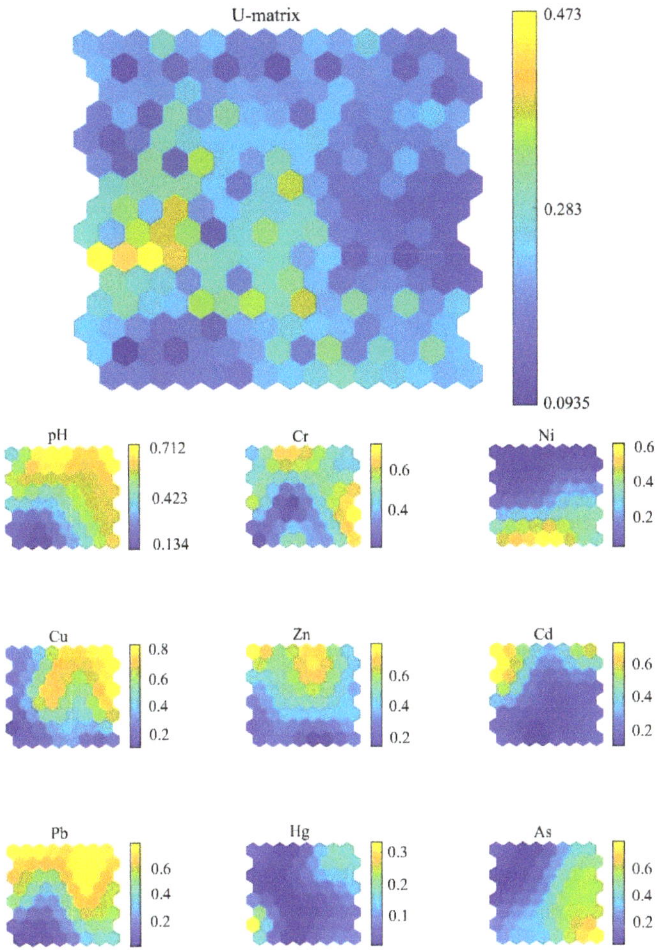

Figure 5. Self-organizing map and K-means cluster for pH and heavy metal concentration.

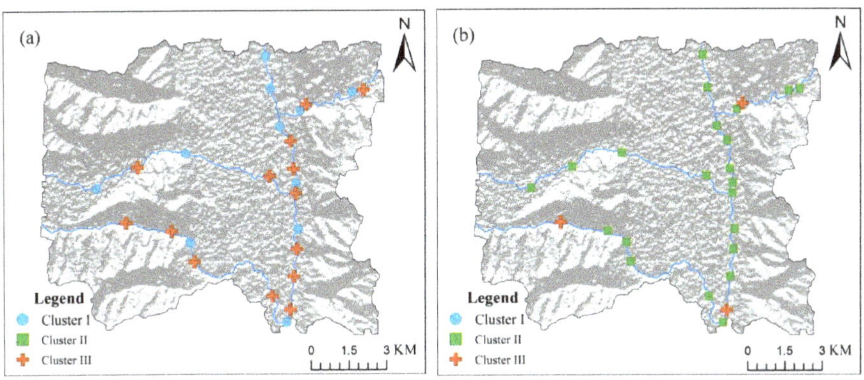

Figure 6. The spatial distribution of the heavy metal concentrations for each cluster. (**a**) flood season; (**b**) dry season.

3.4. Main Factors Influencing Water Quality

Natural and human-driven elements affecting water quality indexes were considered, including elevation, precipitation, temperature, potential evapotranspiration, the NDVI, and the distances from sampling sites to industrial areas, roads, and residential zones. The findings highlight that the NDVI, the distance from the sampling sites to industry, potential evapotranspiration, and key factors including heavy metal levels found in soil and dust are the primary contributors that significantly impact water quality indexes (Figure 7).

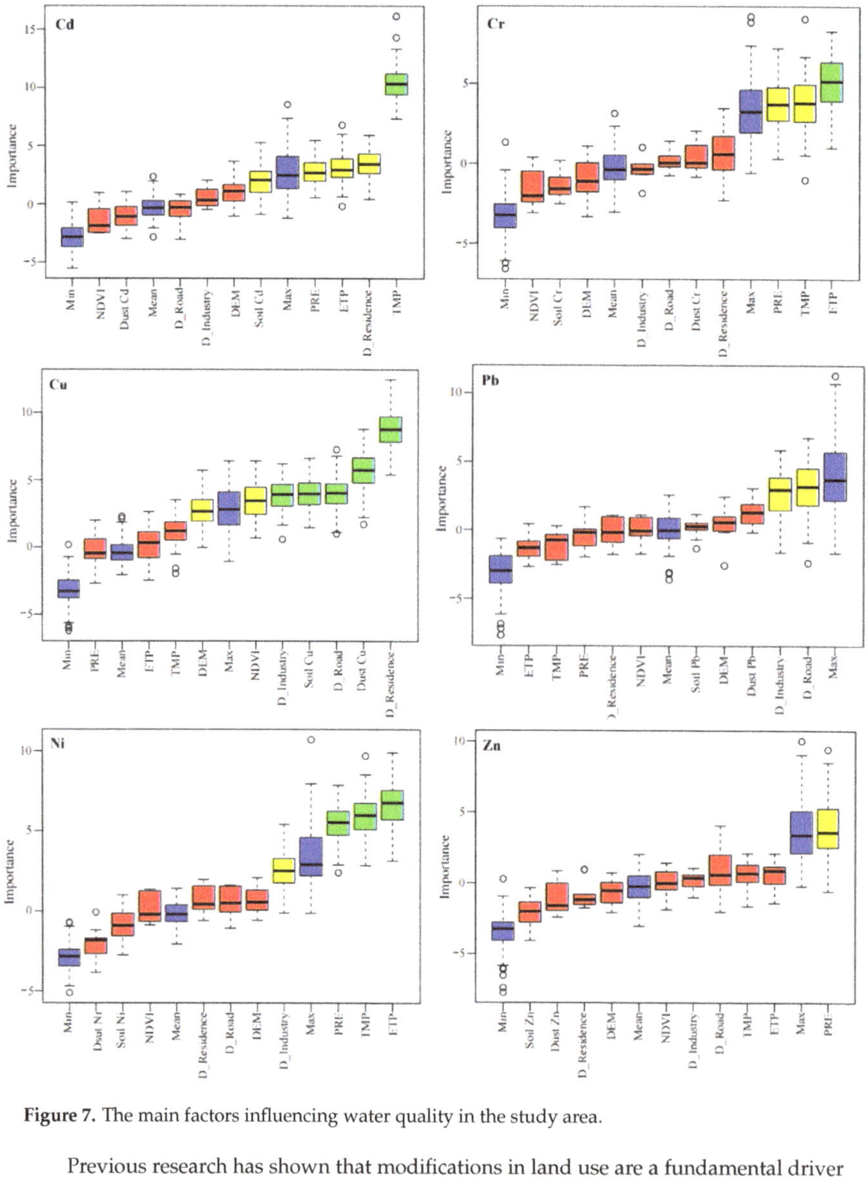

Figure 7. The main factors influencing water quality in the study area.

Previous research has shown that modifications in land use are a fundamental driver behind the decline in water quality, with human activities being the direct catalyst for such changes [10]. Pollution of water resources arising from land use modifications—via processes such as evapotranspiration, groundwater recharge, and surface runoff—constitutes a major challenge to the stability and functionality of aquatic ecosystems. Urban areas are often used as indicators of human activity. In addition, metal processing and electroplating activities release metal sulfides and oxides, which, along with solid waste, flow into rivers through runoff [51]. Moreover, the expansion of cropland has also been linked to inorganic pollution.

Climatic conditions are expected to influence the processes governing the behavior and movement of heavy metals by impacting the basin's hydrology and erosion dynamics [52]. An increasing number of studies have acknowledged the significant effects of climate. Increased precipitation can reduce the time of nitrogen staying in the watershed, thereby reducing the loss of denitrification, storage, and crop utilization. Extreme precipitation events can also cause higher runoff and extreme loading events [53]. In the QTP region, river recharge primarily originates from atmospheric precipitation, along with snow melt, ice melt, and groundwater. Rainfall is generally regarded as the main source of river recharge [54]. Temperature variations can alter the rate at which contaminants degrade naturally in rivers. The average annual temperature caused by the continental climate of the plateau is affected by the topography. In addition, temperature changes could significantly impact the behavior and movement of pesticides through mechanisms like degradation, volatilization, adsorption, and diffusion [55].

Topography affects contaminant transform through the spatial distribution of pollution sources. Furthermore, topographic features shape water quality by affecting pollutant pathways to rivers, including surface runoff velocity and erosion. Research has revealed that physiographic characteristics alone explained 4–15.9% of the variation in water quality in the Hengxi watershed [56]. In the study area, the terrain features a high DEM in the central region and lower DEM values on both sides. This complex topographic structure results in distinctive patterns of water quality distribution across the region.

Artificial and natural wetlands can serve as ecological barriers that intercept and filter agricultural non-point source pollution. Pollutants transported by surface runoff are taken up by vegetation, preventing them from entering water bodies. However, the influence of vegetation on water quality shows significant seasonal variations, likely due to the slower pace of surface runoff during drier periods. On the other hand, heavy rainfall producing substantial surface runoff reduces the regulatory effect of landscape configurations on pollutants during the rainy season and amplifies the role of geomorphic features in determining water quality [57]. At the buffer scale, landscape features provide a more accurate explanation of changes in water quality compared to the subwatershed scale [58,59]. This variation may be due to the greater effectiveness of riparian zones in reducing surface runoff and managing the transport and retention of heavy metals when compared to the broader sub-watershed scale [60,61].

Atmospheric deposition can affect water quality. Earlier studies have indicated that, in the absence of rainfall, pollutants can travel and disperse through the atmosphere, continue migrating toward the ground, and are continuously absorbed by surface layers.

3.5. Water Quality Risk Assessment

The NDVI, the distance from the sampling sites to industry, and potential evapotranspiration and potential evapotranspiration were combined with heavy metal concentrations in dust and soil to evaluate water quality using the IFN-SPA approach [62]. This technique assesses the relative significance of factors through comparative analysis. The approach highlights the significance of each factor by measuring the relative distance between evaluation samples and the defined criteria. The risk levels are categorized into very low risk ($\leq$0.2), low risk ($\leq$0.2–0.4), medium risk ($\leq$0.4–0.6), high risk ($\leq$0.6–0.8), and very high risk ($\leq$0.8–1.0). According to this classification and methodology, the results (Figure 8) revealed that heavy metal risks were lower in the downstream section of the Beichuan River, whereas the upstream section exhibited higher risks. The enterprises in the upper reaches of the basin may have discharged some wastewater and gas polluted with heavy metals. Most of the tributaries of Majuanggou River and Jingyang River showed level II risk. The risk of heavy metal pollution in Sigou River was shown to be level III and level IV. There are indeed some livestock and poultry farms next to the Sigou River, and there will be disorderly discharge and accumulation of feces, and the odor is more intense. Therefore, the risk of heavy metal pollution is higher in the Sigou River.

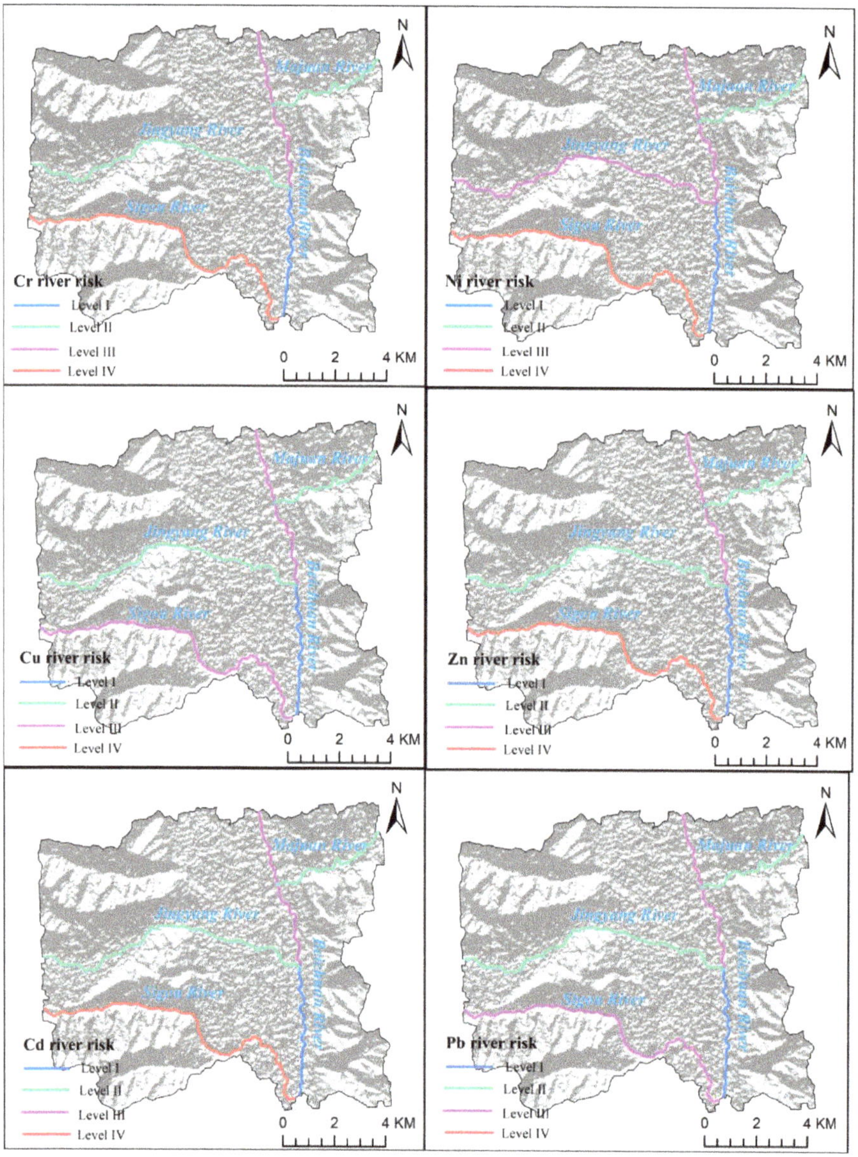

Figure 8. Water quality risk assessment in the study area.

4. Conclusions

The Qinghai–Tibet Plateau is recognized as a critical water resource region for China and plays a vital role in ecological conservation as well as in serving as a source of watersheds. This study focused on examining the spatial and temporal patterns of water quality indicators, exploring key factors that influence water quality and evaluating the associated risks. The results showed that the average concentrations in the flood season and dry season are significantly different. The average value of Cd concentration during the flood season complied with the Class II water quality standard. Heavy metals exhibit varied spatial distribution patterns, and the reason for the difference comes from livestock farms and industrial enterprises. The results of the flood season and dry season were different, which further proved that meteorological factors can influence water quality. The risk of heavy metals in different rivers presents different spatial distribution characteristics, for example, the risk of heavy metals in Sigou River is higher. The findings underline the importance of developing an efficient evaluation framework to address river water pollution management and control in the long term.

Author Contributions: Methodology, D.M.; Software, D.M.; Validation, Q.P.; Investigation, Y.M.; Writing—original draft, X.Z.; Writing—review & editing, Z.Y. All authors have read and agreed to the published version of the manuscript.

Funding: This research was funded by the Basic Research Program of Qinghai Province (2023-ZJ-910M).

Data Availability Statement: The data that support the findings of this study are available from the corresponding author upon reasonable request.

Acknowledgments: The authors thank the editors and reviewers for their helpful and insightful comments.

Conflicts of Interest: The authors declare no conflicts of interest.

References

1. Zheng, N.; Wang, Q.C.; Zheng, D.M. Health risk of Hg, Pb, Cd, Zn, and Cu to the inhabitants around Huludao Zinc Plant in China via consumption of vegetables. *Sci. Total Environ.* **2007**, *383*, 81–89. [CrossRef] [PubMed]
2. Uddin, G.M.; Nash, S.; Rahman, A.; Olbert, A.I. Assessing optimization techniques for improving water quality model. *J. Clean. Prod.* **2023**, *385*, 135671. [CrossRef]
3. Huang, J.; Zhang, Y.; Bing, H.; Peng, J.; Dong, F.; Gao, J.; Arhonditsis, G.B. Characterizing the river water quality in China: Recent progress and on-going challenges. *Water Res.* **2021**, *201*, 117309. [CrossRef]
4. Mari, M.; Nadal, M.; Schuhmacher, M.; Domingo, J.L. Application of self-organizing maps for PCDD/F pattern recognition of environmental and biological samples to evaluate the impact of a hazardous waste incinerator. *Environ. Sci. Technol.* **2010**, *44*, 3162–3168. [CrossRef]
5. Feng, Z.; Xu, C.; Zuo, Y.; Luo, X.; Wang, L.; Chen, H.; Xie, X.; Yan, D.; Liang, T. Analysis of water quality indexes and their relationships with vegetation using self-organizing map and geographically and temporally weighted regression. *Environ. Res.* **2023**, *216*, 114587. [CrossRef] [PubMed]
6. Li, C.; Sun, L.; Jia, J.; Cai, Y.; Wang, X. Risk assessment of water pollution sources based on an integrated k-means clustering and set pair analysis method in the region of Shiyan. *China Sci. Total Environ.* **2016**, *557–558*, 307–316. [CrossRef] [PubMed]
7. Lyu, H.M.; Shen, S.L.; Zhou, A. The development of IFN-SPA: A new risk assessment method of urban water quality and its application in Shanghai. *J. Clean. Prod.* **2021**, *282*, 124542. [CrossRef]
8. Lyu, H.M.; Shen, S.L.; Zhou, A.; Yin, Z.Y. Assessment of safety status of shield tunnelling using operational parameters with enhanced SPA. *Tunn. Undergr. Space Technol.* **2022**, *123*, 104428. [CrossRef]
9. de Paul Obade, V.; Moore, R. Synthesizing water quality indicators from standardized geospatial information to remedy water security challenges: A review. *Environ. Int.* **2018**, *119*, 220–231. [CrossRef]
10. Serpa, D.; Nunes, J.P.; Keizer, J.J.; Abrantes, N. Impacts of climate and land use changes on the water quality of a small Mediterranean catchment with intensive viticulture. *Environ. Pollut.* **2017**, *224*, 454–465. [CrossRef]
11. Chapra, S.C.; Boehlert, B.; Fant, C.; Bierman, V.J.; Henderson, J.; Mills, D.; Mas, D.M.L.; Rennels, L.; Jantarasami, L.; Martinich, J.; et al. Climate change impacts on harmful algal blooms in U.S. freshwaters: A screening-level assessment. *Environ. Sci. Technol.* **2017**, *51*, 8933–8943. [CrossRef]

12. Michalak, A.M. Study role of climate change in extreme threats to water quality. *Nature* **2016**, *535*, 349–350. [CrossRef] [PubMed]
13. Tian, Y.; Yu, C.; Zha, X.; Wu, J.; Gao, X.; Feng, C.; Luo, K. Distribution and potential health risks of arsenic, selenium, and fluorine in natural waters in Tibet, China. *Water* **2016**, *8*, 568. [CrossRef]
14. *GB3838-2002*; Environmental Quality Standards for Surface Water. Ministry of Ecology and Environment of the People's Republic of China: Beijing, China; General Administration of Quality Supervision, Inspection and Quarantine of the People's Republic of China: Beijing, China, 2002.
15. Kohonen, T. Self-organized formation of topologically correct feature maps. *Biol. Cybern.* **1982**, *43*, 59–69. [CrossRef]
16. Lu, R.; Lo, S. Diagnosing reservoir water quality using self-organizing maps and fuzzy theory. *Water Res.* **2022**, *36*, 2265–2274. [CrossRef]
17. Rahman, A.T.M.S.; Kono, Y.; Hosono, T. Self-organizing map improves understanding on the hydrochemical processes in aquifer systems. *Sci. Total Environ.* **2022**, *846*, 157281. [CrossRef] [PubMed]
18. Şahin, M. Impact of weather on COVID-19 pandemic in Turkey. *Sci. Total Environ.* **2020**, *728*, 138810. [CrossRef]
19. Subbiah, S.; Anbananthen, K.S.M.; Thangaraj, S.; Kannan, S.; Chelliah, D. Intrusion detection technique in wireless sensor network using grid search random forest with Boruta feature selection algorithm. *J. Commun. Netw.* **2022**, *24*, 264–273. [CrossRef]
20. Kursa, M.B.; Rudnicki, W.R. Feature selection with the Boruta package. *J. Stat. Softw.* **2010**, *36*, 1–13. [CrossRef]
21. Gauthier, T.D. Detecting trends using Spearman's rank correlation coefficient. *Environ. Forensics* **2001**, *2*, 359–362. [CrossRef]
22. Ho, J.; Afan, H.A.; El-Shafie, A.H.; Koting, S.B.; Mohd, N.S.; Jaafar, W.Z.B.; Hin, L.S.; Malek, M.A.; Ahmed, A.N.; Mohtar, W.H.M.W.; et al. Towards a time and cost effective approach to water quality index class prediction. *J. Hydrol.* **2019**, *575*, 148–165. [CrossRef]
23. Bai, Y.; Sun, W.; Li, N.; Gu, Q. Spatial distribution and pollution characteristics of heavy metals in soil of Diaojiang River basin based on geostatistics. *J. Min. Sci. Technol.* **2017**, *2*, 409–415.
24. Wu, W.; Qu, S.; Nel, W.; Ji, J. The impact of natural weathering and mining on heavy metal accumulation in the karst areas of the Pearl River Basin, China. *Sci. Total Environ.* **2020**, *734*, 139480. [CrossRef] [PubMed]
25. Morsy, A.; Ebeid, M.; Soliman, A.; Halim, A.A.; Ali, A.E.; Fahmy, M. Evaluation of the water quality and the eutrophication risk in Mediterranean sea area: A case study of the Port Said Harbour. *Egypt. Environ. Chall.* **2022**, *7*, 100484. [CrossRef]
26. Choudhury, T.R.; Islam, T.; Md Towfiqul Islam, A.R.; Hasanuzzaman, M.; Idris, A.M.; Rahman, M.S.; Alam, E.; Chowdhury, A.M.S. Multi-media compartments for assessing ecological and health risks from concurrent exposure to multiple contaminants on Bhola Island, Bangladesh. *Emerg. Contam.* **2022**, *8*, 134–150. [CrossRef]
27. Bai, B.; Nie, Q.; Zhang, Y.; Wang, X.; Hu, W. Cotransport of heavy metals and SiO_2 particles at different temperatures by seepage. *J. Hydrol.* **2021**, *597*, 125771. [CrossRef]
28. Zhao, X.; Liu, X.; Xing, Y.; Wang, L.; Wang, Y. Evaluation of water quality using a Takagi-Sugeno fuzzy neural network and determination of heavy metal pollution index in a typical site upstream of the Yellow River. *Environ. Res.* **2022**, *211*, 113058. [CrossRef] [PubMed]
29. Sun, W.; Xu, X.; Lv, Z.; Mao, H.; Wu, J. Environmental impact assessment of wastewater discharge with multi pollutants from iron and steel industry. *J. Environ. Manag.* **2019**, *245*, 210–215. [CrossRef]
30. Haghnazar, H.; Johannesson, K.H.; Gonz'alez-Pinz'on, R.; Pourakbar, M.; Aghayani, E.; Rajabi, A.; Hashemi, A.A. Groundwater geochemistry, quality, and pollution of the largest lake basin in the Middle East: Comparison of PMF and PCA-MLR receptor models and application of the source-oriented HHRA approach. *Chemosphere* **2022**, *288*, 132489. [CrossRef]
31. Ezoe, K.; Ohyama, S.; Hashem, M.A.; Ohira, S.I.; Toda, K. Automated determinations of selenium in thermal power plant wastewater by sequential hydride generation and chemiluminescence detection. *Talanta* **2016**, *148*, 609–616. [CrossRef]
32. Wan, K.; Huang, L.; Yan, J.; Ma, B.; Huang, X.; Luo, Z.; Zhang, H.; Xiao, T. Removal of fluoride from industrial wastewater by using different adsorbents: A review. *Sci. Total Environ.* **2021**, *773*, 145535. [CrossRef] [PubMed]
33. Ding, J.; Cao, J.; Xu, Q.; Xi, B.; Su, J.; Gao, R.; Huo, S.; Liu, H. Spatial heterogeneity of lake eutrophication caused by physiogeographic conditions: An analysis of 143 lakes in China. *J. Environ. Sci.* **2015**, *30*, 140–147. [CrossRef] [PubMed]
34. Choudhury, T.R.; Acter, T.; Alam, M.A.; Sowrav, S.F.F.; Rahman, M.S.; Chowdhury, A.M.S.; Quraishi, S.B. Appraisal of heavy metal contamination and their source apportionment identification in five river water systems of the coastal areas in Bangladesh. *Reg. Stud. Mar. Sci.* **2024**, *70*, 103378. [CrossRef]
35. Xu, W.; Yu, W.; Jing, S.; Zhang, G.; Huang, J. Debris flow susceptibility assessment by GIS and information value model in a large-scale region, Sichuan Province (China). Nat. *Hazards* **2013**, *65*, 1379–1392. [CrossRef]
36. Gong, S.; Bai, X.; Luo, G.; Li, C.; Wu, L.; Chen, F.; Ran, C.; Xi, H.; Zhang, S. Climate change has enhanced the positive contribution of rock weathering to the major ions in riverine transport. *Glob. Planet. Change* **2023**, *228*, 104203. [CrossRef]
37. Abyaneh, Z. Evaluation of multivariate linear regression and artificial neural networks in prediction of water quality parameters. *J. Environ. Health Sci. Eng.* **2014**, *12*, 1–8.
38. Ahmed, A.M.; Shah, S.M.A. Application of adaptive neuro-fuzzy inference system (ANFIS) to estimate the biochemical oxygen demand (BOD) of Surma River. *J. King Saud Univ. Eng. Sci.* **2017**, *29*, 237–243. [CrossRef]
39. Hern'andez, F.; Ib'anez, M.; Portol'es, T.; Cervera, M.I.; Sancho, J.V.; L'opez, F.J. Advancing towards universal screening for organic pollutants in waters. *J. Hazard. Mater.* **2015**, *282*, 86–95. [CrossRef] [PubMed]

40. Lee, J.; Lee, S.; Yu, S.; Rhew, D. Relationships between water quality parameters in rivers and lakes: BOD5, COD, NBOPs, and TOC. *Environ. Monit. Assess.* **2016**, *188*, 1–8. [CrossRef]
41. Rashed, M.N. Adsorption technique for the removal of organic pollutants from water and wastewater. *Organ. Pollut.-Monitor. Risk Treat.* **2013**, *7*, 167–194.
42. Hossain, M.A.; Im, S.; Nasly, M.A. Water quality index: An indicator of surface water pollution in eastern part of Peninsular Malaysia. *Res. J. Recent Sci.* **2013**, *2*, 10–17.
43. Ndiokwere, C.L. A study of heavy metal pollution from motor vehicle emissions and its effect on roadside soil, vegetation and crops in Nigeria. Environ. *Pollut. Ser. B Chem. Phys.* **1984**, *7*, 35–42. [CrossRef]
44. Yao, X.; Yu, X.; Wang, L.; Zeng, Y.; Mao, L.; Liu, S.; Xie, H.; He, G.; Huang, Z.; Liu, Z. Preparation of cinnamic hydroxamic acid collector and study on flotation characteristics and mechanism of scheelite. *Int. J. Min. Sci. Technol.* **2023**, *33*, 773–781. [CrossRef]
45. Faroon, O.; Ashizawa, A.; Wright, S.; Tucker, P.; Jenkins, K.; Ingerman, L.; Rudisill, C. *Toxicological Profile for Cadmium*; Agency for Toxic Substances and Disease Registry (US): Atlanta, GA, USA, 2012. [PubMed]
46. Ashari, I.F.; Banjarnahor, R.; Farida, D.R.; Aisyah, S.P.; Dewi, A.P.; Humaya, N. Application of data mining with the K-means clustering method and Davies bouldin index for grouping IMDB movies. *J. Appl. Informat. Comput.* **2022**, *6*, 7–15. [CrossRef]
47. Rabiaa, E.; Noura, B.; Adnene, C. Improvements in LEACH based on K-means and Gauss algorithms. *Procedia Comput. Sci.* **2015**, *73*, 460–467. [CrossRef]
48. Wang, Z.; Xiao, J.; Wang, L.; Liang, T.; Guo, Q.; Guan, Y.; Rinklebe, J. Elucidating the differentiation of soil heavy metals under different land uses with geographically weighted regression and self-organizing map. *Environ. Pollut.* **2020**, *260*, 114065. [CrossRef] [PubMed]
49. Yang, L.; Chen, L.; Sun, R. River ecosystems and their self-purification capability: Research status and challenges. *Acta Ecol. Sin.* **2009**, *29*, 5066–5075.
50. Terrado, M.; Tauler, R.; Bennett, E.M. Landscape and local factors influence water purification in the Monteregian agroecosystem in Qu'ebec, Canada. *Reg. Environ. Change* **2015**, *15*, 1743–1755. [CrossRef]
51. Xiang, Z.; Wu, S.; Zhu, L.; Yang, K.; Lin, D. Pollution characteristics and source apportionment of heavy metal(loid)s in soil and groundwater of a retired industrial park. *J. Environ. Sci.* **2024**, *143*, 23–34. [CrossRef] [PubMed]
52. Xiong, Q.; Xiao, Y.; Liang, P.; Li, L.; Zhang, L.; Li, T.; Pan, K.; Liu, C. Trends in climate change and human interventions indicate grassland productivity on the Qinghai-Tibetan Plateau from 1980 to 2015. *Ecol. Indic.* **2021**, *129*, 108010. [CrossRef]
53. Ballard, T.C.; Sinha, E.; Michalak, A.M. Long-term changes in precipitation and temperature have already impacted nitrogen loading. *Environ. Sci. Technol.* **2019**, *53*, 5080–5090. [CrossRef]
54. Wang, W.; Liang, T.; Wang, L.; Liu, Y.; Wang, Y.; Zhang, C. The effects of fertilizer applications on runoff loss of phosphorus. *Environ. Earth Sci.* **2013**, *68*, 1313–1319. [CrossRef]
55. Mararakanye, N.; Le Roux, J.J.; Franke, A.C. Long-term water quality assessments under changing land use in a large semi-arid catchment in South Africa. *Sci. Total Environ.* **2022**, *818*, 151670. [CrossRef]
56. Wu, J.; Lu, J. Spatial scale effects of landscape metrics on stream water quality and their seasonal changes. *Water Res.* **2021**, *191*, 116811. [CrossRef] [PubMed]
57. Yu, S.; Xu, Z.; Wu, W.; Zuo, D. Effect of land use types on stream water quality under seasonal variation and topographic characteristics in the Wei River basin, China. *Ecol. Indic.* **2016**, *60*, 202–212. [CrossRef]
58. Hurley, T.; Mazumder, A. Spatial scale of land-use impacts on riverine drinking source water quality. *Water Resour. Res.* **2013**, *49*, 1591–1601. [CrossRef]
59. Mainali, J.; Chang, H. Landscape and anthropogenic factors affecting spatial patterns of water quality trends in a large river basin. *South Korea J. Hydrol.* **2018**, *564*, 26–40. [CrossRef]
60. Fernandes, J.D.F.; de Souza, A.L.; Tanaka, M.O. Can the structure of a riparian forest remnant influence stream water quality? A tropical case study. *Hydrobiologia* **2014**, *724*, 175–185. [CrossRef]
61. Casson, N.J.; Eimers, M.C.; Watmough, S.A.; Richardson, M.C. The role of wetland coverage within the near-stream zone in predicting of seasonal stream export chemistry from forested headwater catchments. *Hydrol. Process.* **2019**, *33*, 1465–1475. [CrossRef]
62. Huang, J.; Zhang, Y.; Arhonditsis, G.B.; Gao, J.; Chen, Q.; Wu, N.; Dong, F.; Shi, W. How successful are the restoration efforts of China's lakes and reservoirs? *Environ. Int.* **2019**, *123*, 96–103. [CrossRef] [PubMed]

water

MDPI

Article

Ecological Compensation Based on the Ecosystem Service Value: A Case Study of the Xin'an River Basin in China

Yuanhua Chen [1,2,*], Qinglian Wu [1] and Liang Guo [1]

1 State Key Laboratory of Urban Water Resource and Environment, Harbin Institute of Technology, Harbin 150090, China; wuqinglian@hit.edu.cn (Q.W.)

2 China Energy Conservation and Environmental Protection Group (CECEP), Eco-Product Development Research Center Co., Ltd., Beijing 100082, China

* Correspondence: chenyuanhua@cecep.cn

Abstract: To establish a sound ecological compensation (EC) mechanism in the Xin'an River Basin, this study suggested utilizing ecosystem service valuation to determine the compensation amount. In this study, the first step was to establish a reasonable watershed EC model using the ecological compensation supply coefficient (ECSC) based on the value spillover theory (VST) of the ecosystem services and the ecological compensation demand coefficient (ECDC). The second step was to classify the ecosystem services of the Xin'an River Basin into three categories, including supply service, regulating service, and cultural service, with 14 specific functions to determine the ecological compensation standard accounting scope in these services. Then, a case study on the Xin'an River Basin for EC standards was presented. The total ecosystem service value (ESV) in the Xin'an River Basin was estimated to be CNY 70.271 billion, with supply service accounting for 22.7%, regulating service accounting for 24.6%, and cultural service accounting for 52.7%. Based on the compensation scope, the ecosystem service values for the upper and lower limits of the EC were calculated as CNY 57.779 billion and CNY 17.292 billion. Combined with the results of the ECSC and ECDC, the upper and lower limits of the EC standard in the Xin'an River Basin were computed to be CNY 4.085 billion and CNY 1.438 billion, respectively. Therefore, the ESV-based EC model for the Xin'an River Basin can effectively address the challenge of inadequate EC in the watershed. It also facilitates balanced regional development and serves as a theoretical foundation and empirical evidence for the government to establish a unified national policy on cross-border river basin ecological compensation.

Keywords: ecological compensation; ecosystem services value; value spillover theory; river basin; Xin'an River Basin

1. Introduction

Watersheds are essential components of natural ecosystems that provide habitats for humans and various organisms. However, due to the rapid development of the social economy, water pollution in watersheds has become inevitable. Additionally, disputes over water resources between the upstream and downstream regions of watersheds continue to arise, threatening the sustainability of watersheds [1]. Since the 1980s, China has implemented an ecological compensation (EC) mechanism to protect the ecological environment by regulating the relationship between stakeholders and casualties of the ecological environment [2]. The basic principle of the Chinese government's policy on EC is to balance the interests of ecosystem service providers and demanders by requiring that beneficiaries and polluters pay for the compensation fund [3,4]. By using EC, ecosystem service providers can protect or restore ecosystems, thereby promoting ecological sustainability [5]. For water resource and eco-environment protection, establishing an EC mechanism in the river basin has become inevitable. However, despite the potential of

EC to achieve both ecological protection and economic development [6,7], practical issues remain in defining the subject and object of EC, standardizing compensation criteria, and allocating compensation funds [8].

Taking China's first trans-provincial watershed ecological compensation pilot project as an example, the Xin'an River Basin has obtained significant achievements in watershed protection and development, which has strengthened the Chinese government's confidence in replicating its successful experience via the subsequent implementations of other large watersheds, such as the Yangtze and Huang Rivers. During the 1990s, the water quality of Qiandao Lake downstream of the Xin'an River Basin showed an increasing trend toward eutrophication. This situation intensified the conflict between ecological conservation and economic expansion in the upstream and downstream regions of the basin, which hindered the sustainable development of the socio-economy within the Xin'an River Basin [9–11]. In 2012, the governments of the Anhui and Zhejiang provinces signed the Agreement on Water Environmental Compensation for Xin'an River Basin, which officially implemented the first trans-provincial EC mechanism in China. Under the collaborative efforts of the Anhui and Zhejiang provinces, three rounds of pilot programs were launched from 2012 to 2014, 2015 to 2017, and 2018 to 2020, respectively, with the goal of establishing an operating mechanism based on the principle of "beneficiaries pay and protectors receive compensation" [12]. Since the establishment of the ecological compensation pilot program in the Xin'an River Basin, there has been a significant improvement in water quality in the basin compared to the past. At present, cross-provincial upstream and downstream horizontal EC pilot projects have been expanded to other rivers, such as the Jiuzhou River, Tingjiang-Hanjiang River, Dongjiang River, Huaihe River, Yangtze River Economic Belt, and the Yellow River Basin [13–16]. However, Huangshan City in Anhui Province received a total of CNY 2.2 billion in ecological compensation funds based on the three-round ecological compensation agreement in the Xin'an River Basin, as shown in Figure 1, with an average of CNY 0.244 billion per year, which is significantly different from the actual investment in ecological protection in Huangshan City [9,17]. In addition, according to existing research, the opportunity cost of Huangshan City's loss due to the implementation of basin compensation during 2013–2017 was between CNY 2.456 billion and CNY 5.327 billion, which led to a significant gap in economic development between the upstream and downstream of the basin. In addition, the compensation efforts were clearly insufficient, making it difficult to enhance the enthusiasm for strengthening the protection in Huangshan City.

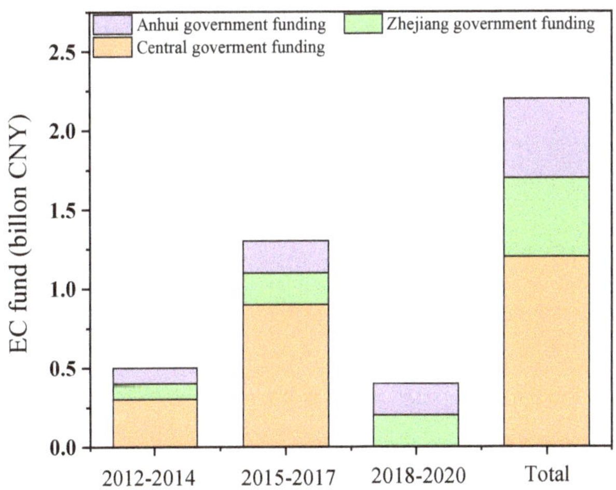

Figure 1. Three rounds of watershed EC agreement between Huangshan and Hangzhou.

Therefore, the Chinese government encourages the implementation of market EC in upstream and downstream watersheds, but a unified and feasible mechanism for watershed EC has yet to be established [18,19]. In this instance, establishing compensation standards that are mutually recognized by both upstream and downstream areas has become the top priority of watershed EC work. To solve this problem, several studies on calculating the watershed EC standards have been conducted in recent years, mainly based on the theory of ecosystem service [20].

According to the Millennium Ecosystem Assessment, there are four main types of ecosystem services, as follows: support services, regulation services, provision services, and cultural services [21,22]. Watershed ecosystems provide vital functions such as flood control, drought resistance, climate regulation, water conservation, flood regulation, land reclamation, biodiversity protection, and tourism, which generate significant ecological, economic, and social benefits for humans. The ecological compensation mechanism based on the ESV is a novel approach that integrates the economic value of ecosystem services into compensation standards to promote a win–win situation for ecological protection and economic development [23].

Based on previous research, a reasonable calculation of the ecosystem service value is a prerequisite for determining the upper limit of EC standards, and it would be unreasonable to directly use the ESV as a standard for EC due to its excessively high value [24]. Dai et al. (2021) indicated that the total value of ecosystem services in the Xin'an River Basin from 2013 to 2017 was between CNY 8.828 billion and CNY 9.088 billion [25]. Yang, N (2019) showed that the total value of water ecosystem services in the Xin'an River Basin in 2020 was CNY 41.409 billion [26]. In these studies, there was an excessive focus on upstream enthusiasm to protect the ecological environment while ignoring downstream willingness to pay (WTP). Once the profit balance of EC is broken, it will cause the EC rules and regulations to lose their effectiveness. Therefore, extensive research and revision of the value of ecosystem services and compensation standards are necessary to ensure their rationality and feasibility [27].

Consequently, this study aims to establish a new EC scheme by exploring a reasonable revision of the ESV or the amount of EC to increase the applicability of Xin'an River Basin EC and achieve a better compensation effect. The formulation of compensation standards will consider multiple factors, such as the actual value of ecosystem services, the rights and interests of compensation subjects and objects, and the source of compensation funds. In this paper, we will propose an EC standard model by employing the ecological compensation supply coefficient (ECSC) based on the value spillover theory and the ecological compensation demand coefficient (ECDC) to adjust the ESV after identifying the types and functions of the main watershed ecosystem services in Huangshan City. Finally, a case study will be conducted to verify the reasonable range of the Xin'an River Basin EC standard based on the established model.

2. Data and Methods

2.1. Studied Area

The Xin'an River originates from Xiuning County, Huangshan City, and flows through the Anhui and Zhejiang provinces (Figure 2). It is the third-largest water system in Anhui Province, after the Yangtze and Huaihe Rivers, and is also the largest river that flows into Qiandao Lake in Zhejiang Province [28].

The Xin'an River Basin covers approximately 11,452.5 square kilometers, of which 5569.75 square kilometers are located in Huangshan City, accounting for 56.79% of the city's total area. In Anhui Province, the Xin'an River's annual average outflow exceeds 6 billion cubic meters, which represents over 60% of the yearly inflow of Qiandao Lake, making it an essential strategic water source in the downstream region of Zhejiang Province [12].

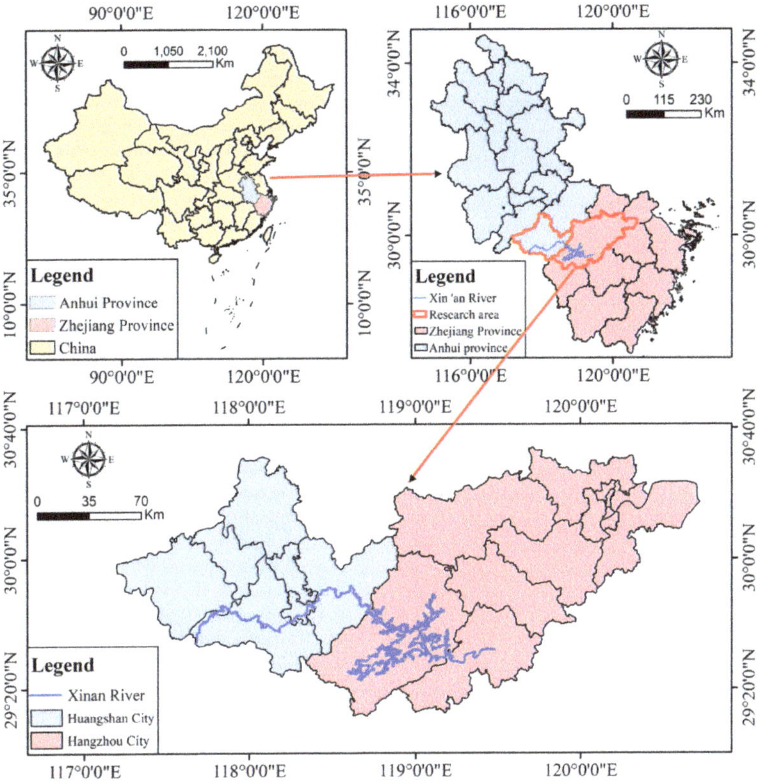

Figure 2. Map of Xin'an River Basin.

2.2. Data Sources

Empirical calculations were carried out to determine the EC standards for the Xin'an River Basin. To obtain the necessary data, remote sensing data were utilized to gather administrative boundary data, first-level ecological system classification data for 2020, the net primary productivity, the vegetation normalized index, soil physical properties, vegetation evaporation, and the digital elevation grid. Other data sources included the "Anhui Statistical Yearbook", "Huangshan Statistical Yearbook", "Hangzhou Statistical Yearbook", "Huangshan Environmental Quality Bulletin", "Hangzhou Environmental Quality Bulletin", "Anhui Water Resources Bulletin", "Anhui Environmental Statistical Yearbook", "Zhejiang Water Resources Bulletin", "Compilation of China's Urban Sewage Treatment Plants", and official websites of the Huangshan Ecological Environment Bureau and Huangshan Water Conservancy Bureau. The data collected encompassed the ecological environment, economic development, population, and other relevant information.

2.3. Methodology

The main idea of this study is illustrated in Figure 3. The first step was to classify the ecosystem services in Huangshan City's upstream area of the Xin'an River. Second, the value of different ecosystem services was calculated using various methods, including shadow engineering, market comparison, and alternative cost. Third, the value of ecosystem services was adjusted using a value correction coefficient. Finally, based on the ecological compensation coefficient and the ecological service spillover theory, this study estimated the EC that Huangshan City should receive in the river basin.

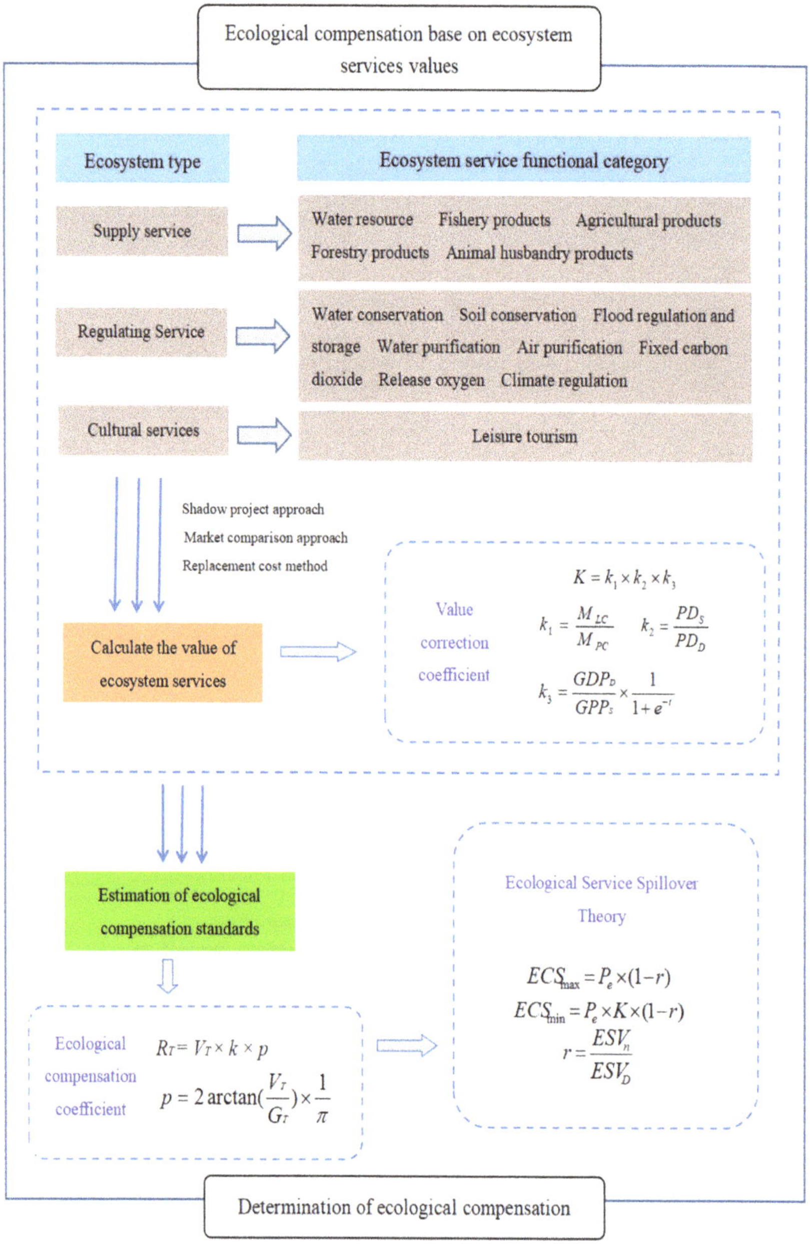

Figure 3. Flow graph of the basin eco-compensation framework based on ecosystem service values.

2.3.1. Value Measurement of Main Ecosystem Services

This study classifies the basin ecosystem services into the following three main categories: supply services, regulating services, and cultural services, including 14 subcategories presented in Table 1. The calculation method for gross ecosystem services value is as follows:

$$V_{total} = \sum_{i=1}^{14} V_i \tag{1}$$

In the formula, V_{total} is the total value of ecosystem service; V_i is the ecosystem service value of the *i*th ecosystem type; and *i* is the ecosystem type, including the water resource products (V_1), fishery products (V_2), agricultural products (V_3), forestry products (V_4), animal husbandry products (V_5), water conservation (V_6), soil conservation (V_7), flood regulation and storage (V_8), water purification (V_9), air purification (V_{10}), fixed carbon dioxide (V_{11}), release (V_{12}), climate regulation (V_{13}), and leisure tourism values (V_{14}), which are calculated and presented in Table 2.

Table 1. Classification and function of ecosystem services in Huangshan City.

Eco-Services Category	Eco-Service Project	Explain
Supply service	Water resource products	Mainly including local water supply and watershed water supply.
	Fishery products	The total amount of various fish products in the watershed.
	Agricultural products	Mainly including grains, oilseeds, cotton, cocoons, tobacco, and tea.
	Forestry products	Mainly including various garden fruits.
	Animal husbandry products	Mainly including meat products, dairy products, and poultry egg products.
Regulating Service	Water conservation	Storage and retention of water in forests and wetlands.
	Soil conservation	Mainly including reducing non-point source pollution and reducing sediment deposition.
	Flood regulation and storage	Mainly including the construction and operation of the reservoir.
	Water purification	Industrial treatment cost of atmospheric pollutants.
	Air purification	The value of purified air is evaluated using the cost of industrial treatment of water pollutants.
	Fixed carbon dioxide	The cost of fixed carbon dioxide in the market.
	Release oxygen	The cost of producing oxygen in the market.
	Climate regulation	The electricity cost required for manually adjusting temperature and humidity.
Cultural service	Leisure tourism	Total tourism revenue of the city.

Table 2. Calculation of ecosystem service values for various ecosystem service types.

Eco-Service Project	Formula	Illustrate
Water resource products	$V_1 = V_L + V_W$ $V_L = P_L \times W_L$ $V_W = P_W \times W_W$	V_1 is the water resource products value (CNY); P_L is the local water consumption of Huangshan City (m^3); W_L is the current water price in Huangshan City (CNY/m^3); P_W is the cost of water pollution control (CNY); and W_W is the net water supply of the watershed (m^3).
Fishery products	$V_2 = P_2 \times W_2$	V_2 is the fishery products value (CNY); P_2 is the price of fishery products (CNY/kg); and W_2 is the total production of fishery products in Huangshan City (kg).
Agricultural products	$V_3 = \sum_{i=1}^{6} P_i \times W_i$	V_3 is the fishery products value (CNY); P_i is the *i*-th price of agricultural products (CNY/kg); W_i is the *i*-th production of agricultural products in Huangshan City (kg); and *i* is the agricultural products type (*i* = 1 to 6).
Forestry products	$V_4 = P_4 \times W_4$	V_4 is the forestry products value (CNY); P_4 is the average price of fruit products (CNY/kg); and W_4 is the total production of fruit products in Huangshan City (kg).

Table 2. *Cont.*

Eco-Service Project	Formula	Illustrate
Animal husbandry products	$V_5 = \sum_{i=1}^{3} P_j \times W_j$	V_5 is the animal husbandry products value (CNY); P_j is the j-th price of animal husbandry products (CNY/kg); W_j is the j-th production of animal husbandry products in Huangshan City (kg); and j is the animal husbandry products type (j = 1 to 3).
Water conservation	$V_6 \times Q_6 \times P_6$	V_6 represents the value of water conservation (CNY); Q_6 is the amount of water conservation (m^3); and P_6 is the market price of water resources (CNY/m^3).
Soil conservation	$V_7 = V_s + V_D$ $V_s = \lambda \times \left(\frac{Q_D}{\rho}\right) \times c$ $V_D = \sum_{i=1}^{2} Q_D \times r_i \times P_i$	V_7 represents the soil conservation value (CNY); Vs represents the value of reducing siltation (CNY); V_D represents the value of reducing non-point source pollution (CNY); λ represents the sedimentation coefficient; Q_D represents the soil conservation quantity (t); ρ represents the soil bulk density (t/m^3); c represents per unit cost of reservoir desilting project(CNY/m^3); r_i represents the purity of the ith pollutant (such as nitrogen or phosphorus) in the soil (%), where i represents the number of nutrient substances in the soil; and P_i represents the cost of treating the i-th pollutant (i =1 to 2).
Flood regulation and storage	$V_8 = Q_8 \times C_w$	V_8 represents the flood storage value (CNY); Q_8 represents the amount of flood storage (m^3); and Cw represents the engineering cost and maintenance cost per unit capacity of the reservoir (CNY).
Water purification	$V_9 = \sum_{i=1}^{n} Q_{9,i} \times C_i$	V_9 represents the total value of water purification (CNY); $Q_{9,i}$ represents the purification amount of the i-th water pollutant (t); Ci represents the treatment cost of the i-th water pollutant (CNY); and i is the water pollutant (i = 1 to n).
Air purification	$V_{10} = \sum_{i=1}^{n} Q_{10,i} \times C_i$	V_{10} represents the total value of air purification (CNY); $Q_{10,i}$ represents the purification amount of the i-th air pollutant (t); Ci represents the treatment cost of the i-th air pollutant (CNY); and i is the air pollutant (i =1 to n).
Fixed carbon dioxide	$V_{11} = Q_{11} \times C_C$	V_{11} is the value of fixed carbon dioxide (CNY); Q_{11} is the total amount of fixed carbon dioxide (t); and C_C is the price of industrial carbon capture (CNY/t).
Release oxygen	$V_{12} = Q_{12} \times C_O$	V_{12} is the value of release oxygen (CNY); Q_{12} is the total amount of release oxygen (t); and C_O is the price for industrial oxygen production (CNY/t).
Climate regulation	$V_{13} = E_{13} \times P_E$	V_{13} is the value of climate regulation (CNY); E_{13} is the total energy consumed by ecosystem transpiration and evaporation (kW·h); and P_E is the electricity price (CNY/kW·h).
Leisure tourism	$V_{14} = C_t \times N$	V_{14} is the value of leisure tourism; C_T is the average travel cost for tourists (sampling survey); and N represents the total number of tourists.

2.3.2. Ecological Compensation Supply Coefficient

The value spillover theory is an extension of the energy value analysis and the water ecological footprint theory in the field of ecological economics. The theory posits that after the primary service providers in an ecological economic system eliminate their own consumption value, they can provide surplus value to other areas. Consequently, only the region that has a spillover value can be worthy of corresponding compensation [29]. Therefore, the reference value for calculating the EC standard in a watershed should not be determined based on all ecosystem service values generated in the upstream areas. Instead, it should be determined by considering the reasonable scope, which encompasses the supply or consumption value of eco-products after deducting the portion required to meet the comfortable living standards of the residents. This is known as the ecological service VST, thereby avoiding overestimating the EC standard based on the value of ecosystem services. In this study, it is assumed that (1) the national per capita value of ecosystem services is the basis for judging the spillover effect of ecological services in a region, defined as the VST coefficient (VSTC) and (2) the region's willingness to accept compensation is substituted by the local level of economic development.

The formula is as follows:

$$ECSC = VSTC_{target} \times WTA \tag{2}$$

$$VSTC_{target} = \frac{ESVnation/P_{nation}}{ESV_{target}/P_{target}} \tag{3}$$

$$WTA = 2\arctan(\frac{ESV_{target}}{GDP_{target}}) \times \frac{1}{\pi} \tag{4}$$

In the formula, $VSTC_{target}$ represents the level of per capita ecosystem service value spillover in the target area compared with the national per capita ecosystem service value (ESV_{nation}/P_{nation}); if $VSTC_{target} < 1$, it indicates that the region is a supplier of ecological products, if $VSTC_{target} > 1$, it indicates that the region is a demander of ecological products, and if $VSTC_{target} = 1$, it indicates that the region is in a self-sufficient state and cannot provide supply ecological products. WTA represents the region's willingness to accept compensation; GDP_{target} represents the total GDP within the region.

2.3.3. Ecological Compensation Demand Coefficient

In order to establish a long-term operating EC mechanism, the perspectives of the compensators must be considered, such as willingness to pay, ability to pay, and scarcity of ecosystem services. In this study, we constructed an ecological compensation demand coefficient, denoted as $ECDC$. The formula for $ECDC$ is as follows:

$$ECDC = k_1 \times k_2 \times k_3 \tag{5}$$

$$k_1 = \frac{M_{LC}}{M_{PC}} \tag{6}$$

$$k_2 = \frac{PD_S}{PD_D} \tag{7}$$

$$k_3 = \frac{GDP_D}{GDP_S} \times \frac{1}{1+e^{-t}},\ t = \frac{1}{En} - 3 \tag{8}$$

In the formula, k_1 represents the adjustment coefficient of payment willingness, which is represented by the ratio of the compensation amount to the actual pollution control investment, reflecting the consumption willingness and preference of the consumption area for ecological products; M_{LC} represents the compensation amount of the previous year; and M_{PC} represents the pollution control investment amount of the previous year. k_2 is the inverse of the ecological product scarcity, which is related to the scarcity of resources. PD_S is the population density of the supply area, and PD_D is the population density of the consumption area. The adjustment coefficient k_3 is related to the consumer's socio-economic development level, and the ratio of the demand area GDP_D to the supply area GDP_S is selected as the indicator of regional economic strength. e is the natural constant; t is the social and economic development level; and En is the Engel coefficient.

2.3.4. Ecological Compensation Standard Model

In this study, the purpose of the EC standard model is to enhance the result practicability and acceptability expected to be adopted in the EC pilot to change the current fixed compensation standards. To establish the new model, it is assumed that (1) the EC standard is calculated on the basis of the region's ecosystem services value; (2) the EC standard is determined through the comprehensive consideration of the supply side and demand side;

(3) the ecosystem service value within the scope of ecological compensation is the basis of EC standard calculation; and (4) the executive EC standard is a matter for negotiation between the supplier and demander, the output from this model is an interval range including the upper and lower limits of the compensation standard, and the difference is between including the supply service value and the cultural service value.

The EC standard model is established as follows:

$$ECS_{max} = ESVC_{target} \times ECSC \times ECDC \quad (9)$$

$$ECS_{min} = (ESVC_{target} - ESSVC_{target} - ECSVC_{target}) \times ECSC \times ECDC \quad (10)$$

In the formula, ECS_{max} represents the upper limit of the compensation standard; ECS_{min} represents the lower limit of the compensation standard; $ESVC_{target}$ represents the ecosystem service value within the compensation scope in the target era; $ESSVC_{target}$ represents the supply service value within the compensation scope in the target area; and $ECSVC_{target}$ represents the cultural service value within the compensation scope in the target area.

3. Results

3.1. Value of Main Ecosystem Services

According to formulas in Table 2, the ecosystem service values of the Xin'an River Basin are calculated as shown in Table 3. According to Formula (1), the total value of the ecological system in the Xin'an River Basin is calculated as CNY 70.271 billion, of which the supply services account for 22.7%, the regulatory services account for 24.6%, and the cultural services account for 52.7%.

Table 3. Total value of ecosystem services in Huangshan City.

Eco-Services Category	Eco-Service Project	Method of Calculation	Value (Billion)
Supply service	Local water use	Shadow project approach	1.19
	Basin water supply	Shadow project approach	3.468
	Fishery products	Market comparison approach	0.281
	Agricultural products	Market comparison approach	6.019
	Forestry products	Market comparison approach	2.254
	Animal husbandry products	Market comparison approach	2.748
Regulating Service	Water conservation	Shadow project approach	0.655
	Soil conservation	Replacement cost method	0.561
	Flood regulation and storage	Shadow project approach	7.84
	Water purification	Replacement cost method	0.363
	Air purification	Replacement cost method	0.011
	Fixed carbon dioxide	Replacement cost method	4.914
	Release oxygen	Replacement cost method	1.323
	Climate regulation	Replacement cost method	1.625
Cultural services	Leisure tourism	Market comparison approach	37.019
Total			70.271

3.2. Scope of Basin Ecological Compensation

There is a significant difference in economic development between the upstream and downstream areas of the Xin'an River Basin, and regional EC will involve disputes and contradictions among various stakeholders. By analyzing the service functions of the Xin'an River ecosystem, the compensation accounting scope of downstream to upstream is determined, as shown in Table 4. Considering that the value of supply services in ecosystem services is transformed into monetary value in the market, it is impossible to allocate compensation responsibility for them, so they cannot be included in the final value compensation. As a result, the primary focus of EC in the Xin'an River Basin is on 10 service functions, including watershed water supply, water conservation, soil conservation, flood control, water purification, air purification, carbon fixation, oxygen release, climate regulation, and recreation and tourism. After calculation, the total value of the ecosystem services within the scope of EC ($ESVC_{target}$) is CNY 57.779 billion, the $ESSVC_{target}$ is CNY 3.468 billion, and the $ECSVC_{target}$ is CNY 37.019 billion.

Table 4. Scope of basin ecological compensation in Huangshan City.

Eco-Services Category	Eco-Service Project	Beneficiaries of Water Ecosystem Services				In the Compensation Range
		Global	Nationwide	This City	Downstream City	
Supply service	Local water use			√		No
	Basin water supply			√	√	Yes
	Fishery products			√		No
	Agricultural products			√		No
	Forestry products			√		No
	Animal husbandry products			√		No
Regulating service	Water conservation		√	√	√	Yes
	Soil conservation		√	√	√	Yes
	Flood regulation and storage		√	√	√	Yes
	Water purification	√	√	√	√	Yes
	Air purification	√	√	√	√	Yes
	Fixed carbon dioxide		√	√	√	Yes
	Release oxygen	√	√	√	√	Yes
	Climate regulation	√	√	√	√	Yes
Cultural services	Leisure tourism		√	√	√	Yes

Note: √ means the area can benefit from this water ecosystem service.

3.3. Correction of Ecological Compensation

First, calculate the ecological compensation supply coefficient. According to the "Biodiversity and Ecosystem Service Economics" research results in China and the "2020 Statistical Bulletin of National Economic and Social Development of Huangshan City", the total value of ecosystem services in China is CNY 7.8 trillion, and the population of China is 1.383 billion, so the per capita value of ecosystem services in China is CNY 5639.913. The GDP of Huangshan City was CNY 8.504 billion, with a registered population of 1.331 million in a 9807-square kilometer area, so the per capita value of the ecosystem services in Huangshan is CNY 52,795.642. According to Formulas (2)–(4), the $VSTC_{target}$ is calculated as 0.107 and the WTA is 0.923, so the $ECSC$ is 0.099.

Then, calculate the ecological compensation demand coefficient. According to the "2020 Statistical Bulletin of National Economic and Social Development of Zhejiang Province", the GDP of Hangzhou City was CNY 1610.6 billion, with a resident population of 11.965 million at the end of 2020 in a 16,596-square kilometer area, and the Engel coefficient of urban residents was 32.3%. According to Formulas (5)–(8), $k_1 = 0.143$, $k_2 = 5.312$, and $k_3 = 0.524$; thus, $ECDC = k_1 \times k_2 \times k_3 = 0.840$.

3.4. Theoretical Total Amount of Ecological Compensation

According to the EC standard model, the $ECS_{max} = 57.779 \times 0.099 \times 0.840 = 4.805$, and the $ECS_{min} = (57.779 - 3.468 - 37.019) \times 0.099 \times 0.840 = 1.438$. Therefore, the upper limit of the EC standard for the Xin'an River Basin is calculated as CNY 4.805 billion, and the lower limit of the EC standard is calculated as CNY 1.438 billion.

4. Discussion

Firstly, we categorized watershed ecosystem services into the following three categories in this study: provisioning services, regulating services, and cultural services. The calculated total value of the ecosystem in the Xin'an River Basin is CNY 70.271 billion. The result shows that the value of cultural services (recreation and tourism) was 37.019, accounting for 52.7% of the total value of ecosystem services. As we all know, Huangshan City is famous for its spectacular Huangshan Scenic Area and is one of China's well-known tourist destinations. The tourism industry in Huangshan City is highly developed, attracting a large number of domestic and foreign tourists for sightseeing and travel every year [30]. Therefore, cultural and tourism services play an important role in the ecosystem services of the Xin'an River Basin. Secondly, the computed ecosystem service value is higher than that reported by Dai et al. (2021) and Yang, N (2019), obviously indicating that the upstream city of Huangshan has made great efforts to protect the watershed ecological environment so the downstream cities can enjoy greater benefits from the watershed ecosystem [25,26]. Consequently, the downstream cities have a greater responsibility for watershed ecological compensation, and the ecosystem service value offered by the watershed ecosystem to downstream cities can serve as a basis for determining the compensation fund. However, using this value directly for the ecological compensation fund clearly exceeds the capacity of the downstream city of Hangzhou. Therefore, this study established the EC standard model using the ECSC and ECDC to adjust the ecosystem service values within the compensation scope.

Finally, this new framework used the ESV overflow from the upstream cities of the Xin'an River as the standard for ecological compensation in the downstream cities, removing the ecosystem services not affecting the downstream cities, such as local water use, fishery products, agricultural products, forestry products, and animal husbandry products, which should not be included in the compensation scope. The EC standard value based on ECSC and ECDC revision in the Xin'an River Basin was calculated, as shown in Figure 4. The results show that the upper limit of the compensation amount in Huangshan City according to the supply–demand relationship is CNY 4.805 billion, and the lower limit of the EC amount after correction according to the compensation scope is CNY 1.438 billion, forming a compensation standard interval, which can be used as a reference for cooperation and decision-making between the compensation subject and object.

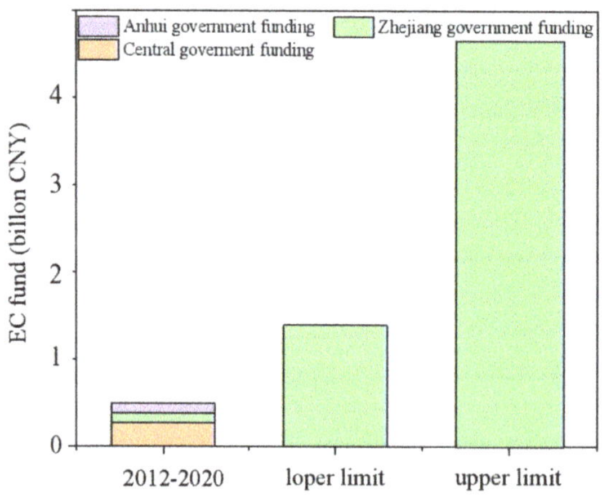

Figure 4. The upper and lower limits of actual annual and theoretical calculation of EC in Xin'an River Basin.

It should be noted that (1) the watershed ecological compensation standard model established in this article relies on the accounting results of the ecosystem service value in the watershed. However, due to the lack of uniformity in the conceptual connotation, scope definition, and value accounting methods of ecosystem services, there are significant differences in the results of different accounting studies in the same region, which to some extent, affects the accuracy and recognition of the model's calculation results in practical applications. Therefore, in the future, we should optimize the algorithm for calculating the ecosystem value in the model, reduce the model's dependence on the value of ecosystem services, and gradually improve the guidance of the watershed ecological compensation model. (2) The model established in this article is primarily used for calculating ecological compensation standards at the macro level. As China's emphasis on ecological protection in river basins increases, there will be more and more compensation between cities and counties in the future. Therefore, in subsequent research, the model established in this article should be further improved by enriching calculation indicators, optimizing calculation methods, and refining application scenarios to continuously enhance the model's applicability.

5. Conclusions

This study shows that Huangshan City can receive compensation of no less than CNY 1.438 billion in market value, and the upper limit of compensation of CNY 4.805 billion based on the compensation model is established, accounting for only 2.29% of Hangzhou's fiscal revenue in 2020 (CNY 209.34 billion), which will not cause significant financial pressure and partially compensate for the problem of insufficient funds in the compensation process, effectively improving the existing deficiencies in compensation and providing a good reference for future basin compensation, and even cross-regional EC. Therefore, the proposed method can be used to provide reasonable suggestions for solving the water resource disputes between upstream and downstream cities, the contradiction between economic development and protection, and promoting the ecological protection of the river basin, thereby offering valuable insights for potential future watershed compensation and cross-regional ecological compensation efforts.

Author Contributions: Y.C.: materials and method, writing—original draft, writing—review and editing. Q.W.: introduction and results, writing—original draft, and writing—review. L.G.: writing—original draft, results and discussion. All authors have read and agreed to the published version of the manuscript.

Funding: This work was financially supported by the ADB Loan Xin'an River Basin Ecological Protection and Green Development Project: Xin'an River Ecological Compensation Standard and Evaluation system (Contract No.: Sery-HSC-1), the State Key Laboratory of Urban Water Resource and Environment (Harbin Institute of Technology) (No. 2023DX06), and the Fundamental Research Funds for the Central Universities.

Data Availability Statement: The datasets obtained and analyzed in this study are available from the corresponding authors upon reasonable request.

Conflicts of Interest: Author Yuan-Hua Chen was employed by the company China Energy Conservation and Environmental Protection Group (CECEP), Eco-Product Development Research Center Co., Ltd. The remaining authors declare that the research was conducted in the absence of any commercial or financial relationships that could be construed as a potential conflict of interest.

References

1. Ni, Q.; Xu, T.; Li, X.; Liu, J.; Zhao, M. Transboundary River basin ecological compensation standard accounting: Based on cost-benefit perspective. *Resour. Environ. Yangtze Basin* **2021**, *30*, 97–110. (In Chinese)
2. Wunder, S.; Engel, S.; Pagiola, S. Taking stock: A comparative analysis of payments for environmental services programs in developed and developing countries. *Ecol. Econ.* **2008**, *65*, 834–852. [CrossRef]
3. Wang, K.; Ou, M.; Wolde, Z. Regional differences in ecological compensation for cultivated land protection: An analysis of Chengdu, Sichuan Province, China. *Int. J. Environ. Res. Public Health* **2020**, *7*, 8242. [CrossRef] [PubMed]
4. Zhai, T.; Wang, J.; Jin, Z.; Qi, Y.; Fang, Y.; Liu, J. Did improvements of ecosystem services supply-demand imbalance change environmental spatial injustices? *Ecol. Indic.* **2020**, *111*, 106068. [CrossRef]
5. He, J.; Wan, Y.; Tang, Z.; Zhu, X.; Wen, C. A developed framework for the multi-district ecological compensation standards integrating ecosystem service zoning in an urban area in China. *Sustainability* **2019**, *11*, 4876. [CrossRef]
6. Fang, Z.; Chen, J.; Liu, G.; Wang, H.; Alatalo, J.M.; Yang, Z.; Mu, E.; Bai, Y. Framework of basin eco-compensation standard valuation for cross-regional water supply-A case study in northern China. *J. Clean Prod.* **2021**, *279*, 123630. [CrossRef]
7. Gao, X.; Shen, J.; He, W.; Zhao, X.; Li, Z.; Hu, W.; Wang, J.; Ren, Y.; Zhang, X. Spatial-temporal analysis of ecosystem services value and research on ecological compensation in Taihu Lake Basin of Jiangsu Province in China from 2005 to 2018. *J. Clean Prod.* **2021**, *317*, 128241. [CrossRef]

8. Peng, Z.; Wu, H.; Ding, M.; Li, M.; Huang, X.; Zheng, R.; Xu, L. Ecological compensation standard of a water-receiving area in an Inter-Basin Water Diversion based on ecosystem service value and public willingness: A case study of Beijing. *Sustainability* **2021**, *13*, 5236. [CrossRef]
9. Ren, Y.; Lu, L.; Zhang, H.; Chen, H.; Zhu, D. Residents' willingness to pay for ecosystem services and its influencing factors: A study of the Xin'an River basin. *J. Clean Prod.* **2020**, *268*, 122301. [CrossRef]
10. Xinhua News Agency. *Xin'an River Basin Eco-Compensation Pilot Makes Multi Party Win-Win*; Xinhua News Agency: Beijing, China, 2019.
11. Sheng, J.; Han, X. Practicing policy mobility of payment for ecosystem services through assemblage and performativity: Lessons from China's Xin'an River Basin Eco-compensation Pilot. *Ecol. Econ.* **2022**, *191*, 107234. [CrossRef]
12. Sheng, J.; Cheng, Q.; Wu, Y. Payment for watershed services and the coordination of interests in transboundary rivers: China's Xin'an River Basin Eco- compensation pilot. *J. Environ. Manag.* **2023**, *328*, 116670. [CrossRef] [PubMed]
13. Zhang, C.; Li, J.; Zhou, X.; Sun, Y. Application of ecosystem service flows model in water security assessment: A case study in Weihe River Basin, China. *Ecol. Indic.* **2021**, *120*, 106974. [CrossRef]
14. Sheng, J.; Qiu, W.; Han, X. China's PES-like horizontal eco-compensation program: Combining marketoriented mechanisms and government interventions. *Ecosyst. Serv.* **2020**, *45*, 101164. [CrossRef]
15. Niu, H.; An, R.; Xiao, D.; Liu, M.; Zhao, X. Estimation of ecosystem services value at a basin scale based on modified equivalent coefficient: A case study of the Yellow River Basin (Henan Section), China. *Int. J. Environ. Res. Public Health* **2022**, *19*, 16648. [CrossRef]
16. Wu, C.; Ma, G.; Yang, W.; Zhou, Y.; Peng, F.; Wang, J.; Yu, F. Assessment of ecosystem service value and its differences in the Yellow River Basin and Yangtze River Basin. *Sustainability* **2021**, *13*, 3822. [CrossRef]
17. Deng, C.; Zhang, S.; Lu, Y.; Li, Q. Determining the ecological compensation standard based on forest multifunction evaluation and financial net present value analysis: A case study in southwestern Guangxi, China. *J. Sustain. Forest.* **2020**, *39*, 730–749. [CrossRef]
18. Dong, J.; Wu, D. An evaluation of the impact of ecological compensation on the cross-section efficiency using SFA and DEA: A case study of Xin'an River Basin. *Sustainability* **2020**, *12*, 7966. [CrossRef]
19. Gao, X.; Shen, J.; He, W.; Sun, F.; Zhang, Z.; Zhang, X.; Yuan, L.; An, M. Multilevel governments' decision-making process and its influencing factors in watershed ecological compensation. *Sustainability* **2019**, *11*, 1990. [CrossRef]
20. Costanza, R.; Arge, R.D.; Groot, S.F. The value of the world's ecosystem services and natural capital. *Nature* **1997**, *387*, 253–260. [CrossRef]
21. Millennium Ecosystem Assessment. *Ecosystems and Human Well-Being: Biodiversity Synthesis*; World Resources Institute: Washington, DC, USA, 2005. Available online: http://www.millenniumassessment.org/en/Synthesis.html (accessed on 23 July 2024).
22. Wang, L.Y.; Chen, C.; Xie, F.; Hu, Z.J.; Zhang, Z.L.; Chen, H.X.; He, X.Y.; Chu, Y.L. Estimation of the value of regional ecosystem services of an archipelago using satellite remote sensing technology: A case study of Zhoushan Archipelago, China. *Int. J. Appl. Earth Obs.* **2021**, *105*, 102616. [CrossRef]
23. Du, H.; Zhao, L.; Zhang, P.; Li, J.; Yu, S. Ecological compensation in the Beijing-Tianjin-Hebei region based on ecosystem services flow. *J. Environ. Manag.* **2023**, *331*, 117230. [CrossRef] [PubMed]
24. Fu, Y.; Cui, X.; Zhao, J.; Zhang, C.; Yan, L. Estimation of ecological compensation standard based on ecological service value calculation. *IOP Conf. Ser. Earth Environ. Sci.* **2021**, *647*, 012161. [CrossRef]
25. Dai, J.; Hu, S.; Zhang, B.; Yang, Q. Study on ecological compensation standard based on opportunity cost and ecosystem service value accounting: A case study of Xin'an river basin. *Hubei Agri. Sci.* **2021**, *60*, 152–157. (In Chinese)
26. Yang, N. *Recahrch on County of Ecological Compensation in Yunnan Province Base on Gross Ecosystem Product*; Yunnan University: Kunming, China, 2019.
27. Zhou, Y.; Zhou, J.; Liu, H.; Xia, M. Study on eco-compensation standard for adjacent administrative districts based on the maximum entropy production. *J Clean Prod.* **2019**, *221*, 644–655. [CrossRef]
28. Fu, L.; Ren, Y.; Lu, L.; Chen, H. Relationship between ecosystem services and rural residential well-being in the Xin'an river Basin, China. *Ecol. Indic.* **2022**, *140*, 108997. [CrossRef]
29. Xu, L.; Li, B.; Yuan, Y.; Gao, X.; Zhang, T.A. Study on eco-compensation based on eco-service assessment in 14 contiguous destitute areas of China. *J. Geo-Inform. Sci.* **2016**, *18*, 286–297. (In Chinese)
30. Lv, C.; Xu, X.; Guo, X.; Feng, J.; Yan, D. Basin water ecological compensation interval accounting based on dual perspectives of supply and consumption: Taking Qingyi River Basin as an example. *J. Clean Prod.* **2023**, *385*, 135610. [CrossRef]

water

MDPI

Article

Antibiotics in Wastewater Treatment Plants in Tangshan: Perspectives on Temporal Variation, Residents' Use and Ecological Risk Assessment

Zhuo Dong [1], Jian Hu [2,*], Pengjie Wang [3], Gengtao Han [1] and Zheng Jia [1]

1 Pharmacy Teaching and Research Office, Medical Department, Tangshan Vocational and Technical College, Tangshan 063000, China; dongzhuo@tsvtc.edu.cn (Z.D.); hgt2608171739@163.com (G.H.); jiazh123@sina.com (Z.J.)

2 Research Center for Eco-Environmental Sciences, Chinese Academy of Sciences, No. 18 Shuangqing Road, Beijing 100085, China

3 Heilongjiang Province Environmental Monitoring Center, Harbin 150090, China; aishu715@163.com

* Correspondence: jianhu@rcees.a.c.cn

Abstract: In 2023, this study monitored nine types of antibiotics in the influent and effluent of wastewater treatment plants (WWTPs) in the urban and suburban areas of Tangshan. The total antibiotics concentration detected in influent WWTPs was highest in winter, followed by spring, summer, and autumn. The antibiotics concentration in influent and effluent urban WWTPs was higher than that in the suburban WWTPs in spring, summer, and winter, while the trend was reversed in autumn. Roxithromycin and oxytetracycline had a risk quotient (RQ) value of $\geq$0.1 in the effluent of WWTPs in winter, indicating that they are medium-risk antibiotics that pose a risk to the aquatic ecosystem after discharge. In the study area, the per capita pollution load of antibiotics was highest in spring, summer, and autumn for sulfamethoxazole, while it was highest in winter for ofloxacin. In the urban area, the use of roxithromycin, sulfamethoxazole, sulfamethoxazole, and ofloxacin was highest in spring, summer, autumn, and winter, respectively, while in suburban areas, the use of sulfamethoxazole, norfloxacin, sulfamethoxazole, and ofloxacin was highest during the same period. The use of antibiotics in the urban area was one order of magnitude higher than that in suburban areas, indicating a possible overuse of antibiotics in urban environments.

Keywords: antibiotics; wastewater treatment plant; temporal variation; use; ecological risk

1. Introduction

Antibiotics play a crucial role in human health and are usually prioritized in drug management [1–3]. However, due to non-compliance with the principles of using antibiotics, there have been cases of inappropriate use and other forms of antibiotic abuse [4–6]. Currently, the misuse of antibiotics is a pressing public health and ecological security issue across various sectors, including medical health, food hygiene, livestock and poultry breeding, and ecological governance, not only in China but also globally [7–9]. China produces about 210,000 tons of antibiotic raw materials every year. Excluding the export of raw materials (about 30,000 tons), the remaining 180,000 tons are used domestically (including medical and agricultural use), with an annual per capita consumption of about 138 g [10]. As veterinary antibiotics, tetracycline antibiotics are the most commonly used, accounting for 40.5% of the total, followed by sulfonamides and macrolides. Quinolones are highly used in hospitals due to the high incidence of respiratory tract infections and mycoplasmal pneumonia in spring, autumn, and winter. Therefore, it is expected that antibiotics are widely present in urban sewage and agricultural wastewater, entering the water ecosystem through various pathways [11–13]. Due to the low metabolic rate of biological organisms against antibiotics, a large amount of antibiotics is excreted with urine and feces, collected by sewage networks, and entered WWTPs.

At present, the main purpose of WWTPs is to remove suspended solids, COD (chemical oxygen demand), nitrogen, phosphorus, and other substances in sewage, but they generally do not have the ability to efficiently remove antibiotics, resulting in high antibiotic concentrations in effluent wastewater [13,14]. It should be noted that sewage treatment technologies like advanced chemical oxidation, chemical precipitation, ultrafiltration, nanofiltration, and ion exchange are not effective at removing antibiotics and may be better suited for industrial wastewater treatment. In contrast, suburban wastewater treatment primarily relies on biological processes, with advanced oxidation and other sewage treatment techniques being less common, resulting in a lower capacity for treating antibiotics in wastewater. The discharge of antibiotics from WWTPs into natural water bodies can lead to a decrease in the self-purification capacity and pollution load of the receiving water bodies, affecting the ecological health of river water environments [15]. Although the mass concentration of antibiotics in surface water is generally between ng/L and mg/L, they are sufficient to have harmful effects on exposed ecosystems or organisms [16]. It is necessary to conduct an assessment of their discharge volume and ecological risks [17]. In order to better understand the temporal variation in antibiotics in WWTPs and their removal rate and potential ecological risks, this study investigated the following: (1) the temporal variation in four types of nine antibiotics in influent and effluent urban and suburban WWTPs; (2) estimating the usage and annual discharge of antibiotics in the urban and suburban environment based on per capita pollution load.

2. Materials and Methods

2.1. Experimental Reagents and Instruments

Ultra-high purity compounds (>99%) of nine antibiotics, including roxithromycin (macrolides), ofloxacin (quinolones), norfloxacin(quinolones), ciprofloxacin (quinolones), tetracycline (tetracyclines), chlortetracycline (tetracyclines), oxytetracycline (tetracyclines), sulfadiazine(sulfonamides) and sulfamethoxazole (sulfonamides), were bought from Sigma-Aldrich (St. Louis, MO, USA). The standard concentrations of each antibiotic were 1000 μg/mL, with a purity of >99% (Tianjin Alta Technology Co., Ltd., Tianjin, China). Acetonitrile, methanol, and formic acid were chromatographically pure (Thermo Fisher Corporation, MA, USA), and anhydrous sodium sulfate, sodium chloride, sodium dihydrogen phosphate dodecahydrate, disodium acetate tetraacetate, etc., were all chemically pure (Sinopharm Chemical Reagent Co., Ltd., Beijing, China). All solutions were prepared using Milli-Q water.

2.2. Sample Collection and Processing

Seasonal sampling campaigns were conducted in 2023 [January to March (winter), April to May (spring), July to August (summer), and October (autumn)] in Tangshan. Samples were collected from influent and effluent WWTPs in the urban ($n = 2$) and suburban areas ($n = 2$). The flow scheme of WWTPs is shown in Figure 1. To reduce experimental errors, instantaneous water samples were collected every 2 h for a total of 4 times within 1 day. The collected water samples were mixed evenly and stored in brown glass bottles to avoid light. They were transported back to the laboratory in an ice bath within 24 h. After filtration with a 0.45 μm glass fiber membrane, 500 mL was accurately measured, 0.25 g Na_2EDTA was added, and the pH was adjusted to about 3.0 with H_3PO_4. The samples were stored at 4 °C, and solid phase extraction was completed within 48 h.

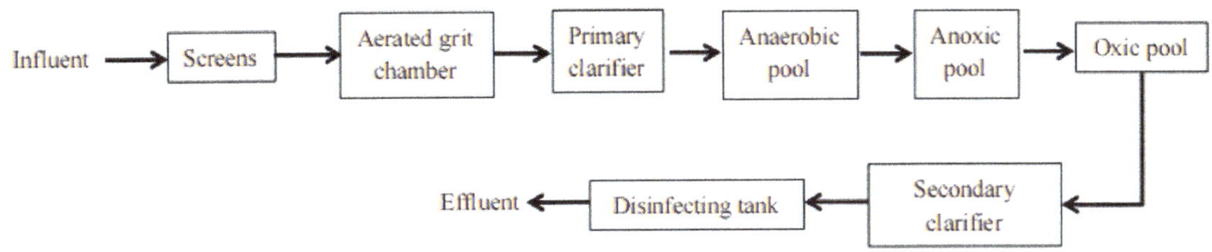

Figure 1. Schematic diagram of wastewater treatment plants (WWTPs) in Tangshan.

2.3. Qualitative and Quantitative Analysis

Take 1 L of the water sample and use a vacuum filtration device to pass it through a 0.45 μm filter membrane. Use a fully automated solid phase extraction instrument and an HLB extraction column to complete the preliminary extraction and concentration. First, activate the extraction column with 10 mL of methanol and 10 mL of ultrapure water in sequence; then extract 1 L of water sample at a rate of 10 mL/min; after completion, use high-purity nitrogen to dry the extraction column for 20 min; finally, wash the extraction column with 5 mL of dichloromethane, 5 mL of ethyl acetate, 5 mL of n-hexane, and 5 mL of methanol in sequence, repeating twice. After extraction, use a rotary evaporator to concentrate the eluent to about 1 mL and transfer it to a test tube; then, use a nitrogen-blowing instrument to blow it nearly dry and dilute it to 1 mL with methanol. The diluted sample passes through a 0.22 μm organic filter membrane and is transferred to a liquid phase vial for testing. The pretreated samples were analyzed using UPLC-MS/MS (QTRAPTM5500 LC/MS/MS system, SCIEX, MA, USA), employing a Waters Cortecs T3 column (2.1 mm × 100 mm, 2.7 μm). The injection volume for liquid chromatography was 2 μL, with a flow rate of 0.3 mL·min^{-1} and a column temperature of 40 °C. The mobile phase was a gradient elution of 0.1% formic acid aqueous solution and acetonitrile. The qualitative and quantitative analysis of antibiotics was carried out using the multiple reaction detection scanning mode (MRM) and electrospray ionization mass spectrometry (ESI/MS) positive and negative ion modes [18,19].

2.4. Quality Control

Using methanol as the solvent, the standard stock solution was diluted to 0.1, 2, 0.5, 1, 2, 5, 10, 20, 50, 100, and 200 μg/L. Linear regression was performed between the concentration of the antimicrobial drug and the corresponding peak area to draw a standard curve for the antibiotics. The standard curve had a good linear correlation within the corresponding linear range (correlation coefficient $R^2 > 0.99$).

2.5. Ecological Risk Assessment Method

The ecological risk assessment is a scientific evaluation of the potential damage of toxic and harmful pollutants to the ecological environment through quantitative characterization methods [20]. In this study, the risk quotient (RQ) was used to evaluate the ecological risk of antibiotics [20,21]. The calculation method of RQ is shown in Equation (1):

$$RQ = MEC(PEC)/PNEC \quad (1)$$

In the formula, MEC is the measured environmental concentration of antibiotics, PEC is the predicted concentration of antibiotics, and PNEC is the predicted no-effect concentration of antibiotics. In this study, the measured concentration of antibiotics, MEC, was used to calculate their risk quotient, and the predicted no-effect concentration (PNEC) was determined using the evaluation factor method. The chronic toxicity data (ChV) of antibiotics came from the Ecological Structure Activity Relationships Program (ECOSAR) predictive analyzer developed by the US Environmental Protection Agency. In this study, the ChV values for roxithromycin, tetracycline, chlortetracycline, oxytetracycline, ciprofloxacin, norfloxacin, ofloxacin, sulfadiazine, and sulfamethoxazole were 0.6, 20, 20, 20, 116, 114, 116, 0.101, and 0.068 mg/L, respectively. An extrapolation factor of 100 was selected to determine the PNEC of each antibiotic [22]. The PNEC values of each antibiotic are shown in Table 1. When RQ < 0.1, it is low risk; when $0.1 \leq$ RQ < 1, it is medium risk; and when RQ $\geq$ 1, it is high risk [23,24].

Table 1. PNEC (μg/L) of the target antibiotic in effluents from wastewater treatment plants (WWTPs).

Target Antibiotic	PNEC
Roxithromycin	1.5
Tetracycline	1.0
Aureomycin	1.0
Oxytetracycline	1.0
Ciprofloxacin	20,000
Norfloxacin	23,000
Ofloxacin	22,000
Sulfadiazine	15
Sulfamethoxazole	6.4

2.6. *Estimation of Use and Emissions*

The daily mass load of antibiotics per capita in the influent of WWTPs [μg/(d·person)] can reflect the use of antibiotics in the service area of WWTPs, as shown in Equation (2) [25]:

$$L_{influent} = (Q \times C_{influent})/P_{total} \quad (2)$$

In the formula, Q is the daily sewage flow of WWTP (m^3/day) (Table S1), $C_{influent}$ is the average concentration of antibiotics detected in the influent of WWTP (ng/L), and P_{total} is the number of residents in the service area of WWTPs (Table S1). P_{total} in the urban and suburban areas of Tangshan were, respectively, provided by the Tangshan Municipal Design Institute. The usage amount of antibiotics (U, kg/year) and the mass load of antibiotics in the effluent of WWTP (M, g/year) are shown in Equations (3) and (4) [26–28]:

$$U = L_{influent} \times P_{total} \times 365 \times 10^{-9} \quad (3)$$

$$M = C_{effluent} \times Q \times 365 \times 10^{-6} \quad (4)$$

In the formula, $L_{influent}$ represents the per capita pollution load of antibiotics [μg/(d·person)], and $C_{effluent}$ represents the average detection concentration of target antibiotics in the effluent of WWTPs (ng/L).

3. Results and Discussion

3.1. *Influent*

The seasonal variation trend of the total antibiotics concentration in the inflow of WWTPs in the Tangshan area was the highest in winter, followed by spring, summer, and autumn (Figures 2 and 3). In spring, summer, and winter, the concentration of antibiotics in the inflow of urban WWTPs was higher than that of suburban WWTPs, while the opposite trend was observed in autumn (Figure 2). China announced that from 8 January 2023, COVID-19 infection will be adjusted from "Class A" to "Class B". The monitoring data released by the National Influenza Center of China shows that since January 2023, the positive rate of influenza virus testing in the southern and northern provinces of China has continued to rise, and various regions have entered a high-incidence season of respiratory infectious diseases, with a significant increase in the number of infected individuals compared to previous years [29].

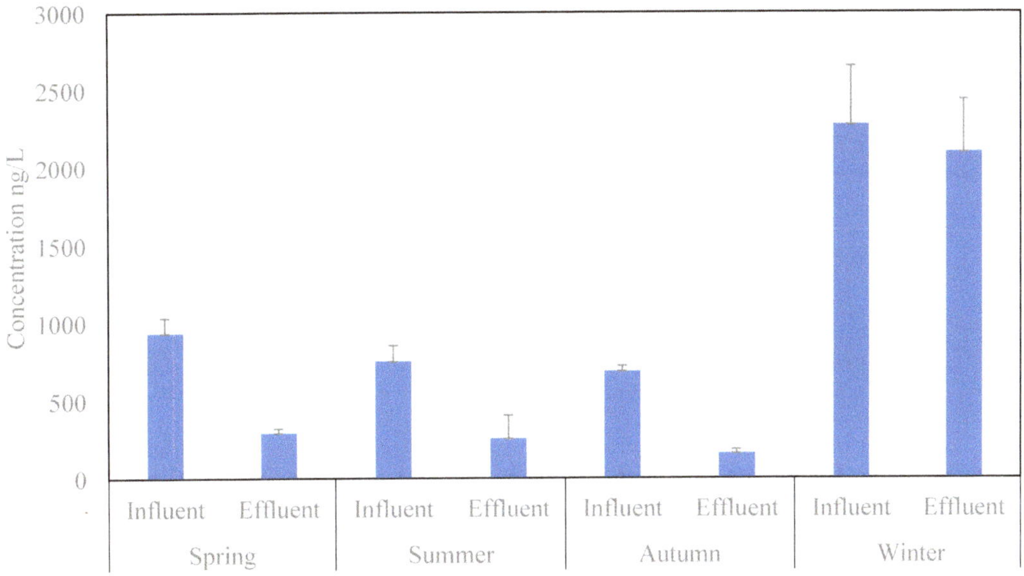

Figure 2. The occurrence of antibiotics in the influent and effluent of wastewater treatment plants (WWTPs) in Tangshan.

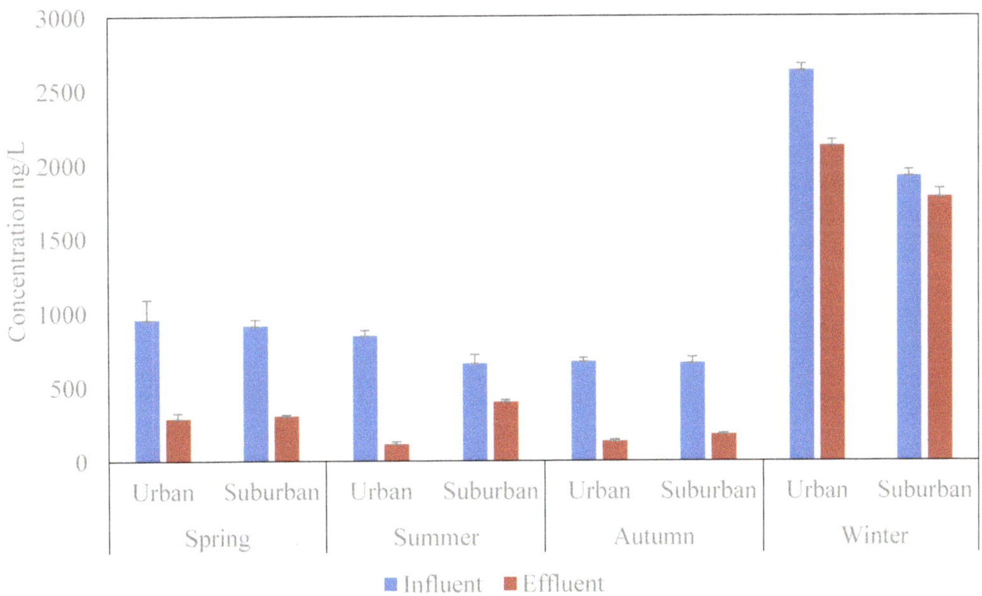

Figure 3. The occurrence of antibiotics in the influent and effluent of urban and suburban WWTPs in Tangshan.

This study selected the winter collection of inlet and outlet water samples from four WWTPs from January to March 2023. It is currently in a period of high incidence of respiratory diseases in the north, as represented by pneumonia. The increase in the use of antibiotic samples by urban populations has led to a much higher total concentration of antibiotics in the inlet water samples of WWTPs than in the other three seasons.

Among the quinolone antibiotics, ofloxacin had the highest concentration detected in the influent water of WWTPs in spring and autumn, followed by norfloxacin and ciprofloxacin. The concentrations of the three quinolone antibiotics detected in the influent water of urban WWTPs were higher than those in the suburban WWTPs (Table 2). In summer and winter, norfloxacin had the highest concentration detected in the influent water of WWTPs, while ofloxacin had the lowest concentration detected. Norfloxacin and ciprofloxacin had the highest concentrations detected in the influent water of urban WWTPs, while ofloxacin showed the opposite trend (Table 2). The concentrations of tetracycline antibiotics detected in the influent water of urban WWTPs were higher than those in the suburban WWTPs in all four seasons, except for oxytetracycline, which had lower concentrations detected in the influent water of urban WWTPs in spring and autumn compared to those in the suburban WWTPs. Among the three tetracycline antibiotics, tetracycline, chlortetracycline, oxytetracycline, and oxytetracycline had the highest concentrations detected in spring, summer, autumn, and winter, respectively, while chlortetracycline (suburban WWTPs), tetracycline (suburban WWTPs), tetracycline (suburban WWTPs), and chlortetracycline (suburban WWTPs) had the lowest concentrations detected (Table 2).

Table 2. The occurrence of antibiotics (ng/L) in the influent of wastewater treatment plants (WWTPs).

Antibiotics		Urban				Suburban			
		Spring	Summer	Autumn	Winter	Spring	Summer	Autumn	Winter
Roxithromycin	Mean	170.40	46.23	78.19	353.33	148.20	34.14	62.42	270.22
	SD	26.78	15.52	3.77	27.34	31.39	1.78	12.55	38.40
Tetracycline	Mean	76.83	18.03	14.72	104.23	61.19	14.17	12.88	71.22
	SD	9.66	4.29	3.89	43.13	12.34	1.75	2.49	17.70
Aureomycin	Mean	50.33	52.98	18.54	96.85	28.21	49.68	16.12	85.15
	SD	13.67	21.24	2.91	51.27	21.58	10.48	3.42	63.47
Oxytetracycline	Mean	40.41	44.87	36.22	256.70	42.30	36.62	62.33	101.67
	SD	15.46	1.56	6.86	17.96	1.41	7.79	33.01	10.80
Ciprofloxacin	Mean	99.83	123.96	79.68	451.79	86.73	107.31	68.25	316.84
	SD	15.32	22.69	12.52	19.76	18.37	31.82	13.77	9.56
Norfloxacin	Mean	122.10	135.17	97.50	458.49	105.76	116.56	84.89	321.42
	SD	17.91	17.68	15.32	20.05	22.41	24.70	18.04	9.87
Ofloxacin	Mean	125.37	47.87	122.62	491.53	106.53	87.72	89.93	422.56
	SD	64.02	1.18	21.57	14.12	53.25	2.19	4.76	93.96
Sulfadiazine	Mean	96.48	133.34	87.67	145.70	117.63	103.14	142.60	110.25
	SD	21.47	15.57	19.43	17.01	11.37	20.24	42.62	20.06
Sulfamethoxazole	Mean	171.60	235.80	140.55	275.04	213.65	121.98	164.26	217.34
	SD	36.39	27.53	49.70	32.12	25.01	16.23	53.80	46.54

During the study period, the macrolide antibiotic drug, roxithromycin, had the highest concentration detected in influent urban WWTPs, which was 1.15 times higher (spring), 1.35 times higher (summer), 1.25 times higher (autumn), and 1.31 times higher (winter) than that in suburban WWTPs (Table 2). The concentrations of sulfa antibiotics, sulfadiazine, and sulfamethoxazole, detected in influent suburban WWTPs were higher than that in urban WWTPs in spring and autumn while showing an opposite trend in summer and winter (Table 2). Sulfadiazine and sulfamethoxazole are antibiotics shared by humans and animals [30–32]. The breeding industry in suburban Tangshan is concentrated, and the use of veterinary antibiotics is high. Most veterinary antibiotics are excreted in the form of raw drugs or metabolites through animal feces and urine after administration and eventually enter the urban drainage system after sewage treatment [33–35]. Despite the legitimate reasons for their use, the current standards for the dosage of various veterinary antibiotics are inconsistent and imprecise, leading to the potential overuse of these drugs in livestock farming. This, in turn, raises the concentration of antibiotics in influent wastewater treatment plants within the farming region.

In influent WWTPs, (1) the concentration ranges of the nine antibiotics selected in this study, except for aureomycin, oxytetracycline, and ciprofloxacin, were much lower than those in Beijing (2018) in winter [36]; (2) the concentration of tetracycline antibiotics in summer was lower than existing research data, while the concentration of aureomycin and oxytetracycline in summer was higher than existing research data (except for Jiulongjiang River Basin), and the concentration of ciprofloxacin in quinolone antibiotics in summer and winter was higher than existing research data (except for Urumqi and Shihezi). In summer, the concentration of norfloxacin surpassed that of the Jiulongjiang River Basin yet remained lower than that of Urumqi and Shihezi at its peak. In summer, the concentration of ofloxacin was lower than that of the Jiulongjiang River Basin yet surpassed that of Yibin, Urumqi, and Shihezi [37–40]. (3) Among the sulfa antibiotics, sulfadiazine's concentration in summer and autumn surpassed that of the Jiulongjiang River Basin yet remained lower than that of Urumqi and Shihezi. The concentration of sulfamethoxazole in summer exceeded that of the Jiulongjiang River Basin, though its peak value was lower than that observed in Beijing (2019) [38–40] (Table S2).

3.2. *Effluent*

The removal effect of antibiotics in WWTPs in different seasons is closely related to treatment processes, operating parameters, influent properties, and types of antibiotics. Currently, most WWTPs employ biological treatment processes to degrade organic matter, including antibiotics. These processes primarily involve microorganisms in activated sludge attaching to the cell surface through adsorption and absorption. Different types of microorganisms utilize their metabolic capabilities to decompose and transform antibiotics. Ultimately, these microorganisms break down the molecular structure of the antibiotics into smaller organic compounds or CO_2 through enzyme production and oxidation, releasing corresponding metabolites. The seasonal variation characteristics of the detected concentration of antibiotics in the effluent of both urban and suburban WWTPs are shown in Figure 3. The total antibiotics concentration in effluent WWTPs in the winter was the highest, followed by spring, autumn, and summer. In winter, the concentration of antibiotics detected in the wastewater from urban WWTPs was higher than that from suburban WWTPs, while the trend was reversed in the other three seasons (Figure 3).

Currently, the WWTPs in Tangshan mainly use the A^2O (anaerobic–anoxic–aerobic) process for sewage treatment. The A^2O method is widely used in the sewage treatment system of northern China [41]. However, northern China experiences lengthy cold seasons, making it challenging for small-scale sewage biochemical treatment processes to operate stably [42]. Low temperatures decrease the activity of nitrifying and denitrifying bacteria, leading to a decline in the nitrogen removal efficiency of the A^2O process and challenges in its stable operation [43,44]. Previous studies have found that temperature has a significant impact on the nitrogen removal efficiency of the A^2O process. Nitrification reactions occur at 20–30 °C and almost stop at temperatures below 5 °C; denitrification reactions occur at 20–40 °C and rapidly decrease at temperatures below 15 °C. The winter temperatures in Tangshan and the surrounding areas are low, with average temperatures below 0 °C from January to February, which is not conducive to the degradation of antibiotics by microorganisms in activated sludge. Therefore, due to the impact of low temperatures on the efficiency of the A^2O process, the total concentration of antibiotics in the effluent of WWTPs in winter is one order of magnitude higher than that in spring, autumn, and summer (Figure 2).

In the effluent of both urban and suburban WWTPs, macrolide antibiotics, such as roxithromycin, have the highest detection concentration in spring, while this corresponds with quinolone antibiotics in autumn and winter (Table 3). In the effluent samples of suburban WWTPs in summer, the concentration of sulfa antibiotics in the effluent is higher than that in the influent (Table 3) [45]. Previous studies have also observed a similar phenomenon, where the concentration of sulfa antibiotics in the effluent after treatment by activated sludge processes has increased. This phenomenon may be due to the following reasons: (1) antibiotics adsorbed in activated sludge are released into the water, resulting in an increase in the concentration of these drugs in the effluent from WWTPs. (2) During the A^2O process, sulfa antibiotics are converted into other substances in the aerobic stage, and these substances are converted back into sulfa antibiotics in the anaerobic stage, resulting in an increase in the concentration of sulfa antibiotics in the effluent. In this study, the removal rates of nine antibiotics in urban and suburban WWTPs increased from 8.18% and 7.30% in winter to 70.14% and 66.82% in spring, and 79.58% and 73.91% in autumn. The removal rates of tetracyclines, quinolones, and sulfonamides in urban WWTPs were higher than those of macrolides in all four seasons, while the suburban WWTPs only followed the same trend in spring, autumn, and winter.

Table 3. Occurrence of antibiotics (ng/L) in the effluent of wastewater treatment plants (WWTPs).

Antibiotics		Urban				Suburban			
		Spring	Summer	Autumn	Winter	Spring	Summer	Autumn	Winter
Roxithromycin	Mean	107.23	22.82	35.68	322.58	104.93	15.80	24.75	248.95
	SD	3.17	9.78	6.62	15.06	13.54	2.64	1.91	38.02
Tetracycline	Mean	9.53	ND	1.85	93.60	7.51	2.98	ND	64.32
	SD	1.60	ND	2.62	39.69	4.22	0.87	ND	15.72
Aureomycin	Mean	5.63	6.13	ND	89.74	6.11	5.27	ND	78.82
	SD	0.95	0.71	ND	50.21	3.07	1.13	ND	60.22
Oxytetracycline	Mean	5.48	5.77	2.36	234.56	6.50	4.90	9.12	91.94
	SD	0.60	0.43	3.34	14.46	2.19	0.90	5.49	10.61
Ciprofloxacin	Mean	26.28	16.16	14.65	418.72	20.27	31.22	21.11	297.48
	SD	4.35	1.85	4.17	18.31	1.09	3.80	9.56	2.77
Norfloxacin	Mean	29.71	20.38	21.59	424.86	27.51	33.13	25.88	295.44
	SD	3.51	1.07	3.16	18.58	4.25	8.25	10.14	8.82
Ofloxacin	Mean	29.10	9.03	20.78	442.08	34.32	26.27	28.48	394.54
	SD	12.45	0.85	4.22	13.32	20.34	2.49	3.28	81.54
Sulfadiazine	Mean	34.22	12.23	139.90	135.16	42.88	133.59	28.73	101.28
	SD	7.38	0.75	1.70	15.78	5.16	14.88	12.17	17.56
Sulfamethoxazole	Mean	37.18	19.60	21.13	256.67	52.20	145.73	39.51	201.87
	SD	7.90	4.53	14.20	34.42	6.84	24.76	6.86	44.41

Note: ND: not detected.

In the effluent of WWTPs, (1) the nine antibiotics selected in this study (excluding tetracycline, aureomycin, oxytetracycline, and ciprofloxacin) exhibited lower winter concentration ranges in comparison to Beijing (2018) [36]; (2) the roxithromycin concentration in summer was lower than that in the Zijiang River Basin of central Hunan, while in spring and autumn, they were an order of magnitude higher than those in Shenyang [36,39,46,47]; (3) Quinolone and tetracycline concentrations in spring and autumn were an order of magnitude higher than those in Shenyang, while in summer, they were an order of magnitude lower than concentrations in Yibin; (4) Sulfa antibiotic concentrations in summer and autumn were higher than those reported by Beijing (2019) and Shenyang, and in summer, they were an order of magnitude higher than those in Yibin [36,46,47] (Table S3). The variation in the spatial and temporal distribution of antibiotic concentrations was significant. This variation is primarily attributed to a complex interplay of factors, including the treatment processes and surface temperatures employed by sewage treatment plants across different regions, the sources and composition of sewage within the service area, and the size of the population served by these WWTPs.

3.3. Ecological Risk Assessment

In the effluent of WWTPs in winter, the RQ values of roxithromycin, tetracycline, chlortetracycline, and oxytetracycline in urban areas were 0.22, 0.1 (0.09), 0.1 (0.08), and 0.23, while those corresponding to roxithromycin and oxytetracycline were 0.17 and 0.1 (0.09) in suburban areas. This indicates that these macrolides and tetracyclines are medium-risk antibiotics. In the other three seasons, the four categories of nine antibiotics had RQ values of ≤ 0.1 in the effluent of WWTPs and showed low-risk antibiotics. It is worth noting that macrolides, including roxithromycin, are medium-risk antibiotics in the effluent of urban and suburban WWTPs in winter, and there is a possibility of the overuse of these drugs by residents in Tangshan. Chen et al. used RQs to assess the ecological risks of antimicrobial drugs. The results showed that erythromycin, roxithromycin, tetracycline, chlortetracycline, sulfamethoxazole, and norfloxacin were high-risk pollutants in water bodies in China, accounting for 20.9% [48]. In winter, various respiratory diseases, including mycoplasma pneumoniae, influenza, adenovirus, and respiratory syncytial virus infections, are highly prevalent. The peak season for mycoplasma pneumoniae infection occurs from August to February of the subsequent year, with the highest incidence of around December to January of the following year [49]. Macrolide antibiotics, such as roxithromycin and clarithromycin, are stable to acidic and have a long half-life (35–48 h), a broad antibacterial spectrum, high bioavailability, are widely distributed in the body, and have significant efficacy, with minimal gastrointestinal irritation [50]. They have become the first choice for treating mycoplasma pneumoniae infection [51,52]. At present, the resistance rate of Mycoplasma pneumoniae to macrolides has been on the rise worldwide [53]. East Asia is the region with the most serious resistance to macrolide drugs for Mycoplasma pneumoniae in the world. Studies have shown that the resistance rate in some areas of China has reached over 90%.

3.4. Estimation of Usage and Sewage Discharge

The per capita pollution load, annual usage, and annual emissions of antibiotics in Tangshan are presented in Tables 4 and 5. The per capita pollution load of antibiotics was highest in spring, summer, and autumn for sulfamethoxazole, while the highest load in winter was for ofloxacin. From spring to winter, the per capita pollution load of antibiotics for urban residents was 9.63–13.74 times that of suburban residents, suggesting that urban residents may be at risk of antibiotic abuse (Table 4). In urban areas, the usage of roxithromycin (5.87 kg/a in spring), sulfamethoxazole (8.96 kg/a in summer), sulfamethoxazole (5.77 kg/a in autumn), and ofloxacin (17.76 kg/a in winter) significantly surpasses that of other antimicrobial agents. In contrast, the usage levels in suburban areas are as follows: sulfamethoxazole (0.17 kg/ain spring), norfloxacin (0.10 kg/ain summer), sulfamethoxazole (0.12 kg/ain autumn), and ofloxacin (0.32 kg/ain winter) (Table 4). It should be noted that the usage of antibiotics in urban areas was one order of magnitude higher than that in suburban areas (Table 5). After treatment by the A^2O process in the WWTPs, the four types of nine antibiotics selected in this study, including roxithromycin (3.87 kg/a), norfloxacin (0.73 kg/a), roxithromycin (1.39 kg/a), and ofloxacin (15.96 kg/a), were the highest in terms of emissions from urban WWTPs in spring, summer, autumn, and winter, respectively. Roxithromycin, sulfamethoxazole, sulfamethoxazole, and ofloxacin were the highest in terms of emissions from suburban WWTPs during the corresponding periods, respectively. The total usage of the nine antibiotics in urban and suburban WWTPs in 2023 was 32.55 and 0.75 kg/a, 30.11 and 0.58 kg/a, 24.29 and 0.57 kg/a, and 96.05 and 1.59 kg/a, respectively, while the total emissions were 9.86 and 0.25 kg/a, 3.83 and 0.33 kg/a, 4.84 and 0.15 kg/a, and 88.38 and 1.48 kg/a, respectively.

Table 4. Estimates of per capita pollution load of antibiotics [μg/(d·person)] in Tangshan.

	Urban				Suburban			
Antibiotics	Spring	Summer	Autumn	Winter	Spring	Summer	Autumn	Winter
Roxithromycin	63.11	17.12	28.96	130.86	5.49	1.26	2.31	10.01
Tetracycline	28.46	6.68	5.45	38.60	2.27	0.52	0.48	2.64
Aureomycin	18.64	19.62	6.87	35.87	1.04	1.84	0.60	3.15
Oxytetracycline	14.96	16.62	13.41	95.07	1.57	1.36	2.31	3.77
Ciprofloxacin	36.98	45.91	29.51	167.30	3.21	3.97	2.53	11.73
Norfloxacin	45.19	50.06	36.11	169.81	3.92	4.32	3.14	11.90
Ofloxacin	46.43	17.73	45.41	182.05	3.95	3.25	3.26	15.65
Sulfadiazine	35.73	49.39	32.47	53.96	4.36	3.82	5.28	4.08
Sulfamethoxazole	63.56	87.33	52.06	101.87	7.91	4.52	6.08	8.05
Total	353.06	310.46	250.25	975.40	33.71	24.86	25.99	70.99

Table 5. Estimates of antibiotics use and emissions (kg/a) in Tangshan.

	Urban				Suburban			
	Spring	Summer	Autumn	Winter	Spring	Summer	Autumn	Winter
				Use				
Roxithromycin	5.87	1.49	2.90	12.54	0.13	0.03	0.06	0.23
Tetracycline	2.93	0.60	0.49	3.25	0.05	0.01	0.01	0.05
Aureomycin	2.01	1.66	0.64	4.20	0.03	0.04	0.01	0.09
Oxytetracycline	1.28	1.66	1.23	9.60	0.04	0.03	0.06	0.09
Ciprofloxacin	3.45	4.23	2.75	16.23	0.08	0.10	0.06	0.26
Norfloxacin	4.22	4.71	3.36	16.48	0.09	0.10	0.08	0.26
Ofloxacin	3.75	1.73	4.20	17.76	0.07	0.07	0.07	0.32
Sulfadiazine	3.24	5.07	2.95	5.54	0.09	0.09	0.10	0.10
Sulfamethoxazole	5.79	8.96	5.77	10.45	0.17	0.09	0.12	0.19
				Emissions				
Roxithromycin	3.87	0.71	1.39	11.58	0.09	0.01	0.02	0.22
Tetracycline	0.33	0.00	0.03	2.90	0.01	0.00	0.00	0.05
Aureomycin	0.22	0.21	0.00	3.92	0.01	0.00	0.00	0.08
Oxytetracycline	0.19	0.20	0.04	8.75	0.01	0.00	0.01	0.08
Ciprofloxacin	0.90	0.57	0.48	15.05	0.02	0.03	0.02	0.24
Norfloxacin	1.04	0.73	0.75	15.27	0.02	0.03	0.02	0.24
Ofloxacin	0.90	0.32	0.70	15.96	0.02	0.02	0.02	0.30
Sulfadiazine	1.15	0.44	0.49	5.14	0.03	0.11	0.02	0.09
Sulfamethoxazole	1.26	0.66	0.95	9.81	0.04	0.11	0.03	0.18

4. Conclusions

Due to the high incidence of respiratory diseases, the use of antibiotics has increased, resulting in the highest concentration of antibiotics in the winter for both influent urban and suburban WWTPs. However, due to the low-temperature environment, the removal rate of antimicrobial drugs by the A^2O process in WWTPs is the lowest in winter. Based on the RQ method for evaluating the ecological risk of antibiotics, it was found that the RQ values of roxithromycin, tetracycline, aureomycin, and oxytetracycline in the winter effluent samples from urban WWTPs were 0.22, 0.1 (0.09), 0.1 (0.08), and 0.23, respectively, identifying them as medium-risk antibiotics. The RQ values of roxithromycin and oxytetracycline in the winter effluent samples from suburban WWTPs were 0.17 and 0.1 (0.09), respectively, identifying them as medium-risk antibiotics. In the other three seasons, the four categories of nine antibiotics selected in this study had RQ values ≤ 0.1 in the effluent of WWTPs, all of which were low-risk pollutants. In the study area of Tangshan, the per capita pollution load of antibiotics was highest in spring, summer, and autumn for sulfamethoxazole while the highest in winter for ofloxacin. The highest use and emissions of antibiotics in the urban in spring, summer, autumn, and winter were roxithromycin (5.87 and 3.87 kg/a), sulfamethoxazole (8.96 kg/a), and norfloxacin (0.73 kg/a), while the highest use and emissions in suburban were sulfamethoxazole (0.17 kg/a) and roxithromycin (0.09 kg/a), norfloxacin (0.10 kg/a) and sulfamethoxazole (0.11 kg/a), sulfamethoxazole (0.12 and 0.03 kg/a), and ofloxacin (0.32 and 0.30 kg/a) in the same seasons.

This study focuses solely on the temporal distribution of target antibiotics in influent and effluent WWTPs. Certain antibiotics are susceptible to hydrolysis and removal in aquatic environments. Research indicates that macrolide antibiotics are prone to hydrolysis. Tetracycline antibiotics are not stable in water; for instance, the hydrolysis rate of oxytetracycline increases with deviations from neutral pH (pH = 7) and rising temperatures, whereas sulfonamides and fluoroquinolones are resistant to hydrolysis. pH and temperature are significant factors influencing hydrolysis. Consequently, it is imperative to further investigate the impact of seasonal variations in pH and temperature at the end of the drainage systems and within the treatment units of WWTPs on the temporal distribution of antibiotics to elucidate the driving factors behind any temporal trends observed in these agents' distribution in the influent and effluent. Moreover, the adsorption of antibiotics by sewage plant sludge is a significant factor in enhancing the removal rate of these antibiotics. For antibiotics primarily removed through sludge adsorption (such as fluoroquinolones and sulfonamides), an extension in sludge retention time concurrently enhances their removal efficiency. However, the removal of certain antibiotics may not be impacted by sludge retention time. Hence, there is an urgent need for a comprehensive assessment of the physical adsorption and biodegradation of antibiotics in WWTPs.

Supplementary Materials: The following supporting information can be downloaded at: https://www.mdpi.com/article/10.3390/w16111627/s1, Table S1: Waste Water Treatment Plant (WWTP); Table S2: Comparison of the concentrations of target antibiotics in effluents from WWTPs in other cities; Table S3: Comparison of the concentrations of target antibiotics in influents from WWTPs in other cities.

Author Contributions: Conceptualization, Z.D. and J.H.; Methodology, G.H.; Formal analysis, J.H.; Investigation, Z.D., P.W. and Z.J.; Resources, Z.D. and P.W. All authors have read and agreed to the published version of the manuscript.

Funding: This research received no external funding.

Data Availability Statement: Data is contained within the article.

Conflicts of Interest: The authors declare no conflict of interest.

References

1. Wang, S.Z.; Wang, J.L. Single atom cobalt catalyst derived from co-pyrolysis of vitamin B12 and graphitic carbon nitride for PMS activation to degrade emerging pollutants. *Appl. Catal. B Environ.* **2023**, *321*, 122051. [CrossRef]
2. Wang, S.Z.; Xu, L.J.; Wang, J.L. Iron-based dual active site-mediated peroxymonosulfate activation for the degradation of emerging organic pollutants. *Environ. Sci. Technol.* **2021**, *55*, 15412–15422. [CrossRef]
3. Zhuang, S.T.; Wang, J.L. Magnetic COFs as catalyst for Fenton-like degradation of sulfamethazine. *Chemosphere* **2021**, *264*, 128561. [CrossRef]
4. Atif, M.; Sadeeqa, S.; Afzal, H.; Latif, S. Knowledge, Attitude and Practices regarding Antibiotics Use among Parents for their Children. *Int. J. Pharm. Sci. Res.* **2018**, *9*, 2140–2148. [CrossRef]
5. Li, S.N.; Ondon, B.S.; Ho, S.H.; Jiang, J.W.; Li, F.X. Antibiotic resistant bacteria and genes in wastewater treatment plants: From occurrence to treatment strategies. *Sci. Total Environ.* **2022**, *838*, 156544. [CrossRef] [PubMed]
6. Kovalakova, P.; Cizmas, L.; McDonald, T.J.; Marsalek, B.; Feng, M.; Sharma, V.K. Occurrence and Toxicity of Antibiotics in the Aquatic Environment: A Review. *Chemosphere* **2020**, *251*, 126351. [CrossRef]

7. Qi, L.H.; Fan, W.H.; Li, J.; Cui, H.F.; Xu, J.X.; Gu, D.M.; Meng, J.J.; Liu, J. Persistent Nocardia beijingensis infection in a patient with postoperative abscess and misuse of antibiotics in China. *Infect. Med.* **2023**, *2*, 343–348. [CrossRef]
8. Shao, Y.T.; Wang, Y.P.; Yuan, Y.W.; Xie, Y.J. A systematic review on antibiotics misuse in livestock and aquaculture and regulation implications in China. *Sci. Total Environ.* **2021**, *798*, 149205. [CrossRef]
9. Luo, X.Z.; Han, S.; Wang, Y.; Du, P.; Li, X.Q.; Thai, P.K. Significant differences in usage of antibiotics in three Chinese cities measured by wastewater-based epidemiology. *Water Res.* **2024**, *254*, 121335. [CrossRef] [PubMed]
10. Xiao, Y.H.; Li, L.J. Legislation of clinical antibiotic use in China. *Lancet Infect. Dis.* **2013**, *13*, 189–191. [CrossRef]
11. Lenart-Boron, A.; Prajsnar, J.; Guzik, M.; Boron, P.; Chmiel, M. How much of antibiotics can enter surface water with treated wastewater and how it affects the resistance of waterborne bacteria: A case study of the Białka river sewage treatment plant. *Environ. Res.* **2020**, *191*, 110037. [CrossRef] [PubMed]
12. Huang, H.W.; Zeng, S.Y.; Dong, X.; Li, D.; Zhang, Y.; He, M.; Du, P.F. Diverse and abundant antibiotics and antibiotic resistance genes in an urban water system. *J. Environ. Manag.* **2019**, *231*, 494–503. [CrossRef] [PubMed]
13. Perez-Bou, L.; Gonzalez-Martinez, A.; Gonzalez-Lopez, J.; Correa-Galeote, D. Promising bioprocesses for the efficient removal of antibiotics and antibiotic-resistance genes from urban and hospital wastewaters: Potentialities of aerobic granular systems. *Environ. Pollut.* **2024**, *342*, 123115. [CrossRef] [PubMed]
14. Zou, M.Y.; Tian, W.J.; Zhao, J.; Chu, M.L.; Song, T.T. Quinolone antibiotics in WWTPs with activated sludge treatment processes: A review on source, concentration and removal. *Process Saf. Environ.* **2022**, *160*, 116–129. [CrossRef]
15. Yin, S.Y.; Gao, L.; Fan, X.M.; Gao, S.H.; Zhou, X.; Jin, W.B.; He, Z.Q.; Wang, Q.L. Performance of sewage sludge treatment for the removal of antibiotic resistance genes: Status and prospects. *Sci. Total Environ.* **2024**, *907*, 167862. [CrossRef] [PubMed]
16. Silva, C.; Almeida, C.M.M.; Rodrigues, J.A.; Silva, S.; Coelho, M.D.; Martins, A.; Lourinho, R.; Cardoso, E.; Cardoso, V.V.; Benoliel, M.J.; et al. Improving the control of pharmaceutical compounds in activated sludge wastewater treatment plants: Key operating conditions and monitoring parameters. *J. Water Process Eng.* **2023**, *54*, 103985. [CrossRef]
17. Nasir, A.; Saleh, M.; Aminzai, M.T.; Alary, R.; Dizge, N.; Yabalak, E. Adverse effects of veterinary drugs, removal processes and mechanisms: A review. *J. Environ. Chem. Eng.* **2024**, *12*, 111880. [CrossRef]
18. Omar, T.F.T.; Aris, A.Z.; Yusoff, F.M.; Mustafa, S. An improved SPE-LC-MS/MS method for multiclass endocrine disrupting compound determination in tropical estuarine sediments. *Talanta* **2017**, *173*, 51–59. [CrossRef] [PubMed]
19. Samaras, V.G.; Thomaidis, N.S.; Stasinakis, A.S.; Lekkas, T.D. An analytical method for the simultaneous trace determination of acidic pharmaceuticals and phenolic endocrine disrupting chemicals in wastewater and sewage sludge by gas chromatography-mass spectrometry. *Anal. Bioanal. Chem.* **2011**, *399*, 2549–2561. [CrossRef]
20. European Commission Joint Research Centre (EC-JRC). *Technical Guidance Document on Risk Assessment in Support of Commission Directive 93/67/EEC on Risk Assessment for New Notified Substances, Commission Regulation (EC) No 1488/94 on Risk Assessment for Existing Substances, and Directive 98/8/EC of the European Parliament and of the Council Concerning the Placing of Biocidal Products on the Market*; Part I–IV, European Chemicals Bureau (ECB), JRC-ISPRA (VA), Italy, April 2003; Part II. EUR; Institute for Health and Consumer Protection: Ispra, Italy, 2003.
21. Shams, D.F.; Izaz, M.; Khan, W.; Nayab, S.; Tawab, A.; Baig, S.A. Occurrence of selected antibiotics in urban rivers in northwest Pakistan and assessment of ecotoxicological and antimicrobial resistance risks. *Chemosphere* **2024**, *352*, 141357. [CrossRef]
22. Letsinger, S.; Kay, P. Comparison of prioritisation schemes for human pharmaceuticals in the aquatic environment. *Environ. Sci. Pollut. Res.* **2019**, *26*, 3479–3491. [CrossRef] [PubMed]
23. Verlicchi, P.; Al Aukidy, M.; Zambello, E. Occurrence of pharmaceutical compounds in urban wastewater: Removal, mass load and environmental risk after a secondary treatment-a review. *Sci. Total Environ.* **2012**, *429*, 123–155. [CrossRef] [PubMed]
24. Rodriguez-Mozaz, S.; Chamorro, S.; Marti, E.; Huerta, B.; Gros, M.; Sànchez-Melsió, A.; Borrego, C.M.; Barceló, D.; Balcázar, J.L. Occurrence of antibiotics and antibiotic resistance genes in hospital and urban wastewaters and their impact on the receiving river. *Water Res.* **2015**, *69*, 234–242. [CrossRef] [PubMed]
25. Gan, X.M.; Yan, Q.; Gao, X.; Zhang, Y.X.; Zi, C.F.; Peng, X.Y.; Guo, J.S. Occurrence and Fate of Typical Antibiotics in a Wastewater Treatment Plant in Southwest China. *Environ. Sci.* **2014**, *35*, 1817–1823. (In Chinese)
26. Liu, W.R.; Yang, Y.Y.; Liu, Y.S.; Zhang, L.J.; Zhao, J.L.; Zhang, Q.Q.; Zhang, M.; Zhang, J.N.; Jiang, Y.X.; Ying, G.G. Biocides in wastewater treatment plants: Mass balance analysis and pollution load estimation. *J. Hazard. Mater.* **2017**, *329*, 310–320. [CrossRef] [PubMed]
27. Huang, Z.; Zhao, J.L.; Yang, Y.Y.; Jia, Y.W.; Zhang, Q.Q.; Chen, C.E.; Liu, Y.S.; Yang, B.; Xie, L.T.; Ying, G.G. Occurrence, mass loads and risks of bisphenol analogues in the Pearl River Delta region, South China: Urban rainfall runoff as a potential source for receiving rivers. *Environ. Pollut.* **2020**, *263*, 114361. [CrossRef] [PubMed]
28. Lei, H.J.; Yang, B.; Ye, P.; Yang, Y.Y.; Zhao, J.L.; Liu, Y.S.; Xie, L.T.; Ying, G.G. Occurrence, fate and mass loading of benzodiazepines and their transformation products in eleven wastewater treatment plants in Guangdong province, China. *Sci. Total Environ.* **2021**, *755*, 142648. [CrossRef] [PubMed]
29. Zhu, K.Y.; Tu, W.X.; Feng, Y.N.; Bai, W.Q.; Xie, Y.R.; Zhang, Q.; Ren, J.H.; Shi, G.Q.; Xiang, N.J.; Meng, L. Risk assessment of public health emergencies concerned in China, December 2023. *Dis. Surveill.* **2023**, *38*, 1421–1424.
30. Zheng, H.S.; Zhang, Y.F.; Li, S.; Feng, X.C.; Wu, Q.L.; Leong, Y.K.; Chang, J.S. Antibiotic sulfadiazine degradation by persulfate oxidation: Intermediates dependence of ecotoxicity and the induction of antibiotic resistance genes. *Bioresour. Technol.* **2023**, *368*, 128306. [CrossRef]

31. Hu, L.; Zhang, G.; Wang, Q.; Wang, X.; Wang, P. Effect of Microwave Heating on Persulfate Activation for Rapid Degradation and Mineralization of p-Nitrophenol. *ACS Sustain. Chem. Eng.* **2019**, *7*, 11662–11671. [CrossRef]
32. Duan, W.; Cui, H.; Jia, X.; Huang, X. Occurrence and ecotoxicity of sulfonamides in the aquatic environment: A review. *Sci. Total Environ.* **2022**, *820*, 153178. [CrossRef] [PubMed]
33. Daghrir, R.; Drogui, P. Tetracycline antibiotics in the environment: A review. *Environ. Chem. Lett.* **2013**, *11*, 20–227. [CrossRef]
34. Xu, L.Y.; Zhang, H.; Xiong, P.; Zhu, Q.Q.; Liao, C.Y.; Jiang, G.B. Occurrence, fate, and risk assessment of typical tetracycline antibiotics in the aquatic environment: A review. *Sci. Total Environ.* **2020**, *753*, 141975. [CrossRef] [PubMed]
35. Leichtweis, J.; Vieira, Y.; Welter, N.; Silvestri, S.; Dotto, G.L.; Carissimi, E. A review of the occurrence, disposal, determination, toxicity and remediation technologies of the tetracycline antibiotic. *Process Saf. Environ.* **2022**, *160*, 25–40. [CrossRef]
36. Liu, X.H.; Zhang, G.D.; Liu, Y.; Lu, S.Y.; Qin, P.; Guo, X.C.; Bi, B.; Wang, L.; Xi, B.D.; Wu, F.C.; et al. Occurrence and fate of antibiotics and antibiotic resistance genes in typical urban water of Beijing, China. *Environ. Pollut.* **2019**, *246*, 163–173. [CrossRef] [PubMed]
37. Li, Z.; Xu, H.B.; Qu, J.; Qu, L.L.; Wang, S.; Wang, N.; Zheng, X.B.; Zhang, X. Characteristics of Typical Antibiotics in Effluent from Shenyang Sewage Treatment Plant and Its Ecological Risk Assessment. *Sci. Technol. Innov.* **2023**, *468*, 87–91. (In Chinese)
38. Li, Y.; Wang, J.; Lin, C.Y.; Lian, M.S.; He, M.C.; Liu, X.T.; Ouyang, W. Occurrence, removal efficiency, and emission of antibiotics in the sewage treatment plants of a low-urbanized basin in China and their impact on the receiving water. *Sci. Total Environ.* **2024**, *921*, 171134. [CrossRef] [PubMed]
39. Tang, T.T.; Zhao, Z.Y.; Wang, Y.; Qiao, X.J.; Gu, B.C.; Wang, W.Q. Antibiotic Pollution Level and Ecological Risk Assessment of Township Wastewater Treatment Plants. *Environ. Sci.* **2024**. [CrossRef]
40. Wang, L.; Zhu, D.; Cao, Y.X.; Yu, X.D.; Hui, Y.M.; Li, W.C.; Wang, D.H. Seasonal changes and ecological risk assessment of pharmaceutical and personal care products in the effluents of wastewater treatment plants in Beijing. *Acta Sci. Circumstantiae* **2021**, *41*, 2922–2932. [CrossRef]
41. Lan, Z.H.; Zhang, Y.P.; Liang, R.L.; Wang, Z.Q.; Sun, J.; Lu, X.W.; He, Y.; Wang, Y.J. Comprehensive comparison of integrated fixed-film activated sludge (IFAS) and AAO activated sludge methods: Influence of different operational parameters. *Chemosphere* **2024**, *357*, 142068. [CrossRef]
42. Pu, M.; Ailijiang, N.; Mamat, A.W.; Chang, J.L.; Zhang, Q.F.; Liu, Y.F.; Li, N.X. Occurrence of antibiotics in the different biological treatment processes, reclaimed wastewater treatment plants and effluent-irrigated soils. *J. Environ. Chem. Eng.* **2022**, *10*, 107715. [CrossRef]
43. Wang, R.M.; Ji, M.; Zhai, H.Y.; Guo, Y.J.; Liu, Y. November Occurrence of antibiotics and antibiotic resistance genes in WWTP effluent-receiving water bodies and reclaimed wastewater treatment plants. *Sci. Total Environ.* **2021**, *796*, 148919. [CrossRef]
44. Li, W.H.; Shi, Y.L.; Gao, L.H.; Liu, J.M.; Cai, Y.Q. Occurrence, distribution and potential affecting factors of antibiotics in sewage sludge of wastewater treatment plants in China. *Sci. Total Environ.* **2013**, *445–446*, 306–313. [CrossRef]
45. Liu, Z.G.; Zhang, Y.; Zhou, W.; Wang, W.; Dai, X.H. Comparison of Nitrogen and Phosphorus Removal between Two Typical Processes under Low Temperature in a Full-Scale Municipal Wastewater Treatment Plant. *Water* **2022**, *14*, 3874. [CrossRef]
46. Zhang, H.; Du, M.M.; Jiang, H.Y.; Zhang, D.D.; Lin, L.F.; Yea, H.; Zhang, X. Occurrence, seasonal variation and removal efficiency of antibiotics and their metabolites in wastewater treatment plants, Jiulongjiang River Basin, South China. *Environ. Sci. Process. Impacts* **2015**, *17*, 225. [CrossRef]
47. Liu, J.; Lu, J.J.; Tong, Y.B.; Li, C. Occurrence and elimination of antibiotics in three sewage treatment plants with different treatment technologies in Urumqi and Shihezi, Xinjiang. *Water Sci. Technol.* **2017**, *75*, 1474–1484. [CrossRef] [PubMed]
48. Chen, L.H.; Cao, Y.; Li, Q.; Meng, T.; Zhang, S. Pollution Characteristics and Ecological Risk Assessment of Typical Antibiotics in Environmental Media in China. *Environ. Sci.* **2023**, *44*, 6894–6908. [CrossRef] [PubMed]
49. Debnath, S.K.; Debnath, M.; Srivastava, R. Opportunistic etiological agents causing lung infections: Emerging need to transform lung-targeted delivery. *Heliyon* **2022**, *8*, e12620. [CrossRef] [PubMed]
50. Belizário, J.; Garay-Malpartida, M.; Faintuch, J. Lung microbiome and origins of the respiratory diseases. *Curr. Res. Immunol.* **2023**, *4*, 100065. [CrossRef]
51. Bebear, C.; Dupon, M.; Renaudin, H.; Debarbeyrac, B. Potential Improvements in Therapeutic Options for Mycoplasmal Respiratory-infections. *Clin. Infect. Dis.* **1993**, *17*, S202–S207. [CrossRef]
52. Tsai, T.A.; Tsai, C.K.; Kuo, K.C.; Yu, H.R. Rational stepwise approach for Mycoplasma pneumoniae pneumonia in children. *J. Microbiol. Immunol.* **2021**, *54*, 557–565. [CrossRef] [PubMed]
53. Rafei, R.; Al Iaali, R.; Osman, M.; Dabboussi, F.; Hamze, M. A global snapshot on the prevalent macrolide-resistant emm types of Group A Streptococcus worldwide, their phenotypes and their resistance marker genotypes during the last two decades: A systematic review. *Infect. Genet. Evol.* **2022**, *99*, 105258. [CrossRef] [PubMed]

water

MDPI

Article

A Comprehensive Assessment of the Ecological State of the Transboundary Irtysh River (Kazakhstan, Central Asia)

Elena Krupa [1,2], Sophia Romanova [1], Aizada Serikova [1] and Larisa Shakhvorostova [1,*]

1 Institute of Zoology of the Republic of Kazakhstan, Almaty 050060, Kazakhstan; elena_krupa@mail.ru (E.K.); sofiyarom@mail.ru (S.R.); serikova.aiz@mail.ru (A.S.)
2 Kazakh Agency for Applied Ecology, Almaty 050010, Kazakhstan
* Correspondence: larisa.shakhvorostova@zool.kz

Abstract: The diverse anthropogenic load on the transboundary Irtysh River necessitates an assessment of its ecological state, which was the goal of this work. We conducted this research in July 2023 in the upper and lower reaches of the Kazakh part of the Irtysh basin. We determined transparency; temperature; pH; salinity (TDS); oxygen, $N\text{-}NO_3$, $N\text{-}NO_2$, $N\text{-}NH_4$, PO_4, Mn, Fe, Si, Cd, Cu, Zn, Pb, Cr, Co, and Hg contents; permanganate index; and zooplankton variables at 27 stations. We assessed the ecological state of the river by comparing the contents of pollutants with their maximum permissible concentrations (MPC_{fw}), Classification Scales, and bioindications. An excess of MPC_{fw} was detected for $N\text{-}NO_2$, Cu, and Fe and locally for Cr and Zn. According to the Classification Scales, most analysed variables corresponded to slightly polluted waters; $N\text{-}NO_2$, Cr, and Zn corresponded to moderately and heavily polluted waters. Zooplankton was represented by 82 species, with an average abundance of 6728 individuals/m^3, biomass of 2.81 mg/m^3, Shannon index of 1.99–2.08 bit, Δ-Shannon of 0.09, and average individual mass of 0.0019 mg. The spatial distribution of abiotic and biotic variables indicated increased organic and toxic pollution downstream in the Irtysh. Potential sources of pollution of the Irtysh basin are discussed.

Keywords: pollution; zooplankton; bioindication; heavy metals; nutrients; water quality; transboundary basin

1. Introduction

The most practical approach to objectively assessing the ecological state of aquatic ecosystems involves a combination of chemical and biological methods (bioindication) [1]. Bioindication is based on the response of biological communities to a set of external factors, which makes it possible to characterise the ecological well-being of a reservoir as a whole [2,3]. Using chemical variables for analysis aims to identify the potential causes of registered environmental problems.

Zooplankton is effectively used to assess the ecological state of lakes and reservoirs [4–9]. The bioindication of rivers is associated with methodological difficulties since the decisive factor in forming planktonic communities is the flow speed and constant removal of individuals by water masse [10]. The hydrological regime of a water body significantly impacts the species composition and quantitative variables of planktonic communities. In fast-flowing rivers, as a rule, rotifers dominate [11–15]. For planktonic crustaceans, favourable conditions develop in water bodies with slow water exchange (ponds, lakes, reservoirs, and slow-flowing rivers). Accordingly, at the same pollution level, the structure of zooplankton communities in rivers, lakes, and reservoirs can vary significantly. This must be considered when assessing the ecological state of rivers using bioindication methods mainly developed for lakes and reservoirs [4].

The transboundary Irtysh River flows through the territory of China, Kazakhstan, and Russia. Flow regulation and intensive use of water resources have given rise to many environmental and social problems throughout the Irtysh basin. The construction of reservoirs [16] has led to the disruption of the floodplain watering regime and the degradation of floodplain ecosystems [17]. Irreversible water intake over the entire river basin area [18] led to the shallowing of the Irtysh. To ensure the safety of vessel traffic in the Pavlodar Irtysh region, dredging work is being carried out, which, together with sand extraction, causes a decrease in water transparency. The unsatisfactory water quality of the Irtysh River [19] is associated with the extraction and processing of minerals, water removal for agricultural and industrial purposes, and the discharge of poorly treated or untreated wastewater into the river and its tributaries [20–22].

Despite the close attention of scientists and the public to the environmental problems of the Irtysh River, a comprehensive assessment of its water resources has not been carried out to date. There are data (2010–2011) on the content of nutrients and some heavy metals in the water of the upper (Black Irtysh) and middle (in the zone of influence of the upper Irtysh cascade of reservoirs) reaches of the Irtysh River [22]. The same work assessed toxic pollution of the right tributaries of the middle reaches of the Irtysh (Ulba, Krasnoyarka, and others) using biotesting methods. An analysis of the long-term dynamics (1986–2011) of the pollutant contents in certain sections of the Irtysh River has been given [18]. The zooplankton of the upper Irtysh cascade of reservoirs has been relatively well studied [23–25], but only one work [26] has been devoted to river zooplankton.

The purpose of this study was to comprehensively assess the ecological state of the Kazakh part of the Irtysh River based on chemical variables and the structure of zooplankton communities. This work partially fills the existing gaps in the hydrochemical, toxicological, and hydrobiological description of the Irtysh River. In territorial terms, it covers the most poorly studied areas of the basin—its upper (Black Irtysh within Kazakhstan) and lower (Pavlodar region) reaches. This work demonstrates some methodological approaches that can be used to monitor water basin studies.

1.1. General Characteristics of the Irtysh River Basin

The Irtysh River begins on the western slopes of the Mongolian Altai in China, flows through Kazakhstan, and empties into the Ob River in Russia. The total length of the Irtysh River is 4248 km; in Kazakhstan, it is 1698 km. The nutrition of the Irtysh River is mixed: in the upper reaches, it is snow–glacial; in the lower reaches, it is snow, rain, and soil. The width of the riverbed reaches 0.2–0.9 km, with an average slope of 3–7 cm/km. The depth varies from 1 to 2 m on the rifts to 3 to 15 m on the reaches. The average current speed in low water is 2.5–3.5 km/h and up to 4.5–5.1 km/h in high water. The bottom is mostly pebble, sometimes sandy. The banks are composed of clay and overgrown with trees and shrubs.

In the upper reaches, from the source to the confluence with Zaisan Lake, the river is called the Black Irtysh (Figure 1). The length of this section in Kazakhstan is about 120 km. The riverbed is winding, breaks into branches, and forms an extensive delta before flowing into Lake Zaisan. The width of the floodplain is from 1.0 to 12.0 km. The Kalzhir River flows into the Black Irtysh (Figure 2a,b), originating in the mountain Markakol Lake at 1447 m above sea level. The left tributary Kenderlik (Figure 2c,d) originates on the northern slopes of the Sauyr ridge at an altitude of 3000–3200 m and flows into the mouth of the Black Irtysh.

Below Zaisan Lake, the Irtysh River is regulated by the Bukhtarminsky, Ust-Kamenogorsky, and Shulbinsky reservoirs (upper Irtysh cascade). Large right tributaries Bukhtarma, Kurchum, Ulba, Uba, and others flow into this section. Here are the cities of Altai, Serebryansk, Ust-Kamenogorsk, Semey, and Ridder. The length of this section of the river is about 820 km, with a height difference from 388 to 159 m above sea level.

There are no tributaries in the lower reaches of the Kazakhstani part of the Irtysh River (below the upper Irtysh cascade of reservoirs). The length of the section is about 760 km. The width of the floodplain varies from 5 to 15 km. The cities of Kurchatov, Aksu, and Pavlodar and numerous villages are located here.

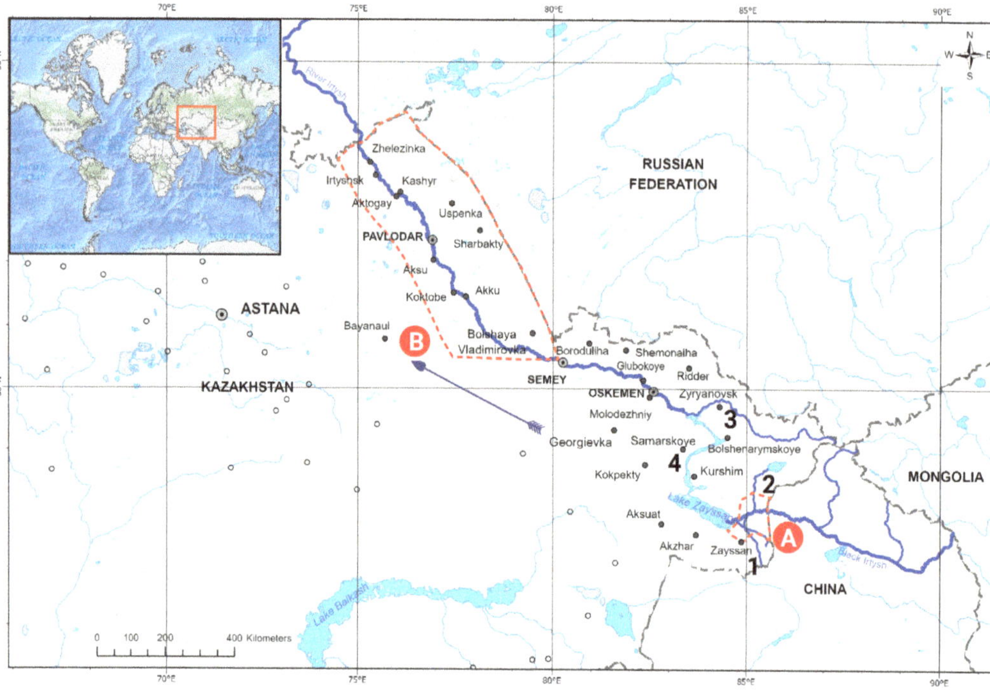

Figure 1. Map scheme of the location of the surveyed parts of the Irtysh River basin, July 2023. A—Black Irtysh River, B—Pavlodar Irtysh region. Arabic numerals: 1—left tributary Kenderlik, 2—right tributary Kalzhir, 3—right tributary Bukhtarma, 4—upper Irtysh cascade of reservoirs. The arrow shows the direction of flow of the Irtysh River, determining left-bank or right-bank tributaries from up to downstream (the river flow direction).

Figure 2. Tributaries of the Black Irtysh River, July 2023: (**a**) right tributary Kalzhir, mountainous part; (**b**) right tributary Kalzhir, before flowing into the Black Irtysh River; (**c**,**d**) left tributary Kenderlik. Photo by E. Krupa.

1.2. Potential Sources of Pollution of the Irtysh River

A significant part of the Irtysh basin is located in the zone of geochemical anomalies. There are deposits of gold [27] and polymetallic (Au, Ag, Mo, Cu, Fe, Pb, and Zn) ores [28]. In the Pavlodar region (lower reaches), 142 mineral deposits are known. Due to the geological features of the territory, the ratio of metals to water for the Irtysh basin is Zn > Cu > Mn > Pb > Mo > Cr > Cd > Co [29,30]. Industrial enterprises in the middle and lower reaches of the Kazakh part of the Irtysh basin carry out ore processing. The total number of industrial enterprises in East Kazakhstan is 133 [31]. Most of them are concentrated in the cities of Ust-Kamenogorsk (61), Ridder (13), Glubokovsky (24), and Altai (14) (middle reaches). There are 185 industrial enterprises in the Pavlodar region (downstream), a significant part of which are located in the cities of Pavlodar (59), Aksu (23), and their surrounding areas (46) [32]. Near Pavlodar city, in the riverbed of the Irtysh River, non-metallic building materials (sands of various sizes and gravel–pebble deposits) are mined (Figure 3a).

Figure 3. The Irtysh River in the Pavlodar region, July 2023: (**a**) sand mining in the riverbed (below Pavlodar City); (**b**) wastewater discharge into the Irtysh River (Terenkol village); (**c**) water intake in the area of Zhelezinka village; (**d**) water intake in the area of Terenkol village. Photos by E. Krupa and A. Linnik.

Agriculture and livestock farming are widespread in the Irtysh basin. Within the Pavlodar region, the most significant areas sown with crops are concentrated in the lower part of the basin along both banks of the Irtysh River [32]. In the East Kazakhstan region, the most agriculturally developed territories are located at a considerable distance from the floodplain of the Black Irtysh River. The right bank part of the Black Irtysh drainage basin is represented by a foothill plain and mountainous areas unsuitable for agriculture. Directly adjacent to the left bank of the Black Irtysh floodplain are foothill sandy and alluvial–deltaic plains unsuitable for agriculture.

The environmental problems of the Irtysh basin are associated with the industrial and agricultural development of the region. The water of the Irtysh River is used for drinking water in populated areas (Figure 3c,d) as well as for agriculture, livestock farming, and industries. Wastewater is discharged into the river and tributaries, sometimes without preliminary treatment (Figure 3b). In the Pavlodar region alone, there are 29 enterprises with 49 wastewater outlets [33]. The total wastewater discharge into the Irtysh River and its tributaries is almost 3000 million m^3 annually [18].

2. Materials and Methods

Comprehensive studies of the Kazakh part of the Irtysh River were carried out in the East Kazakhstan and Pavlodar regions in four parts, which differ significantly in the intensity and nature of anthropogenic load (Table 1).

Table 1. Description of the surveyed parts of the Irtysh River, indicating potential sources of pollution.

Part Number	Description	Length, km	Number of Stations	Altitude above Sea Level	Potential Sources of Pollution
I	Black Irtysh River	120	5	390–419	Buran and Ordynka villages, transboundary flow from China
II	Above Pavlodar and Aksu cities	236	7	113–172	Kurchatov City, flow from the upper Irtysh cascade of reservoirs
III	Zone of influence of Pavlodar and Aksu cities	81	7	101–115	Pavlodar and Aksu cities, surface runoff
IV	Below Naberezhnoe village	316	8	79–102	Settlements, surface runoff

Part I includes the Black Irtysh River from the border with the People's Republic of China to its confluence with Zaisan Lake (Figures 1 and 4a). It can be considered relatively unimpacted since the Black Irtysh River is located far from industrial centres and agricultural enterprises. Currently, there is only one relatively large village, Boran, with a population of 1506 (according to 2009 data). The second village of Ordynka (Zhideli) has only 327 residents.

Figure 4. Surveyed parts of the Irtysh River, July 2023: (**a**) Part I (the Black Irtysh River); (**b**) Part II (the Irtysh River above Aksu city); (**c**) Part III (the Irtysh River in the zone of influence of Aksu city); (**d**) Part III (the Irtysh River in the zone of influence of Pavlodar city); (**e**) Part IV (the Irtysh River below Pavlodar city); (**f**) Part IV (the Irtysh River below Pavlodar city, 15 km from the border with Russia). Photos by E. Krupa.

The following three parts (II–IV, Pavlodar Irtysh region) are located below the upper Irtysh cascade of reservoirs, from the eastern border of the Pavlodar region to the border with Russia (Figure 4b–f). The section of the river above the Pavlodar and Aksu cities (Part II) experiences residual pollution coming from the upper reservoirs of the upper Irtysh cascade. Part III is located in the zone of influence of industrial and utility enterprises in the Pavlodar and Aksu cities. The lowest section of the Irtysh River (Part IV) experiences residual pollution from the upstream sections of the river. There are no large cities or industrial enterprises there. Numerous settlements and agricultural fields make an additional contribution to the overall level of pollution in the lower reaches of the Kazakh part of the Irtysh River.

The selection of sites allows us to assess the water quality of the transboundary Irtysh River at the inlet (receipt from the territory of China), the entry of pollutants into the territory of Kazakhstan, and their further transboundary transfer to the Russian part of the basin.

2.1. Field Methods

Sampling was conducted in the Irtysh River at 27 stations in July 2023 (Figure 5).

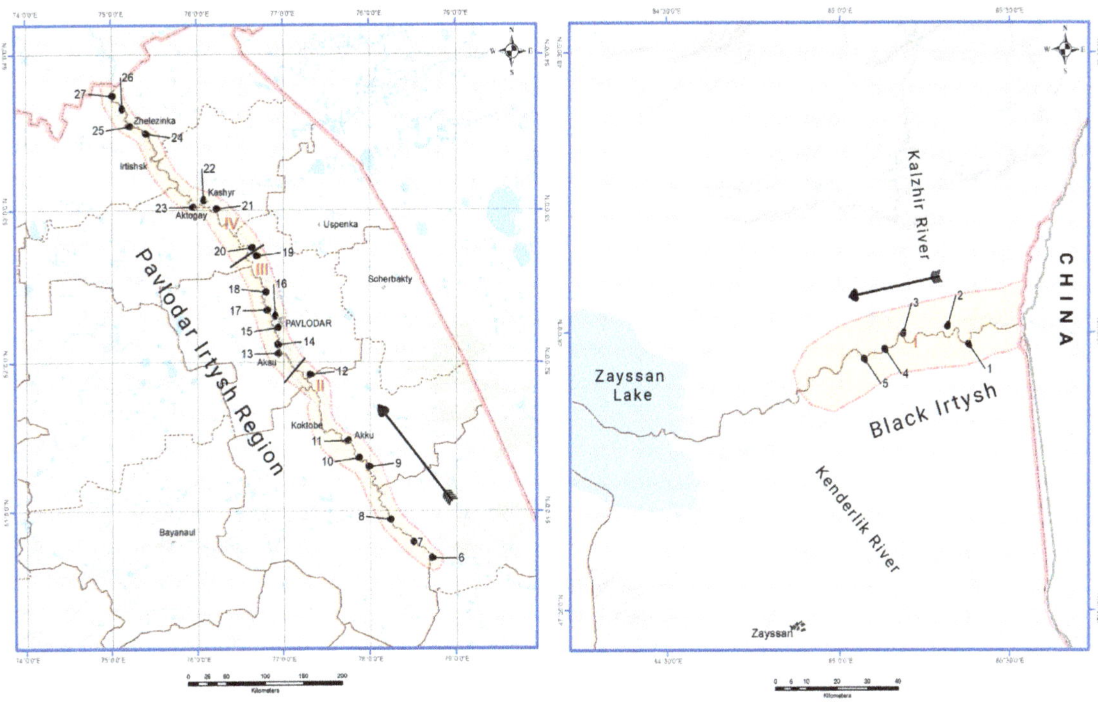

Figure 5. Material sampling stations in the Black Irtysh (Part I) and Pavlodar Irtysh (Parts II–IV) regions, July 2023. Roman numerals indicate the numbers of surveyed parts of the Irtysh River basin. Arabic numerals indicate station numbers. Arrows indicate the direction of flow.

Station coordinates were determined using a Garmin eTrex GPS navigator. Transparency, water temperature, and pH values were determined at each station. The pH value was measured using an AMTAST digital pH meter. Water samples were taken to determine dissolved oxygen, mineralisation (TDS), nitrates ($N\text{-}NO_3$), nitrites ($N\text{-}NO_2$), ammonium nitrogen ($N\text{-}NH_4$), phosphates (PO_4), manganese (Mn), total iron (Fe), silicon (Si), and easily oxidised organic substances (permanganate index PI). Water samples were also taken to determine the content of seven heavy metals, cadmium (Cd), copper (Cu), zinc (Zn), lead

(Pb), chromium (Cr), cobalt (Co), and mercury (Hg), as the most dangerous toxic pollutants in the Irtysh water basin. Samples for the determination of nutrients were fixed with chloroform for the determination of heavy metals with concentrated nitric acid and easily oxidised organic substances with sulphuric acid at a dilution of 1:3. The determination of dissolved oxygen was carried out in field conditions using the Winkler method [34]. This method is based on the ability of manganese hydroxide to oxidise into a compound of higher valency in an alkaline environment. Manganese hydroxide quantitatively binds oxygen dissolved in water and then, in an acidic environment, it is again converted into divalent compounds while oxidising an equivalent amount of iodine. The released iodine was determined by titration with thiosulfate. Zooplankton samples were collected by filtering 50–100 L of water through Apstein's plankton net [35] and fixed with 40% formaldehyde to a final concentration of 4%.

2.2. Laboratory Analysis Methods

The determination of the ionic composition of water, TDS, and contents of Mn, Fe, Si, nitrites, nitrates, ammonium, and phosphates was carried out by the following methods [34,36]. First, the content of ions (Ca^{2+}, Mg^{2+}, $SO_4{}^{2-}$, Cl^-, $HCO_3{}^-$, $CO_3{}^{2-}$, and $Na^+ + K^+$) was determined using traditional hydrochemical methods to calculate TDS further. Hydrocarbonate and carbonate ions were determined by volumetric direct titration. Calcium and magnesium ions and total hardness values were determined by the complexometric method with indicators chrome black murexide ET-00. Sulphates were determined by the gravimetric method. The volumetric argentometric method was used to determine chlorides. The total content of sodium and potassium ions was calculated as the difference between the sum of anions and cations. Finally, we summed the content of all anions and cations to calculate the total mineralisation or TDS. The content of nutrients and manganese was determined by the photocolorimetric method (method sensitivity 0.02 mg/dm^3, error $\pm$2–4%). Depending on the element being determined, we used various indicators: for ammonium ions—Nessler's reagent, for nitrite and nitrate ions—Griess's reagent, for phosphates—ascorbic acid, for iron—sulfosalicylic acid, for silicon—ammonium molybdate, for manganese—formaldoxime.

The determination of heavy metals was carried out at the Institute of Nuclear Physics of the Ministry of Energy of the Republic of Kazakhstan (Almaty, Kazakhstan) by mass spectrometry with inductively coupled plasma on an ELAN 9000 device (Perkin Elmer SCIEX, Shelton, CT, USA) according to Interstate Standard 31 870-2012 [37] with an accuracy of 0.00006–0.0002 μg/dm^3.

We identified the species of planktonic invertebrates using the MCX-300 Orchid LED microscope (MICROS, Hunnenbrunn, Austria) according to the identification keys and figures given in the monographs [38–40]. We processed zooplankton samples according to the methods described by Kiselev [35], with our additions. First, a sample was brought to a particular volume (150–500 cm^3). After thorough mixing, three sub-samples were taken from the sample using a 1 mL stamp pipette. In this sub-sample, all encountered individuals and age stages of certain species (the most numerous) were counted in Bogorov's cell. The sample was concentrated to 125–150 cm^3 in the next step. Three sub-samples were retaken from it, where less abundant age stages or species were counted. The whole procedure was repeated once more while the sample was concentrated to a volume of 50 cm^3. In the end, the sample, with its volume of 20–25 cm^3, was viewed in its entirety for counting large and rare species of planktonic invertebrates. The results of counting individuals of all species and age stages were recalculated per one m^3 using the below formula (separately for each sample dilution):

$$N = \frac{n \times (V1/V2)}{V3} \quad (1)$$

where N—abundance (specimens/m^3), n—number of individuals (specimens) in a portion, V1—dilution volume (cm^3), V2—subsample volume (cm^3), and V3—filtered water volume (m^3). The filtered volume of water (50–100 L) was recalculated per m^3 (0.05–0.10 m^3).

We calculated an individual's mass (including different age stages of planktonic invertebrates) according to formulas specific to each species [41]. The formulas were based on the relationship between an individual's mass and length. We calculated each species' total abundance and biomass by summing the abundance and biomass of their individual age and size stages. Ultimately, we determined zooplankton's total abundance and biomass (per m^3) by summing up the abundance and biomass of all species in the sample.

For each sample, we calculated the Shannon index by abundance (Shannon Ab) and biomass (Shannon Bi) [42], the Δ-Shannon index [43], and the average individual mass of a specimen [4]. Shannon index values were calculated in the Primer 6 programme. Δ-Shannon values were found as the arithmetic difference between two versions of the Shannon index, Shannon Ab and Shannon Bi [43]. The average individual mass of a specimen was found by dividing the total biomass by the total abundance of zooplankton [4].

The calculation of species similarity was performed as the network analysis in JASP 0.9.0.0 (Jeffrey's Amazing Statistics Program, University of Amsterdam, Amsterdam, The Netherlands) with the botnet package in R-Statistica 4.1.1 (R Core Team, Vienna, Austria). JASP plot analysis was created as a calculation result of the 50% similarity; the level was significant only when $p < 0.05$ [44].

2.3. Environmental Assessment

We assessed the ecological state of the Irtysh River by comparing the recorded concentrations of pollutants with (a) their maximum permissible content for water bodies of fishery importance (MPC_{fw}) [45]; (b) the Ecological Classification Scale in the section "Nutrients" [46]; and (c) the Regional Classification Scale in the section "Heavy metals" [47].

Bioindication of the ecological state of the Irtysh River was carried out based on the analysis of the abundance of zooplankton, the composition of dominant species and their environmental preferences, values of the Δ-Shannon index [43], the average individual mass of a specimen [4], and the presence of specimens with morphological deviations in copepod populations [48]. The Δ-Shannon index is a numerical analogue of the graphical index of Clarke's W-statistic [49] and characterises the structure of species dominance in communities. In undisturbed communities, large species dominate, especially in terms of biomass, so Shannon Ab is more than Shannon Bi. As a result, Δ-Shannon values are positive [43]. At eutrophication, the increased dominance of small species in abundance and biomass reflects the inverse ratio of these variables and negative Δ-Shannon values.

3. Results and Discussion

3.1. Hydrophysical, Hydrochemical, and Toxicological Characteristics

The banks of the Irtysh River are composed mainly of clays (Figure 3). Woody vegetation is widespread in the floodplain of the Black Irtysh River. Significant areas in the Pavlodar Irtysh region are characterised by sparse vegetation cover. In the coastal zone, especially in the lower reaches (sections II–IV), common reed *Phragmites australis* Cav. Trin. ex Steud. is widespread (Figure 6d). In shallow waters, the umbrella plant *Butomus umbellatus* L. (Figure 6b) and several species of pondweeds, including *Potamogeton perfoliatus* L. (Figure 6c), are widespread. Filamentous algae develop on mass (Figure 6a). Bottom sediments are represented mainly by pebbles and sand in places with a thin layer of silt.

During the research period, the water temperature reached 23.0–25.2 °C. The transparency of the water was low and did not exceed 5–7 cm. The large number of particles suspended in the water was due to several factors: in the upper reach, these factors were a relatively high flow speed and the destruction of clayey banks; in the lower reach, these factors were the destruction of clayey banks (locally) and dredging and sand mining in the river bed (Figure 2a). The dissolved oxygen content was at a high level (Table 2).

Figure 6. Higher aquatic and semi-aquatic vegetation in the lower reaches of the Irtysh River, July 2023: (**a**) filamentous algae; (**b**) umbrella pine *Butomus umbellatus* L.; (**c**) pondweed *Potamogeton perfoliatus* L.; (**d**) common reed *Phragmites australis* (Cav.) Trin. ex Steud. Photos by E. Krupa.

Table 2. Hydrophysical, hydrochemical, and toxicological characteristics of the surveyed parts of the Irtysh River, July 2023 (mean and standard error).

Variable	[1] The Part of the Irtysh River			
	I	II	III	IV
Temperature, °C	23.00 ± 0.17	24.90 ± 0.26	24.40 ± 0.38	25.20 ± 0.13
pH	7.20 ± 0.06	7.90 ± 0.06	7.60 ± 0.03	7.70 ± 0.05
Oxygen, mg/dm^3	9.04 ± 0.06	8.44 ± 0.40	8.39 ± 0.29	10.07 ± 0.06
TDS, mg/dm^3	100.7 ± 3.4	176.2 ± 3.0	177.7 ± 2.3	175.5 ± 3.2
Hardness, mg.eq./dm^3	1.03 ± 0.04	1.70 ± 0.01	1.72 ± 0.01	1.75 ± 0.01
PI, mgO/dm^3	5.21 ± 0.75	5.59 ± 0.84	3.98 ± 0.14	3.94 ± 0.15
N-NO$_3$, mg/dm^3	0.319 ± 0.169	0.009 ± 0.007	0.018 ± 0.018	0.007 ± 0.004
N-NO$_2$, mg/dm^3	0.027 ± 0.004	0.046 ± 0.005	0.063 ± 0.008	0.043 ± 0.007
N-NH$_4$, mg/dm^3	0.193 ± 0.124	0.384 ± 0.026	0.525 ± 0.058	0.482 ± 0.035
Total N, mg/dm^3	0.539 ± 0.166	0.440 ± 0.026	0.606 ± 0.080	0.532 ± 0.035
PO$_4$, mg/dm^3	0.034 ± 0.010	0.023 ± 0.004	0.021 ± 0.006	0.028 ± 0.002
Si, mg/dm^3	4.02 ± 0.58	3.87 ± 0.46	4.01 ± 0.39	3.97 ± 0.25
Fe, mg/dm^3	0.58 ± 0.07	0.14 ± 0.01	0.28 ± 0.08	0.31 ± 0.04
Mn, µg/dm^3	30.0 ± 3.7	78.6 ± 13.6	85.1 ± 17.2	116.0 ± 11.6
Cd, µg/dm^3	0.05 ± 0.00	0.05 ± 0.00	0.06 ± 0.01	0.06 ± 0.01
Co, µg/dm^3	0.06 ± 0.01	0.16 ± 0.04	0.11 ± 0.03	2.60 ± 2.86
Cr, µg/dm^3	0.93 ± 0.33	2.24 ± 0.03	2.67 ± 0.34	38.00 ± 34.57

Table 2. *Cont.*

Variable	[1] The Part of the Irtysh River			
	I	II	III	IV
Cu, μg/dm^3	0.61 ± 0.21	2.22 ± 0.10	1.80 ± 0.08	3.90 ± 2.30
Pb, μg/dm^3	0.07 ± 0.03	0.04 ± 0.0	0.08 ± 0.04	0.04 ± 0.00
Zn, μg/dm^3	1.00 ± 0.00	15.61 ± 9.61	7.59 ± 4.02	28.61 ± 25.24
Hg, μg/dm^3	below detection limit			

Note: [1] Descriptions of the surveyed parts of the river are given in Table 1.

According to the results of chemical analysis (Table 2), the Irtysh water was an alkaline, fresh, soft, hydrocarbonate class of the calcium group [45]. The contents of easily oxidised organic substances (PI), phosphates, silicon, nitrate, ammonium nitrogen, and heavy metals were low. Mercury was not detected in any water samples. Similarly, low levels of heavy metals, including mercury, were recorded in the river water in 2010–2011 [18]. A comparison of the results obtained with the data of the work cited above showed a slight decrease in the copper content in transboundary runoff from 2010–2011 to 2023.

The water of the Black Irtysh River (Part I) was characterised by lower temperatures, pH values, TDS, and hardness and higher contents of N-NO_3, PO_4, and Fe compared to the lower parts of the river (II–IV, the Pavlodar Irtysh region). In the direction from the upper to the lower reaches of the river, the amount of easily oxidised organic substances (PI) decreased slightly, and the contents of Mn, Co, Cu, Cr, and Zn in the water increased. The distribution of Si, Cd, and Pb along the longitudinal profile of the river was relatively uniform.

Nonparametric correlation analysis showed that changes in temperature; water hardness; and N-NH4, Mn, Cr, Fe, and PI contents along the longitudinal profile of the river were statistically significant (Table 3). Negative statistically significant relationships between Fe and Co, Fe and Cu, and Fe and TDS reflected the asynchronous nature of the spatial variability of these variables across the surveyed areas. Positive connections between Cu and Co, Cu and Zn, and Cr and Mn may indicate a single source of their entry into the river.

Table 3. Spearman correlation coefficients (R) between environmental variables of the Irtysh River, July 2023, at $p < 0.05$.

Pair of Variables	R	Pair of Variables	R	Pair of Variables	R
Altitude–Temperature	−0.490	Altitude–Cr	−0.731	Cu–Co	0.731
Altitude–Hardness	−0.664	Fe–pH	−0.625	Cu–Zn	0.674
Altitude–PI	0.664	Temperature–TDS	0.584	Cu–pH	0.632
Altitude–N-NH_4	−0.586	Fe–Co	−0.607	Cr–Mn	0.693
Altitude–Mn	−0.585	Fe–Cu	−0.595	TDS–Fe	−0.594

3.2. Zooplankton

Eighty-two species were identified among the zooplankton, including 60 rotifers, 12 cladocerans, and 10 copepods (Table 4). The species richness of zooplankton communities linearly doubled from up to downstream. The species composition of zooplankton had a low similarity between sampling stations (Figure 7a). In total, the low similarity in the species composition of planktonic invertebrates was between Parts II and IV of the river (Figure 7b) and high similarity was found for Parts I (the Black Irtysh River) and III (the Irtysh River in the zone of influence of the Pavlodar and Aksu cities). Rotifers *Brachionus angularis*, Bdelloida gen. sp., *Keratella cochlearis*, and *Euchlanis oropha* were found everywhere. Rotifers *Brachionus quadridentatus*, *Cephalodella* sp., *Filinia longiseta*, *Synchaeta stylata*, cladocerans *Alona affinis*, *Bosmina (Bosmina) longirostris*, *Bosminopsis deitersi*,

Macrothrix hirsuticornis, and *Pleuroxus trigonellus* were relatively widespread, mainly in the lower reaches of the river (Parts II, III, IV).

Table 4. Species composition and frequency of occurrence of planktonic invertebrates in the Irtysh River, July 2023.

Taxon Name	Part of the Irtysh River			
	I	II	III	IV
Rotifera				
Asplanchna henrietta (Langhans)	0	0	13	57
Asplanchna intermedia (Hudson)	20	0	13	0
Asplanchna priodonta (Gosse)	0	0	13	0
Asplanchna sieboldi (Leydig)	0	0	0	14
Bdelloida gen. sp.	60	29	63	100
Brachionus angularis (Gosse)	40	29	88	100
Brachionus bennini (Leissling)	20	0	75	71
Brachionus budapestiensis (Daday)	0	0	13	0
Brachionus calyciflorus (Pallas)	0	0	0	29
Brachionus calyciflorus anuraeiformis (Brehm)	0	0	0	29
Brachionus calyciflorus dorcas (Gosse)	0	0	13	57
Brachionus diversicornis (Daday)	0	0	25	71
Brachionus diversicornis homoceros (Wierzejski)	0	0	0	14
Brachionus plicatilis (Muller)	0	0	0	14
Brachionus quadridentatus (Hermann)	40	0	13	29
Brachionus quadridentatus zernovi (Voronkov)	0	29	13	14
Brachionus quadridentatus ancylognathus (Schmarda)	0	0	0	29
Brachionus quadridentatus brevispinus (Ehrenberg)	0	0	13	43
Brachionus variabilis (Hempel)	20	0	0	71
Keratella cochlearis (Gosse)	40	43	50	86
Keratella cochlearis tecta (Gosse)	0	14	0	29
Keratella quadrata (Muller)	20	0	0	0
Keratella quadrata dispersa (Carlin)	40	0	0	0
Cephalodella gibba (Ehrenberg)	0	14	0	0
Cephalodella sp.	20	0	13	14
Euchlanis calpidia (Myers)	0	14	0	0
Euchlanis deflexa (Gosse)	0	14	0	0
Euchlanis lyra (Hudson)	0	14	0	0
Euchlanis oropha (Gosse)	40	43	13	14
Euchlanis sp.	40	29	0	0
Filinia longiseta (Ehrenberg)	40	0	13	100
Hexarthra mira (Hudson)	0	0	0	14
Hexarthra intermedia (Wiszniewski)	0	0	0	14
Lecane (Monostyla) bulla (Gosse)	20	0	0	0
Lecane (Monostyla) sp.	0	14	0	0
Lecane (s.str.) *flexilis* (Gosse)	0	14	0	0

Table 4. *Cont.*

Taxon Name	Part of the Irtysh River			
	I	II	III	IV
Rotifera				
Notholca acuminata (Ehrenberg)	20	0	0	0
Notommatidae gen. sp.	0	0	13	0
Platyias patulus (Muller)	0	0	13	0
Polyarthra dolichoptera (Idelson)	20	0	0	29
Polyarthra luminosa (Kutikova)	0	0	0	14
Polyarthra major (Burchhardt)	0	0	25	14
Polyarthra minor (Voigt)	0	0	0	57
Polyarthra remata (Skorikov)	0	0	0	14
Polyarthra vulgaris (Carlin)	0	0	0	14
Pompholyx sulcata (Hudson)	0	0	0	57
Postclausa hyptopus (Ehrenberg)	60	0	0	0
Synchaeta pectinata (Ehrenberg)	40	0	0	0
Synchaeta stylata (Wierzejski)	40	0	25	71
Testudinella patina (Hermann)	0	14	0	0
Trichocerca (*Diurella*) *bidens* (Lucks)	0	0	0	14
Trichocerca (*Diurella*) sp.	0	0	0	14
Trichocerca (*Diurella*) *myersi* (Hauer)	0	0	0	43
Trichocerca (s.str.) *cylindrica* (Imhof)	0	0	0	57
Trichocerca longiseta (Schrank)	0	0	0	14
Trichocerca sp.	0	0	0	14
Trichotria pocillum (Muller)	0	14	13	0
Trichotria tetractis (Ehrenberg)	0	0	13	0
Trichotria truncata (Whitel.)	0	0	0	43
Rotifera gen. sp.	0	0	0	14
Total Rotifera	19	15	22	39
Cladocera				
Alona affinis (Leydig)	20	14	13	0
Alona rectangula (Sars)	0	0	13	0
Bosmina (Bosmina) kessleri (Uljanin)	0	0	0	14
Bosmina (Bosmina) longirostris (O.F. Muller)	80	0	13	86
Bosminopsis deitersi (Richard)	0	14	50	100
Ceriodaphnia pulchella (Sars)	0	0	0	14
Diaphanosoma sp.	0	14	0	43
Ilyocryptus acutifrons (Sars)	0	0	13	0
Pleuroxus trigonellus (O.F.Muller)	0	14	50	29
Scapholeberis mucronata (O.F.Muller)	0	14	0	0
Sida crystallina (O.F.Muller)	0	0	13	0
Macrothrix hirsuticornis (Norman et Brady)	0	14	13	14
Total Cladocera	2	6	8	7

Table 4. *Cont.*

Taxon Name	Part of the Irtysh River			
	I	II	III	IV
Copepoda				
Acanthocyclops robustus (Sars)	0	0	0	14
Ectocyclops phaleratus (Koch)	0	14	13	0
Eucyclops serrulatus (Lilljeborg)	20	0	0	0
Eucyclopinae gen.sp.	60	29	38	14
Mesocyclops leuckarti (Claus)	100	43	75	100
Microcyclops afghanicus (Lindberg)	0	29	0	0
Paracyclops affinis (Sars)	0	0	13	0
Thermocyclops crassus (Fischer)	0	0	50	43
Cyclopoida gen.sp.	0	29	0	0
Harpacticoida gen.sp.	20	14	38	0
Total Copepoda	4	6	6	4
Total species	25	27	36	50

Notes: I—the Black Irtysh River from the border with China to the confluence with Lake Zaisan, II—the Irtysh River above Pavlodar city, III—the Irtysh River in the zone of influence of the Pavlodar and Aksu cities, IV—the lower part of the Irtysh River to the border with Russia.

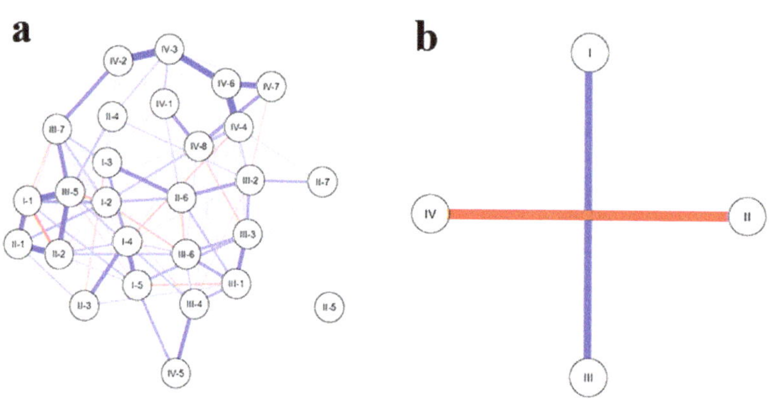

Figure 7. The similarity of the species composition of planktonic invertebrates in the Irtysh River according to JASP analysis, July 2023: (**a**) comparison between individual stations; (**b**) comparison across selected parts of the Irtysh River, where I—the Black Irtysh River from the border with China to its confluence with the Zaisan Lake, II—the Irtysh River above Pavlodar city, III—the Irtysh River in the zone of influence of the Pavlodar and Aksu cities, IV—the lower part of the Irtysh River to the border with Russia. Arabic numerals indicate the numbers of stations within each of the surveyed parts. The colour and line width between points reflect the correlation: blue—positive, red—negative, and the colour saturation—correlation strength.

Quantitative zooplankton variables were very low, except for downstream of the Irtysh River (Table 5). Rotifers dominated. Copepods were subdominant. In the total quantitative variables of zooplankton, the share of rotifers increased from 45.1–52.5% in Part I to 97.6% in Part IV. In the zooplankton of the Black Irtysh River (Part I), the dominant complex was represented by the rotifers Bdelloida gen. sp. and the crustaceans *Bosmina longirostris*, *Eucyclops serrulatus*, and *Mesocyclops leuckarti* (Table 6). The role of the first three species in zooplankton decreased almost linearly downstream, while the share of the rotifer

Brachionus angularis, on the contrary, increased sharply. The cyclops *Mesocyclops leuckarti* and *Thermocyclops crassus* played a significant role in zooplankton locally.

Table 5. Quantitative variables of zooplankton in the Irtysh River (mean with standard error), July 2023.

Part of the Irtysh River	Rotifera	Cladocera	Copepoda	Total
Abundance, specimen/m^3				
I	147 ± 75	16 ± 7	161 ± 92	326 ± 164
II	276 ± 123	109 ± 98	135 ± 69	526 ± 206
III	455 ± 137	91 ± 26	361 ± 80	906 ± 198
IV	23,582 ± 6388	181 ± 20	395 ± 46	24,158 ± 6362
Average abundance	6347 ± 2538	105 ± 28	274 ± 41	6728 ± 2553
Biomass, mg/m^3				
I	0.22 ± 0.09	0.12 ± 0.04	0.27 ± 0.11	0.62 ± 0.18
II	0.35 ± 0.20	1.00 ± 0.94	0.16 ± 0.09	1.54 ± 1.23
III	0.41 ± 0.09	0.66 ± 0.18	0.50 ± 0.19	1.57 ± 0.17
IV	5.04 ± 1.17	0.98 ± 0.25	1.03 ± 0.20	7.05 ± 1.40
Average biomass	1.56 ± 0.50	0.73 ± 0.25	0.51 ± 0.10	2.81 ± 0.68

Table 6. Composition of dominant species in the zooplankton of the Irtysh River, July 2023.

Taxon Name	Part of the Irtysh River							
	I	II	III	IV	I	II	III	IV
	Abundance, %				Biomass, %			
Bdelloida gen. sp.	21.7	4.8	5.1	0.7	19.3	13.5	11.8	3.2
Brachionus angularis	1.2	1.4	31.5	71.9	0.1	0.1	2.4	43.5
Euchlanis oropha	2.5	19.9	0.4	0.02	1.6	4.6	0.1	0.02
Eucyclops serrulatus	27.4	6.2	1.0	4.4	17.0	0.7	0.3	0.1
Scapholeberis mucronata	0.0	2.5	0.0	0.0	0.0	20.9	0.0	0.0
Bosmina longirostris	4.3	0.0	0.7	0.2	15.7	0.0	3.2	3.0
Mesocyclops leuckarti	21.5	1.7	23.1	1.4	27.2	2.0	11.8	11.8
Thermocyclops crassus	0.0	0.0	14.7	0.2	0.0	0.0	17.4	2.5

The average abundance of zooplankton varied widely and reached a maximum in the lower reaches of the Irtysh River (Part IV) (Table 7). According to the Shannon index values, the diversity of zooplankton communities varied from low to medium levels [4]. The spatial dynamics of the Shannon and Δ-Shannon indices indicated a change in zooplankton structure, which was most pronounced downstream (Part IV). Low values of the average individual mass of a specimen were associated with the dominance of small rotifers in zooplankton communities.

According to the Spearman correlation coefficient values, from the upper to the lower reaches, the abundance of rotifers, cladocerans, and copepods; the number of species; and the Shannon Bi index (negative relationship with "Altitude") increased statistically significantly; the average individual mass of a specimen (positive relationship with "Altitude") decreased (Table 8). For rotifers, a negative relationship with PI and positive relationships with oxygen, Cr, N-NH_4, and water hardness were recorded. The abundance of cladocerans and copepods increased slightly with increasing ammonium nitrogen content. The relationship between the abundance of Cladocera and Cr was negative. The values of the average individual mass of a specimen decreased statistically significantly with increasing water temperature and Cr concentrations.

Table 7. Structural variables of zooplankton in the Irtysh River (mean with standard error), July 2023.

Part of the Irtysh River	Average Species Number	Shannon Ab	Shannon Bi	Δ-Shannon	Average Individual Mass, mg
I	9.6 ± 1.9	2.48 ± 0.24	1.91 ± 0.35	0.57 ± 0.24	0.0031 ± 0.0010
II	5.7 ± 1.7	1.82 ± 0.43	1.23 ± 0.36	0.59 ± 0.15	0.0016 ± 0.0010
III	9.5 ± 0.9	2.40 ± 0.14	2.10 ± 0.14	0.35 ± 0.18	0.0028 ± 0.0008
IV	19.7 ± 1.4	1.67 ± 0.11	2.69 ± 0.13	−0.88 ± 0.20	0.0004 ± 0.0001
Average	11.2 ± 1.2	2.08 ± 0.14	1.99 ± 0.16	0.08 ± 0.15	0.0019 ± 0.0004

Table 8. Spearman correlation coefficients (R) between biological and environmental variables of the Irtysh River in July 2023, $p < 0.05$.

Pair Variables	R	Pair Variables	R
Altitude–Rotifera Ab	−0.795	Rotifera Ab–Cr	0.603
Altitude–Cladocera Ab	−0.720	Rotifera Ab–N-NH_4	0.565
Altitude–Copepoda Ab	−0.642	Rotifera Ab–Hardness	0.603
Altitude–Total Ab	−0.852	Cladocera Ab–PI	−0.620
Altitude–Species Number	−0.657	Cladocera Ab–Cr	−0.620
Altitude–Shannon Bi	−0.563	Cladocera Ab–N-NH_4	0.468
Altitude–Average Mass	0.596	Copepda Ab–N-NH_4	0.446
Rotifera Ab–PI	−0.471	Average Mass–Temperature	−0.580
Rotifera Ab–Oxygen	0.544	Average Mass–Cr	−0.614

Thus, the species richness of the zooplankton of the Irtysh River was generally assessed as high, especially considering a single sampling of material in July 2023. For comparison, 104 taxa were recorded in the zooplankton of the Northern Dvina River over eight years (2012–2019) [13]. The predominance of rotifers in the Irtysh zooplankton is generally typical for rivers [11,50], but depends on local conditions. Rotifers dominated the riverbed zone of the regulated Missouri River (USA) [10] and crustaceans dominated in areas between reservoirs. In the Syrdarya River (Southern Kazakhstan) below the Shardara reservoir, the total species richness of crustaceans (34) was higher than that of rotifers (21) [51]. Above the reservoir and in the mouth zone, before the Syr Darya flows into the Aral Sea, rotifers dominated (58–73% of the total species richness) [52].

The spatial dynamics of species composition and quantitative and structural variables of zooplankton reflected the variability of external conditions in the Irtysh River. One of the reasons for the increase in the abundance of planktonic invertebrates was the decrease in the speed of the Irtysh flow in the Pavlodar region. A similar increase in the abundance of planktonic invertebrates was recorded in the lower reaches of the Northern Dvina [13] and Nakdong [53] Rivers. The increase in the abundance of zooplankton communities can also be associated with increased organic pollution in the lower reaches of the Irtysh River, which was confirmed by chemical analysis data (Table 2).

3.3. *Assessment of the Ecological State of the Irtysh River by Chemical Data*

3.3.1. Assessment of Organic Pollution

Comparison with Global Values

In river ecosystems around the globe, the content of ammonium nitrogen varies from 0.005 to 0.04 mg/dm^3 (median 0.015), nitrate nitrogen varies from 0.05 to 0.2 (median 0.10), and phosphates vary from 0.002 to 0.025 (median 0.01) mg/dm^3 [54]. According to the results, in the Irtysh River, the amount of ammonium nitrogen was 12.9–35.2 times and phosphates were 2.1–3.4 times (Table 2) higher than the global median values for rivers. In Part I (Black Irtysh River), the content of nitrate nitrogen exceeded the global median values by 31.9 times; in Part II, it exceeded global values by 1.8 times; in other parts, it was below global values.

Comparison with MPC_{fw}

In the Irtysh River, a widespread excess of MPC_{fw} was detected only for nitrite nitrogen, with a maximum in the zone of influence of the Pavlodar and Aksu cities (Part III) (Table 9). A slight excess of MPC_{fw} for ammonia nitrogen was recorded at the same site. An analysis of published data showed that an excess of MPC_{fw} for nitrite nitrogen was observed in the Irtysh River earlier, in the period from 1990 to 1999 [18]. In July 2023, the average nitrite nitrogen content in the Pavlodar Irtysh region was 1.4 times lower compared to data from previous years.

Table 9. Assessment of the level of pollution of the surveyed parts of the Irtysh River by chemical variables, July 2023.

Variable	I	II	III	IV	[1] MPC_{fw} (mg/dm^3)	I	II	III	IV
	Multiplicity of Exceeding MPC_{fw}					Water Quality Classes			
[2] PI	–	–	–	–	–	2b	2b	2a	2a
[2] $N\text{-}NO_3$	0.04	0.001	0.002	0.0008	9.10	2b	1	1	1
[2] $N\text{-}NO_2$	1.35	2.3	3.15	2.15	0.02	4a	4a	4b	4a
[2] $N\text{-}NH_4$	0.39	0.77	1.05	0.96	0.50	2b	3b	4a	3b
[2] PO_4	0.68	0.46	0.42	0.56	0.05	3a	2b	2b	2b
Fe	5.8	1.4	2.8	3.1	0.10	–	–	–	–
Mn	3.0	7.9	8.5	11.6	0.01	–	–	–	–
[3] Cd	0.10	0.10	0.12	0.12	0.0005	1	1	1	1
[3] Co	0.006	0.016	0.011	0.26	0.01	1	1	1	1
[3] Cr	0.2	0.4	0.5	7.6	0.005	1	1	1	4
[3] Cu	0.6	2.2	1.8	3.9	0.001	1	2	2	2
[3] Pb	0.007	0.004	0.008	0.004	0.01	1	1	1	1
[3] Zn	0.1	1.5	0.8	2.8	0.01	1	3	1	4
[3] Hg	Below detection limit				0.0001	1	1	1	1

Notes: I–IV—parts of the Irtysh River. [1] MPC_{fw} according to Guseva [45]; [2] ranks and classes of water quality according to Romanenko et al. [46]: 1—unpolluted water, 2a—very clean, 2b—quite clean, 3a—fairly clean, 3b—slightly polluted, 4a—moderately polluted, 4b—heavily polluted; [3] water quality classes according to the regional scale developed by Krupa et al. [47]: 1—clean water, 2—slightly polluted, 3—moderately polluted, 4—highly polluted.

Comparison with Classification Scales

Classification Scales [46] make it possible to assess water quality from the point of view of its use for various purposes. Based on the content of easily oxidised organic substances (PI) and nitrate nitrogen, the Irtysh water was assessed as pure of the second quality class (Table 9). The phosphate content was at the level of quite clean (second class) and fairly clean (third class) water. The content of ammonia nitrogen varied from the level of completely clean waters in the Black Irtysh River to slightly and moderately polluted waters in the Pavlodar Irtysh region. To the greatest extent, the Irtysh River was polluted with nitrite nitrogen at the level of quality class 4 (moderate and severe pollution). Deterioration in water quality in terms of nitrite and ammonium nitrogen content was recorded in Part III in the zone of influence of the Pavlodar and Aksu cities.

Assessment Based on the Ratio of Forms of Nitrogen Compounds

Nutrient compounds enter natural ecosystems mainly in the form of ammonium ions, which are oxidised to unstable nitrite ions and then to nitrates [55]. The ratio of the forms of nitrogen compounds allows us to estimate the approximate time of their entry into the aquatic ecosystem [56]. Table 10 shows that persistent nitrate ions predominated in the Black Irtysh (Part I) water, indicating predominantly "old" pollution. As mentioned above, there are currently only two small villages in the Kazakh part of the Black Irtysh floodplain, and their influence on the river ecosystem can be assessed as weak. At the same time, the Chinese part of the Black Irtysh basin is densely populated; industry and agriculture are developed here [57]. Thus, the water quality of this section of the Irtysh River is determined mainly by the transboundary transport of pollutants from the territory of the People's Republic of China. A significant predominance of ammonium ions (an indicator of recent pollution) in the lower reaches of the Irtysh River indicated a constant influx of pollutants into the Pavlodar Irtysh region. Potential sources of nitrogen compounds are the discharge of municipal wastewater (Figure 2b) from the Pavlodar and Aksu cities and numerous villages in the Irtysh River's coastal zone.

Table 10. The ratio of nitrogen forms (percentage of the total amount) in the water of the surveyed parts of the Irtysh River, July 2023 (mean with standard error).

Variable	Part of the Irtysh River			
	I	II	III	IV
N-NO_3	45.0 ± 18.7	2.1 ± 1.6	1.7 ± 1.7	0.3 ± 0.6
N-NO_2	16.7 ± 12.5	10.6 ± 0.9	10.5 ± 0.6	8.3 ± 1.2
N-NH_4	38.3 ± 15.4	87.3 ± 2.3	87.8 ± 1.9	90.4 ± 1.2

3.3.2. Assessment of Toxic Pollution

Comparison with MPC_{fw}

A widespread excess of MPC_{fw} was detected for Fe, Mn, and Cu (for Cu, except for the Black Irtysh River) and locally for Cr and Zn (Table 9). The maximum excess of MPC_{fw} for Fe was recorded in the upper reaches of the river (Part I, Black Irtysh River). The excess of MPC_{fw} for Mn, Cr, and Zn increased almost linearly from the upper to lower sections of the Irtysh River.

Comparison with the Regional Classification Scale

Earlier, we developed the Regional Classification Scale based on the determination of background concentrations of heavy metals and statistical analyses of their distribution in ecologically diverse water bodies of Kazakhstan [47,58]. Rich reserves of ore minerals [59] lead to naturally elevated levels of some heavy metals, especially copper, in water bodies, which was considered when developing a regional water quality classification [47]. For example, the average background copper content in unpolluted lakes of Kazakhstan is 0.0052 mg/dm^3, and in unpolluted rivers, it is 0.0106 mg/dm^3 [60], which is 5.2–10.6 times higher than MPC_{fw} [44].

According to the Regional Classification Scale, Cd, Hg, and Pb content in the Irtysh River was at the level of extremely clean waters of the first quality class. Despite exceeding the MPC_{fw} for copper by 2.2–3.9 times (Table 9), we classified its content in the Irtysh River at the level of clean (quality class 1) and slightly polluted (quality class 2) waters [47]. Regarding Cr and Zn content, the quality of the Irtysh water deteriorated from quality class 1 in the upper reaches (unpolluted waters) to quality class 4 (heavy pollution) in the lower reaches.

Thus, in July 2023, the content of heavy metals in the water of the Irtysh River was generally at a low level, with their spatial distribution being heterogeneous. The low content of all heavy metals (except Fe) in the Black Irtysh River indicated an insignificant contribution of natural factors (leaching of metals from rocks) [59] and transboundary transfer from China to the overall level of toxic pollution of the Kazakh part of the river. As before [18], in 2023, an increased iron content was recorded in the water of the Black Irtysh River.

The deterioration of water quality for almost all variables in the Pavlodar Irtysh region may be due to the influx of wastewater from the cities of Pavlodar and Aksu and numerous villages, polluted surface runoff, and the transit transfer of pollutants from the upstream sections of the river (the middle reaches of the Kazakhstan part). As already mentioned, the Altai, Ust-Kamenogorsk, Semey, Ridder, and Zyryanovsk cities with large industrial enterprises are in the middle reaches. In 2010–2011, in the zone of influence of the city of Ust-Kamenogorsk, the average content of Cu in the river water reached 2.2 MPC_{fw}, Zn 2.4 MPC_{fw} [18]. Downstream, near Semey city, the Cu content varied from 0.9 to 5.9 MPC_{fw}, with a low Zn content.

3.4. Bioindication

3.4.1. Assessment of Organic Pollution

The biological assessment of aquatic ecosystems is based on the nonlinear relationship between the level of organic pollution and the structure of biological communities [4]. With increasing organic pollution (eutrophication), communities become enriched in species, the total abundance of zooplankton increases as well as the dominance of small species, the average individual mass of a specimen decreases [4], and the Δ-Shannon index becomes negative [43]. Eurysaprobic species, which tolerate high organic matter in the ecosystems, dominate the communities. For example, rotifers of the genus Brachionus and *Pompholyx sulcata* and the crustaceans *Bosmina longirostris* and *Thermocyclops crassus* [4] usually inhabit eutrophic waters. In Kazakhstan, *Brachionus angularis* prefers heavily polluted and extremely dirty waters of the 4 and 5 quality classes [48,61]. According to our data [60], with increasing organic pollution of fresh lakes and reservoirs in Kazakhstan, the average individual mass of a specimen decreases from 0.0157–0.0354 to 0.0036–0.0057 mg.

An indicator of increased organic pollution in the Irtysh River was a twofold increase in the total species richness of zooplankton downstream (Table 4). The number of rotifer species increased from 19 in the Black Irtysh River to 39 in the lower reaches, and cladocerans increased from 2 to 7–8. The species richness of rotifers of the genus *Brachionus* increased significantly from 4 to 13. Signs of eutrophication of the river ecosystem in the lower reaches were the appearance of the small-sized rotifer *Pompholyx sulcata* in the zooplankton (Table 4), increased dominance of *Brachionus angularis* and the cyclops *Thermocyclops crassus* (Table 6), negative Δ-Shannon index values, and decreases in the average individual mass of a specimen (Table 7).

There are Classification Scales for lakes and reservoirs [46,60,62,63] that have ranked values of quantitative biological variables for waters of different quality classes. There are no Classification Scales for river ecosystems. River zooplankton communities are predominantly influenced by flow speed and the amount of suspended matter. Substances suspended in water create unfavourable conditions for crustaceans, particularly clogging cladoceran filtration apparatus [64]. The flow velocity and, as a rule, the amount of suspended substances decrease from the upper to the lower sections of rivers, which is one of the reasons for the increase in zooplankton abundance [13,53].

This trend is also true for the rivers of Kazakhstan. We compared the abundance of zooplankton in the transboundary and regulated rivers Irtysh (East Kazakhstan), Syrdarya (South Kazakhstan), and Ili (Southeast Kazakhstan). A comparative analysis showed that in the Irtysh River, the abundance of zooplankton (0.3–24.2 thousand specimens/m^3) was approximately at the same level as in the Ili River (0.5–25.3 thousand specimens/m^3) [52] but lower than in the Syrdarya River (0.8–85.2 thousand specimens/m^3) [51]. In all cases, the abundance of planktonic invertebrates increased from the upper to the lower sections of the rivers. In the Ili and Syrdarya Rivers, an increase in the abundance of zooplankton occurred against the background of an increase in water transparency (from 0.005–0.010 m before reservoirs to 0.3–0.4 m below reservoirs). Considering approximately the same water transparency (0.03–0.04 m) in the Irtysh River, the increase in the abundance of zooplankton in the lower reaches (section IV) by more than 25 times (Table 5) against the background of the changes in its structure can be associated with anthropogenic pollution. This assumption is confirmed by the results of the chemical (Table 2) and statistical analysis (Tables 3 and 8).

3.4.2. Toxic Contamination Assessment

To assess the level of toxic pollution of aquatic ecosystems using zooplankton, a reliable indicator is the presence of copepods with deviations in morphology [60]. In the zooplankton communities of the Irtysh River, only the cyclops *Mesocyclops leuckarti* was widespread. The absence of individuals with deviations in morphology in their populations generally confirmed the low level of toxic pollution of the river water.

Thus, chemical analysis data and the structure of zooplankton communities indicated increased organic pollution and a low level of toxic pollution in the Irtysh River in July 2023. Regarding the content of nitrogen compounds, the most noticeable deterioration in water quality occurred in the zone of influence of the Aksu and Pavlodar cities (Part III). The most pronounced changes in the structure of zooplankton communities downstream (Part IV) indicated a delayed response of planktonic invertebrates to increased organic pollution of the river ecosystem. Local deterioration of water quality in terms of chromium and zinc content downstream (Part IV) did not hurt the structure of zooplankton communities.

With the intense anthropogenic load on the entire Irtysh basin, the low content of heavy metals in the water may be due to their sorption on the surface of suspended particles, bottom sediments, and plant accumulation. Clays can sorbate up to 80% of nickel, zinc, lead, cobalt, aluminium, and iron ions [65,66]. Among the *Potamogeton* species, the highest amounts of heavy metals were recorded in *Potamogeton perfoliatus* (1.88 μg/g for Cd; 13.14 μg/g for Cu; 13.32 μg/g for Pb; 57.96 μg/g for Zn) [67]. *Butomus umbellatus* accumulated chromium in the highest concentrations; pondweed accumulated zinc and chromium [68]. Such results indicate the high self-purifying ability of the Irtysh River ecosystem.

4. Conclusions

A comprehensive assessment of the ecological state of the Kazakh part of the Irtysh River in its upper (Black Irtysh River) and lower (Pavlodar region) reaches was given. The maximum content of easily oxidised organic substances, phosphates, and nitrate nitrogen in the Irtysh River did not exceed the level of completely clean or slightly polluted waters. The maximum content of ammonia nitrogen did not exceed the level of moderately polluted waters. The nitrite nitrogen content was most often at the level of moderately polluted waters; in the zone of influence of the cities of Pavlodar and Aksu, it increased to the level of highly polluted waters. The ratio of the forms of nitrogen compounds indicated a constant influx of fresh pollution into the Irtysh River in the Pavlodar region. Of the nine heavy metals (Fe, Mn, Cd, Co, Cr, Cu, Pb, Zn, Hg), widespread excess of MPC_{fw} was recorded only for Fe, Cu, and Mn and locally for Zn and Cr. Based on the content of heavy metals, the water of the Irtysh River was predominantly assessed as clean and slightly polluted, with a local deterioration in the quality of water in the lower reaches to the level of highly polluted waters. The structure of zooplankton communities also indicated increased organic pollution and a low level of toxic pollution in the Irtysh River. The spatial distribution of biotic and abiotic indicators allows us to draw the following conclusions: (a) in terms of the content of organic substances and heavy metals, water from the territory of the People's Republic of China is of satisfactory quality; (b) the constantly increased iron content in the Black Irtysh River may be due to natural factors, namely, its leaching from underlying rocks; (c) in the lower reaches of the Kazakhstan part of the Irtysh basin, there is a deterioration in water quality in terms of the content of ammonium and nitrite nitrogen, total nitrogen, and manganese; (d) the main contribution to the deterioration of river water quality comes from the discharge of industrial and municipal wastewater in the middle and lower reaches of the Kazakhstan part of the basin; (e) at a high level of anthropogenic load on the Irtysh basin, the distribution of the analysed variables indicated intensive processes of self-purification of the water; (f) self-purification of the water column occurs due to high flow speed and favourable oxygen conditions; sorption of heavy metals by particles and clays suspended in water, which make up the banks of the Irtysh; and absorption of nutritional compounds and heavy metals by aquatic and semi-aquatic vegetation.

Author Contributions: Conceptualisation, methodology, writing, editing, E.K.; investigation, E.K., S.R. and A.S.; research management, E.K. and L.S.; editing, L.S. All authors have read and agreed to the published version of the manuscript.

Funding: This research was funded by the Ministry of Science and Higher Education of the Republic of Kazakhstan, the Scientific Program "Assessment of biological resources of the Kazakh part of the transboundary Irtysh basin in the context of climate change (BR18574062)".

Data Availability Statement: Data are contained within the article.

Acknowledgments: We are grateful to V.A. Kamkin (Pavlodar State University, named after Toraigyrov, Pavlodar, Kazakhstan), who determined species of semi-aquatic and aquatic plants, and D.V. Malakhov (Institute of Zoology, Almaty, Kazakhstan), who made the schematic maps.

Conflicts of Interest: The companies Institute of Zoology and Kazakh Agency employed Author Elena Krupa for Applied Ecology. The remaining authors declare that the research was conducted in the absence of any commercial or financial relationships that could be construed as a potential conflict of interest.

References

1. EC. Directive 2000/60/EC of the European Parliament and of the Council Establishing a Framework for Community Action in the Field of Water Policy. OJ L327, 22.12.2000. Available online: https://www.eea.europa.eu/policy-documents/directive-2000-60-ec-of (accessed on 22 December 2000).
2. Aazami, J.; Sari, A.E.; Abdoli, A.; Sohrabi, H.; Van den Brink, P.J. Assessment of ecological quality of the Tajan river in Iran using a multimetric macroinvertebrate index and species traits. *Environ. Manag.* **2015**, *56*, 260–269. [CrossRef]
3. Dembowska, E.A.; Mieszczankin, T.; Napiórkowski, P. Changes of the phytoplankton community as symptoms of deterioration of water quality in a shallow lake. *Environ. Monit. Assess.* **2018**, *190*, 95. [CrossRef]
4. Andronikova, I.N. *Structural and Functional Organization of Zooplankton of Lake Ecosystems of Various Trophic Types*; Nauka: St. Petersburg, Russia, 1996; 189p. (In Russian)
5. Krupa, E.G. The dominance structure of species in zooplankton communities of water bodies of Kazakhstan as an indicator of their ecological state. In *Bioindication in the Monitoring of Freshwater Ecosystems*; Lubavitch: St. Petersburg, Russia, 2011; pp. 175–180. (In Russian)
6. Ochocka, A.; Pasztaleniec, A. Sensitivity of plankton indices to lake trophic conditions. *Environ. Monit. Assess.* **2016**, *188*, 622. [CrossRef]
7. Svensson, O.; Bellamy, A.S.; Van den Brink, P.J.; Tedengren, M.; Gunnarsson, J.S. Assessing the ecological impact of banana farms on water quality using aquatic macroinvertebrate community composition. *Environ. Sci. Pollut. Res.* **2018**, *25*, 13373–13381. [CrossRef]
8. Krupa, E.G.; Barinova, S.S.; Isbekov, K.B.; Assylbekova, S.Z. The use of zooplankton distribution maps for assessment of ecological status of the Shardara reservoir (Southern Kazakhstan). *Ecohydrol. Hydrobiol.* **2018**, *18*, 52–65. [CrossRef]
9. Krupa, E.G.; Barinova, S.S.; Romanova, S.M. Zooplankton size structure in the Kolsay Mountain Lakes (Kungei Alatau, South-eastern Kazakhstan) and its relationships with environmental factors. *Water Resour.* **2019**, *46*, 403–414. [CrossRef]
10. Havel, J.E.; Medley, K.A.; Dickerson, K.D.; Angradi, T.R.; Bolgrien, D.W.; Paul, A. Effect of main-stem dams on zooplankton communities of the Missouri River (USA). *Hydrobiologia* **2009**, *628*, 121–135. [CrossRef]
11. Krylov, A.V. *Zooplankton of Lowland Small Rivers*; Nauka: Moscow, Russia, 2005; 263p. (In Russian)
12. Yermolaeva, N.I. Zooplankton and water quality of the Ishim River in Northern Kazakhstan. *Arid. Ecosyst.* **2015**, *5*, 176–187. [CrossRef]
13. Imant, E.N.; Novoselov, A.P. Dynamics of Zooplankton Composition in the Lower Northern Dvina River and Some Factors Determining Zooplankton Abundance. *Rus. J. Ecol.* **2021**, *52*, 59–69. [CrossRef] [PubMed]
14. Bolotova, N.; Mukhin, I.; Lopicheva, O. Possibilities of Bioindication of River Ecosystem Water Quality by Plankton Communities. In *INTERAGROMASH 2022: XV International Scientific Conference "INTERAGROMASH 2022"*; Springer: Berlin/Heidelberg, Germany, 2023; Volume 575, pp. 331–339. [CrossRef]

15. Sahu, S.K.; Sarkar, S.D.; Gogoi, P.; Naskar, M. A Geostatistical Framework Predicting Zooplankton Abundance in a Large River: Management Implications towards Potamoplankton Sustainability. *Environ. Manag.* **2023**, *71*, 1037–1051. [CrossRef]
16. Huang, F.; Xia, Z.; Guo, L.; Yang, F. Effects of reservoirs on seasonal discharge of Irtysh River measured by Lepage test. *Water Sci. Eng.* **2014**, *7*, 363–372. [CrossRef]
17. Beisembaeva, M.A.; Dubrovskaya, L.I.; Zemtsov, V.A. Anthropogenic changes in water resources and maximum levels of the Irtysh River in the flat Part of the basin in the Republic of Kazakhstan. *News Tomsk. Polytech. Univ. Georesour. Engin.* **2018**, *329*, 6–15. (In Russian)
18. Burlibaev, M.Z.; Kuts, S.I.; Fashchevsky, B.V.; Tsaregorodtseva, A.G.; Shenberger, I.V.; Burlibaeva, D.M.; Aytureev, A.M. *Flooding of the Ertis Floodplain is the Main Factor in the Sustainable Development of the River Ecosystem*; Kaganat: Almaty, Kazakhstan, 2014; 396p. (In Russian)
19. Kulikova, E.V. Hydrology, hydrochemistry and pollution level of the transboundary section of the Irtysh River. In *Ecology and Hydrofauna of Water Bodies of Transboundary Basins of Kazakhstan*; Bastau: Almaty, Kazakhstan, 2008; pp. 318–327. (In Russian)
20. Vinokurov, Y.I.; Chibilev, A.A.; Krasnoyarova, B.A.; Pavleichik, V.M.; Platonova, S.G.; Sivokhip, Z.T. Regional environmental problems in the transboundary basins of the Ural and Irtysh rivers. *Izv. RAN Geograph. Ser.* **2010**, *3*, 95–104. (In Russian)
21. Zakarkina, N.A.; Tskhai, A.I.; Epifantseva, T.M.; Akulova, G.V. Monitoring wastewater composition from some industrial enterprises in Pavlodar, the Bylkyldak storage lake and groundwater. *News NA Sci. RK Ser. Chem. Tech.* **2011**, *3*, 21–24. (In Russian)
22. Burlibayeva, D.M.; Burlibayev, M.Z.; Opp, C.; Bao, A. Regime dynamics of hydrochemical and toxicological parameters of the Irtysh River in Kazakhstan. *J. Arid. Land* **2016**, *8*, 521–532. [CrossRef]
23. Evseeva, A.A. Characteristics of zooplankton in the Bukhtarma reservoir in 2004–2008. *Bull. Semipalat. St. Univ. Shakarima* **2009**, *2*, 129–134. (In Russian)
24. Evseeva, A.A. Zooplankton and assessment of the ecological state of the Shulbinskoye reservoir. *J. Selevinia* **2010**, 112–116. (In Russian)
25. Evseeva, A.A. Zooplankton of the Ust-Kamenogorsk reservoir. *Bull. Kaz. Nation. Univ. Ecol. Ser.* **2012**, *1*, 165–168. (In Russian)
26. Evseeva, A.A. Characteristics of zooplankton of the Black Irtysh River (transboundary zone with China) in 2005–2007. In *Ecology and Hydrofauna of Water Bodies of Transboundary Basins of Kazakhstan*; Bastau: Almaty, Kazakhstan, 2008; pp. 337–346. (In Russian)
27. Wang, Y.; Xu, J.; Ding, R.; Zhang, H.; Cheng, X.; Bian, C. Ore Forming Fluids of Several Gold Deposits in the Irtysh Gold Belt, Xinjiang, China. *J. Earth Sci.* **2020**, *31*, 298–312. [CrossRef]
28. Abdulin, A.A.; Kayupov, A.K. (Eds.) *Metallogeny of Kazakhstan. Ore Formations. Deposits of Iron and Manganese*; Nauka: Alma-Ata, Kazakhstan, 1982; 208p. (In Russian)
29. Panin, M.S. Technogenic pollution with heavy metals in the catchment area of the Irtysh River basin. In Proceedings of the 2nd International Conference "Ecology, Radiation, Health", Semipalatinsk, Kazakhstan, 18–22 April 1998; p. 190. (In Russian).
30. Panin, M.S. *Ecological and Biogeochemical Assessment of Technogenic Landscapes of Eastern Kazakhstan*; Evero: Almaty, Kazakhstan, 2000; 338p. (In Russian)
31. Stat.gov.kz. Bureau of National Statistics of the Agency for Strategic Planning and Reform of the Republic of Kazakhstan. Statistics of the Region. East Kazakhstan Region. Dynamic Tables. (In Russian). Available online: https://stat.gov.kz/ru/region/vko/dynamic-tables/38/ (accessed on 5 January 2024).
32. Stat.gov.kz. Bureau of National Statistics of the Agency for Strategic Planning and Reform of the Republic of Kazakhstan. Statistics of the Region. Pavlodar Region. Dynamic Tables. (In Russian). Available online: https://stat.gov.kz/ru/region/pavlodar/dynamic-tables/38/ (accessed on 5 January 2024).
33. Khamzina, S.S.; Sharipova, Z.M.; Omarova, G.M. *Water Resources of the Pavlodar Region, Their Protection, and Rational Use: Textbook*; Innovative Eurasian University: Pavlodar, Kazakhstan, 2013; 248p. (In Russian)
34. Semenov, A.D. *Manual for the Chemical Analysis of Terrestrial Surface Waters*; Gidrometeoizdat: Leningrad, Russia, 1977; 541p. (In Russian)
35. Kiselev, I.A. Research methods of plankton. In *Life of the Fresh Water of the USSR*; Pavlovsky, E.N., Zhadin, V.I., Eds.; Academy of Sciences: Moscow, Russia, 1956; pp. 188–253. (In Russian)
36. Fomin, G.S. *Water. Control of Chemical, Bacterial, and Radiation Safety according to International Standards*; NGO Alternative: Moscow, Russia, 1995; 618p. (In Russian)
37. *Interstate Standard 31 870-2012*; Drinking Water. Determination of Element Content by Atomic Spectrometry Methods. Standardinform: Moscow, Russia, 2013. (In Russian)
38. Rylov, V.M. *Fauna of the USSR. Crustaceans. Freshwater Cyclopoida*; Nauka: Moscow, Russia, 1948; 312p. (In Russian)
39. Kutikova, L.A. *Rotifers of the Fauna of the USSR*; Science: Leningrad, Russia, 1964; 744p. (In Russian)
40. Tsalolikhin, S.Y. (Ed.) *Key to Freshwater Invertebrates in Russia and Adjacent Territories. Volume 2. Crustaceans*; Zoological Institute: St. Petersburg, Russia, 1995; 629p. (In Russian)
41. Balushkina, E.V.; Vinberg, G.G. The relationship between the length and body weight of planktonic crustaceans. In *Experimental and Field Studies of the Biological Foundations of Lake Productivity*; Vinberg, G.G., Ed.; Institute of Lake and River Fishery: Leningrad, Russia, 1979; pp. 58–79. (In Russian)
42. Magurran, E. *Ecological Diversity and Its Measurement*; Mir: Moscow, Russia, 1998; 184p. (In Russian)

43. Krupa, E.G.; Barinova, S.S. Environmental variables regulating the phytoplankton structure in high mountain lakes. *Res. J. Pharm. Biol. Chem. Sci.* **2016**, *7*, 1251–1261.
44. Love, J.; Selker, R.; Marsman, M.; Jamil, T.; Dropmann, D.; Verhagen, J.; Ly, A.; Gronau, Q.F.; Šmíra, M.; Epskamp, S.; et al. JASP: Graphical Statistical Software for Common Statistical Designs. *J. Statist. Softw.* **2019**, *88*, 1–17. [CrossRef]
45. Guseva, T.V. *Hydrochemical Variables of the State of the Environment*; Socio-Ecological Union: Moscow, Russia, 2002; 148p. (In Russian)
46. Romanenko, V.D.; Oksyuk, O.P.; Zhukinsky, V.N.; Stolberg, F.V.; Lavrik, V.I. *Environmental Assessment of the Impact of Hydraulic Engineering on Water Bodies*; Monograph; Naukova Dumka: Kyiv, Ukraine, 1990; 256p. (In Russian)
47. Krupa, E.; Barinova, S.; Romanova, S.; Aubakirova, M.; Ainabaeva, N. Heavy Metals in Fresh Waters of Kazakhstan and Methodological Approaches to Developing a Regional Water Quality Classification. *Cent. Asian J. Water Resear.* **2020**, *6*, 19–41. [CrossRef]
48. Krupa, E.G. Water invertebrates. In *Biodiversity of Wetland in the Syrdarya River Delta*; Ospanov, M.O., Stamkulova, K.Z., Eds.; Almaty, Kazakhstan, 2012; pp. 29–32.
49. Clarke, K.R. Comparison of dominance curves. *J. Exp. Mar. Biol. Ecol.* **1990**, *138*, 143–157. [CrossRef]
50. Napiórkowski, P.; Napiórkowska, T. The diversity and longitudinal changes of zooplankton in the lower course of a large, regulated European river (the lower Vistula River, Poland). *Biologia* **2013**, *68*, 1163–1171. [CrossRef]
51. Krupa, E.G. Zooplankton of the Syrdarya River as an indicator of anthropogenic impact. In *Ecology and Hydrofauna of Transboundary Basins of Kazakhstan*; Bastau: Almaty, Kazakhstan, 2008; pp. 92–112. (In Russian)
52. Krupa, E.G. *Zooplankton of Limnic and Lotic Ecosystems of Kazakhstan. Structure, Patterns of Formation*; Palmarium Academic Publishing: Saarbrucken, Germany, 2012; 346p. (In Russian)
53. Kim, H.W.; Joo, G.J. The longitudinal distribution and community dynamics of zooplankton in a regulated large river: A case study of the Nakdong River (Korea). *Hydrobiologia* **2000**, *438*, 171–184. [CrossRef]
54. Meybeck, M.; Helmer, R. The quality of rivers: From pristine stage to global pollution. *Glob. Planet. Chang.* **1989**, *1*, 283–309. [CrossRef]
55. Nikanorov, A.M. *Hydrochemistry*; Gidrometeoizdat: St. Petersburg, Russia, 2001; 444p. (In Russian)
56. Gashkina, N.A. Spatio-Temporal Variability of the Chemical Composition of Small Lakes' Waters in Modern Environmental Change Conditions: Abstract of the Dissertation of the Doctor of Geographical Sciences. Ph.D. Thesis, Moscow State University, St. Petersburg, Russia, 2014; 46p. (In Russian). Available online: https://www.dissercat.com/content/prostranstvenno-vremennaya-izmenchivost-khimicheskogo-sostava-vod-malykh-ozer-v-sovremennykh (accessed on 5 January 2024).
57. Fan, J.; Abudumanan, A.; Wang, L.; Zhou, D.; Wang, Z.; Liu, H. Dynamic Assessment and Sustainability Strategies of Ecological Security in the Irtysh River Basin of Xinjiang, China. *Chin. Geogr. Sci.* **2023**, *33*, 393–409. [CrossRef] [PubMed]
58. Krupa, E.; Barinova, S.; Aubakirova, M. Tracking pollution and its sources in the catchment-lake system of major waterbodies in Kazakhstan. *Lakes Reserv.* **2020**, *25*, 18–30. [CrossRef]
59. Mazurov, A.K. Metallogenic zoning of Kazakhstan. *News Tomsk. Polytech. Univ.* **2005**, *308*, 33–39. (In Russian)
60. Krupa, E.G.; Ainabaeva, N.S.; Aubakirova, M.O. *Methodological Recommendations (Methodological Guidelines) for Assessing the Ecological State of Reservoirs Based on Biological and Chemical Indicators*; MPK Crystal: Almaty, Kazakhstan, 2017; 30p. (In Russian)
61. Krupa, E.G.; Barinova, S.S.; Romanova, S.M.; Khitrova, E.A. *Hydrochemical and Hydrobiological Characteristics of Lakes in the Shchuchinsk-Borovsky Resort Area (Northern Kazakhstan) and the Main Methodological Approaches to Assessing the Ecological State of Small Reservoirs*; IP Volkova E.V.: Almaty, Kazakhstan, 2021; 330p. (In Russian)
62. Zhukinsky, V.N.; Oksiyuk, O.P.; Oleinik, G.N.; Kosheleva, S.I. Principles and experience in constructing an ecological classification of the quality of land surface waters. *Hydrobiol. J.* **1981**, *17*, 38–39. (In Russian)
63. Kitaev, S.P. *Fundamentals of Limnology for Hydrobiologists and Ichthyologists*; Karelian Scientific Center of the RAS: Petrozavodsk, Russia, 2007; 395p. (In Russian)
64. Zhou, J.; Qin, B.; Han, X. The synergetic effects of turbulence and turbidity on the zooplankton community structure in large, shallow Lake Taihu. *Environ. Sci. Pollut. Res.* **2018**, *25*, 1168–1175. [CrossRef] [PubMed]
65. Abollino, O.; Giacomino, A.; Malandrino, M.; Mentasti, E. The Efficiency of Vermiculite as Natural Sorbent for Heavy Metals. Application to a Contaminated Soil. *Water Air Soil Pollut.* **2007**, *181*, 149–160. [CrossRef]
66. Klimov, E.S.; Kalyukova, E.N.; Buzaeva, M.V. Sorption Properties of Natural Sorbent Silica Clay in Relation to Nickel Cations. *Rus. J. Appl. Chem.* **2010**, *83*, 1080–1082. [CrossRef]
67. Matache, M.L.; Marin, C.; Rozylowicz, L.; Tudorache, A. Plants accumulating heavy metals in the Danube River wetlands. *J. Environ. Health Sci. Eng.* **2013**, *11*, 39. [CrossRef]
68. Vlasov, B.P.; Gigevich, G.S. Estimation of pollution of lakes of Belarus under the contents of heavy metals in water plants and bottom sediments. *Limnol. Rev.* **2006**, *6*, 289–294.

water

MDPI

Article

Nitrate Source and Transformation in Groundwater under Urban and Agricultural Arid Environment in the Southeastern Nile Delta, Egypt

Alaa M. Kasem [1,2,3], Zhifang Xu [1,2,4,*], Hao Jiang [5], Wenjing Liu [1,2,4], Jiangyi Zhang [1,2,4] and Ahmed M. Nosair [6]

1 Key Laboratory of Cenozoic Geology and Environment, Institute of Geology and Geophysics, Chinese Academy of Sciences, Beijing 100029, China; alaa.kasem@cu.edu.eg (A.M.K.); liuwenjing@mail.iggcas.ac.cn (W.L.); zhangjiangyi@mail.iggcas.ac.cn (J.Z.)
2 University of Chinese Academy of Sciences, Beijing 100049, China
3 Department of Natural Resources, Faculty of African Postgraduate Studies, Cairo University, Giza 12613, Egypt
4 CAS Center for Excellence in Life and Paleoenvironment, Beijing 100044, China
5 Key Laboratory of Aquatic Botany and Watershed Ecology, Wuhan Botanical Garden, Chinese Academy of Sciences, Wuhan 430074, China; jianghao@wbgcas.cn
6 Environmental Geophysics Lab (ZEGL), Geology Department, Faculty of Science, Zagazig University, Zagazig 44519, Egypt; ahmed_nosair@zu.edu.eg
* Correspondence: zfxu@mail.iggcas.ac.cn; Tel.: +86-010-82998289; Fax: +86-010-62010846

Abstract: With the intensification of human activities, nitrate pollutants in groundwater are receiving increasing attention worldwide. Especially in the arid Nile Delta of Egypt, groundwater is one of the most valuable water resources in the region. Identifying the source of nitrate in groundwater with strong human disturbances is important to effective water resource management. This paper examined the stable isotopes ($\delta^{15}N/\delta^{18}O\text{-}NO_3$ and $\delta^{2}H/\delta^{18}O\text{-}H_2O$) and the hydrogeochemical parameters of the shallow groundwaters in the arid southeast of the Nile Delta to assess the potential sources and transformation processes of nitrate under severe urban and agricultural activities. The results revealed that the groundwaters were recharged by the Nile River. Meanwhile, the infiltration of irrigation water occurred in the west, while the mixing with the deep groundwater occurred in the east regions of the study area. The TDS, SO_4^{2-}, NO_3^-, and Mn^{2+} concentrations of groundwaters (n = 55) exceeded the WHO permissible limit with 34.6%, 23.6%, 23.6%, and 65.5%, respectively. The NO_3^- concentrations in the shallow groundwaters ranged from 0.42 mg/L to 652 mg/L, and the higher levels were observed in the middle region of the study area where the unconfined condition prevailed. It extended to the deep groundwater and eastward of the study area in the groundwater flow direction. The $\delta^{15}N\text{-}NO_3$ and $\delta^{18}O\text{-}NO_3$ values suggested that the groundwater NO_3^- in the west and east regions of semi-confined condition were largely from the nitrification of soil organic nitrogen (SON) and chemical fertilizer (CF). In contrast, wastewater input (e.g., domestic sewage and unlined drains) and prevalent denitrification were identified in the middle region. The denitrification might be tightly coupled with the biogeochemical cycling of manganese. This study provides the first report on the groundwater NO_3^- dynamics in the Nile Delta, which generated valuable clues for effective water resource management in the arid region.

Keywords: nitrate pollution; dual nitrate isotopes; water isotopes; the Nile Delta

1. Introduction

Groundwater nitrate (NO_3^-) pollution resulting from human activities is a pressing global concern [1,2]. Particularly, regions with high population density and intensive land use face increased vulnerability to this contamination [3–8]. This issue is prevalent in arid and semi-arid areas where water scarcity is already a significant challenge [9–11]. High NO_3^- levels in groundwater cause ecological issues such as eutrophication of nearby surface water and hypoxia [1,12–18]. Furthermore, NO_3^- in drinking water has been associated with health risks such as methemoglobinemia and cancer [19–22]. In response to the prevailing health concerns associated with NO_3^- contamination, the World Health Organization (WHO) has set a maximum threshold of 50 mg/L for drinking water [23].

To effectively manage NO_3^- pollution in groundwaters, a comprehensive understanding of its sources and transformation processes is necessary. While NO_3^- stems from various sources, including agricultural practices, wastewater, organic matter in the soil, atmospheric nitrogen deposition [24–26], and complex transformation processes such as denitrification and nitrification [27–29] make it more difficult to determine the exact source of nitrate. The use of isotopic tracer techniques, particularly dual NO_3^- stable isotopes, has been widely employed in various areas to investigate NO_3^- sources and transformations based on the distinct isotopic compositions of NO_3^- sources and the predictable isotopic fractionations [22,30–38]. For instance, studies have utilized the dual NO_3^- isotopes to explore the relationship between NO_3^- contamination in groundwaters and agricultural practices [39] and urban activities [30].

Despite the widespread use of this approach, uncertainties can arise when attempting to identify NO_3^- sources [40–42]. First, isotopic compositions of NO_3^- arising from various sources can exhibit some degree of overlap [38]. Second, the isotope fractionation process may obscure the initial isotopic compositions [38]. Therefore, other isotopes (e.g., $\delta^2H/\delta^{18}O$-H_2O and $\delta^{11}B$), water chemistry (e.g., Cl^- and SO_4^{2-}), and hydrogeochemical parameters were jointly applied to enhance the accuracy of NO_3^- source identification [22,43–48]. The integration of these additional data sets is expected to provide valuable and definitive insights into both water and NO_3^- cycling dynamics.

In the arid Nile Delta, the groundwater is a valuable resource because people rely on it for different purposes. The growing population, rapid urbanization, and increased usage of chemical fertilizers have led to severe NO_3^- pollution in the groundwaters. The authors of [49] highlighted significant urban expansion in the Nile Delta region from 1987 to 2015. It is presumed that this expansion has led to an increase in the discharge of untreated wastewater into water bodies. Furthermore, the two main wastewater unlined drains (Belbies and Kalyobiya) cross the study areas. These pollutants will inevitably infiltrate into the groundwater. High NO_3^- levels in the groundwaters have been reported (210 mg/L); however, the sources of pollution have not yet been identified, and the driving mechanisms still remain to be studied. There has been a lack of isotopic studies on NO_3^- pollution in the groundwaters in this region [50,51]. This seriously hinders the implementation of groundwater nitrate pollution prevention and control work in the region.

Located at the eastern fringe of the Nile Delta, the studied area is characterized with unlined drainage channels crisscrossing, a wide distribution of illegal shallow wells, and the expansion of urban areas, which further deteriorate the problem of groundwater pollution. In this study, for the first time, we use an integrating approach of a multiple isotope tracer ($\delta^{15}N/\delta^{18}O$-$NO_3$ and $\delta^2H/\delta^{18}O$-H_2O) combined with hydrogeochemical parameters to identify the sources of NO_3^- contamination and its transportation and transformation processes in the shallow groundwater of this area. The results are expected to provide insights to the effective management of the water resources in the arid region with intense human disturbance.

2. Materials and Methods

2.1. Study Area

The study area is located in the southeastern Nile Delta, a natural extension of the Nile Delta floodplain stretching toward the east. It extends between longitudes 31°3′22.42″–31°40′25.4″ E and latitudes 31°14′4.87″–30°37′16.87″ N and covers an area of approximately 2050 km^2 (Figure 1a). The study area had a population of roughly 12 million people, and the population density was calculated to be 5800 individuals/km^2 [52]. The land use is dominated by the interrelation of rural–urban and agricultural lands (Figure S2). The cultivation in the area primarily relies on Nile River water and groundwater for irrigation as a water source. The climate is arid and rainless in summer (May to September), with temperatures ranging from 30 to 40 °C, while the winter season (November to February) is relatively mild, with precipitation ranging from 10 to 20 mm/year and temperatures ranging from 10 to 20 °C. The study area is typically characterized by its flat plains forming the Delta landscape, gently sloping toward the north–northeast and east. It exhibits low elevation and is surrounded by a moderately elevated plateau to the south and southeast (Figure 1b).

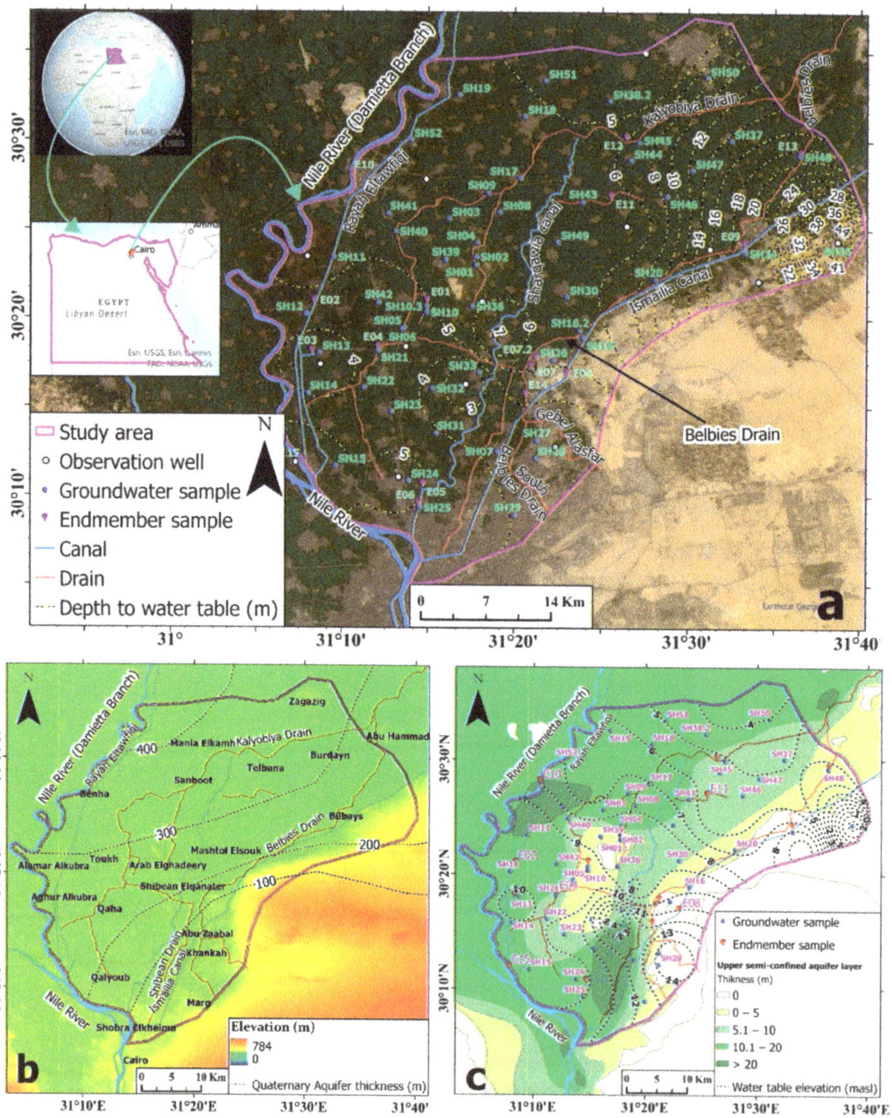

Figure 1. Maps of the study area and the sampling sites. (**a**) Landsat 8 and depth to the water table; (**b**) digital elevation model (DEM) and aquifer thickness; and (**c**) thickness of the upper semi-confined aquifer layer (modified from [53]) and hydraulic head.

The Quaternary aquifer is semi-confined and consists of two main units, the semi-pervious upper Holocene silt, sand, and silty clay unit (upper Holocene layer) (Figures S1 and S3), with thickness ranging from 0 to 20 m (Figure 1c) [53]. It is underlain by unconsolidated Pleistocene sand and gravel with clay lenses, with thickness ranging from a few meters in the south to about 400 m north (Figures 1b and S3). Moreover, the basalt and clay layer underlain the Quaternary aquifer acts as an aquiclude south of the study area, where the aquifer is thin (Figure S4). In contrast, in other areas, the Quaternary is uncomfortably underlain by the Tertiary aquifer, which is hydraulically connected particularly with the pumping increase from the Quaternary aquifer [54,55]. The consequences of the high extraction due to the irrigation of the newly reclaimed areas at the southeast corner of the study area create a cone of depression. At the same time, the increase in wastewater and surface water infiltration to the groundwater south of Shibean Elqanater City has increased the water table (Figure 1c).

The groundwater depth was measured in January 2022; it ranges from 2 m to 47 m, with the shallowest levels found near Khanka city and the deepest levels located in newly reclaimed areas southeast of the study area, at well SH35 (Figure 1a). The elevation of the water table fluctuates between 14 and −1 masl, with the highest values observed in the south and southwestern part of the study area, in the Khanka city and its vicinity. The groundwater flow direction exhibits a southwest-to-northeast pattern in the central, southern, and northern regions and a northwest-to-southeast direction in the southeastern part of the study area (Figure 1c). The groundwater recharge primarily occurs through two main sources: Nile River and irrigation canal seepage and infiltration of excess irrigation water, industrial usage, and surface water bodies (e.g., cesspools and drains) [51,56,57]. On the other hand, groundwater discharge is mainly attributed to pumping withdrawals for various purposes, such as agricultural irrigation, drinking water supply, and industrial usage.

2.2. Sample Collection

A total of 71 samples (16 surface water and 55 shallow groundwater) were collected between December 2022 and January 2023. The surface water samples were collocated as an endmember to represent the potential pollution sources for the shallow groundwater, not for water quality purposes. There are 4 samples from two different wastewater drains (Bilbies and Kalyobiya drains), 2 from Brackish lakes, 2 from agricultural drains, and 8 from surface freshwater. The majority of wells were established for private domestic purposes, with depths ranging from 9 to 43 m below the ground surface, and the average depth is 29.7 m. Each well was pumped 5 to 10 min before sampling to eliminate any stagnant water in the pipes. Of the 55 groundwater samples, 3 were from relatively deep agricultural wells (well depths of more than 50 m), and 52 were from shallow domestic wells. All samples were passed through 0.45 μm cellulose membranes and then collected into pre-washed high-density polyethylene (HDPE) bottles without headspace to avoid evaporation for the analysis of dissolved ions, heavy metals, $\delta^2H/\delta^{18}O$-H_2O, and $\delta^{15}N/\delta^{18}O$-$NO_3$. The samples were acidified for major cation and heavy metal analysis by adding a few drops of ultrapure nitric acid to lower the pH below 2 and stored in 60 mL HDPE bottles. Another 125 mL was stored in HDPE bottles for anion, $\delta^{15}N$-NO_3, and $\delta^{18}O$-NO_3 analyses. In addition, two 15 mL bottles were prepared for other anions and δ^2H-H_2O and $\delta^{18}O$-H_2O, respectively. The collected samples were carefully stored in a cold and dark environment until they can be transported back to the laboratory for analysis.

2.3. Analytical Methods

In situ measurements of water temperature (T), electrical conductivity (EC), and pH were conducted using portable pH (JENCO-6010N, USA) and EC (JENCO-EC3175, USA) meters. In addition, the HCO_3^- concentration was determined by titration within 24 h after sampling using HCl and Methyl Orange as the indicator [58]. The concentrations of Cl^-, SO_4^{2-}, F^-, SiO_2, NO_3^-, and NO_2^- were analyzed through ion chromatography (IC, DIONEX, ICS-1500, USA), with analytical precisions of ±5%. The concentration of cations

(K^+, Na^+, Mg^{2+}, and Ca^{2+}) and trace elements (As^{3+}, B^{3+}, Li^+, Sr^{3+}, Al^{3+}, Mn^{2+}, Cr^{3+}, Fe^{2+}, Co^{2+}, Cu^{2+}, Zn^{2+}, Ba^{2+}, and Ni^{2+}) were measured by an inductively coupled plasma optical emission spectrometer (ICP-OES, IRIS Intrepid II XSP, USA) with analytical precisions of ±5%. All the analyses were conducted at the Key Laboratory of Cenozoic Geology and Environment, Institute of Geology and Geophysics, Chinese Academy of Sciences, Beijing. δ^2H-H_2O and $\delta^{18}O$-H_2O were analyzed using an IRMS at the Institute of Resources and Environment, Henan Polytechnic University, and the isotope data were reported in per mil (‰) relative to the ratios of the Vienna Standard Mean Ocean Water (VSMOW). The analytical precisions were ±0.5‰ for δ^2H and ±0.1‰ for $\delta^{18}O$. The $\delta^{15}N$-NO_3 and $\delta^{18}O$-NO_3 analyses were conducted with the bacterial denitrification method [59,60]. This method examines the isotopic analysis of nitrous oxide (N_2O) reduced from NO_3^- through the activity of denitrifying bacteria that lack N_2O reductase activity, which is analyzed using an IRMS at the Environmental Stable Isotope Lab of the Chinese Academy of Agricultural Sciences. The isotope data were reported in per mil (‰) relative to atmospheric N_2 (AIR) and VSMOW. The analytical uncertainties were 0.3‰ for the $\delta^{15}N$-NO_3^- and 0.1‰ for the $\delta^{18}O$-NO_3^-.

3. Results and Discussion

3.1. Spatial Distribution of NO_3^- in the Shallow Groundwaters

Significant variability in NO_3^- concentrations across the studied area was observed (Figure 2a). The NO_3^- concentrations mainly ranged from 0.42 mg/L to approximately 350 mg/L, except in well SH26 (9 m depth, with a value of 651.8 mg/L). The presence of basalt at the bottom of this well acts as a NO_3^- trapping mechanism, forming a concentrated NO_3^- pool. Overall, 23.6% of the groundwater samples exceeded the permissible limit of 50 mg/L established by WHO [23]. The high NO_3^- levels were governed by unconfined conditions and the thickness of the aquifer, resulting in higher concentrations in the Abu Zaabal area (i.e., vicinity of Khanka city) at the central of the study area, extending toward the northwest and east regions following the groundwater flow path (Figure 2a). The study area was characterized into three regions based on the NO_3^- concentrations: the west, middle, and east regions. The west region exhibited low NO_3^- concentrations, which can be attributed to the dilution process from recharging by the Nile River (Damietta branch) or the presence of high thickness of the upper Holocene unit that regulates the infiltration of NO_3^- into the shallow groundwater. The middle region was generally characterized by unconfined conditions (Figure 1c), along with a small thickness of the aquifer, making it highly vulnerable to the infiltration of surface pollutants. Moreover, due to the point pollution sources (e.g., cesspools and drains), NO_3^- exhibited a wide concentration range [57,61]. Finally, the east region showed little impact from surface contaminants due to the increasing thickness of the upper Holocene layer.

3.2. Spatial Distribution of Hydrogeochemical Parameters

The hydrogeochemical parameters in the shallow groundwater aquifer exhibited a distinct spatial distribution pattern (Figures 2 and S5). The pH values ranged from 6.2 to 7.85, with a mean value of 7.08, indicating the groundwater's slightly acidic to neutral nature. The EC varied from 700 μS/cm to 3650 μS/cm, with an average of 1476 μS/cm (Table 1). The total dissolved solids (TDS), a critical indicator of dissolved chemical concentrations, exhibited a range of 455 mg/L and 2373 mg/L, with a mean value of 960 mg/L. The waters displayed elevated salinity levels, primarily attributed to the seepage of surface pollutants, particularly in the middle region. Furthermore, the increased extraction from well SH35 in the southeast part contributed to more significant mixing with the deep saline aquifer (Miocene aquifer) (Figure 2b). These observations aligned with previous studies by [55,57,62]. The SO_4^{2-} and Cl^- concentrations ranged from 2.5 mg/L to 827.2 mg/L (mean: 200.7 mg/L) and from 33.3 mg/L to 604 mg/L (mean: 170.3 mg/L), respectively. Compared with the previously reported values [61], the current study showed higher

SO_4^{2-} and Cl^- concentrations, which can be attributed to the enhanced seepage of surface pollutants, particularly notable in the middle region (Figure 2c,d).

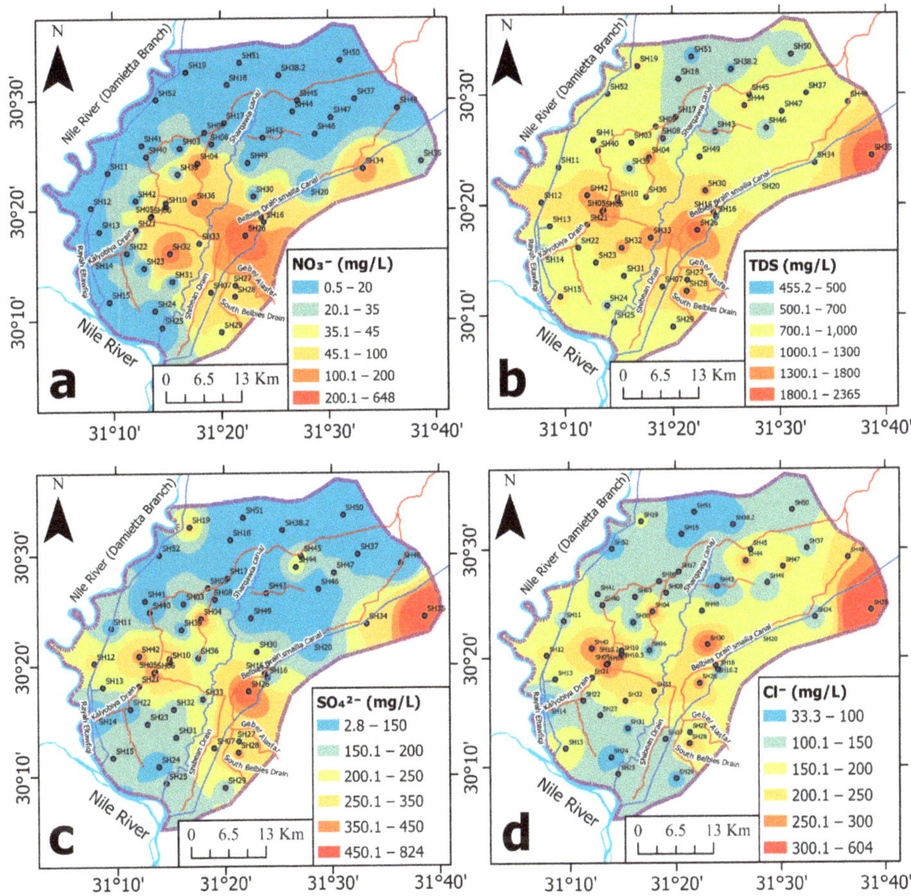

Figure 2. Spatial distribution maps of hydrogeochemical parameters of the groundwater in the study area: (**a**) TDS; (**b**) NO_3^-; (**c**) SO_4^{2-}; and (**d**) Cl^-.

The high SO_4^{2-} and Cl^- concentrations in the middle region of the study area were consistent with the high NO_3^- concentrations ($R^2 = 0.2$; $p = 0.3$ and $R^2 = 0.33$; $p < 0.01$, respectively). As SO_4^{2-} and Cl^- are typically enriched in wastewater, the correlations strengthened the scenario of increased recharge from the surface contaminant source. However, the insignificant Cl^- and NO_3^- correlation can be attributed to biochemical mediated NO_3^- concentration attenuation. In contrast, well SH35 in the southeast part records high SO_4^{2-} and Cl^- values, which can be mainly due to the mixing with the saline water from the underlain Miocene aquifer. In the west and east regions, the levels of these anions were not high, likely due to the thick upper Holocene layer in these regions and the dilution process in the west (Figure 2c,d). The HCO_3^- concentration showed high values in the west and southwest areas and decreased from the Damietta branch toward the east, with values ranging from 165.2 mg/L to 753.4 mg/L (mean: 390.9 mg/L).

Table 1. Descriptive statistics of analyzed parameters of shallow groundwater and endmember samples in the study area.

Parameter	Fresh Surface Water (n = 7)				Agricultural Drain (n = 2)				Polluted Lake (n = 2)			
	Mean	SD	Min	Max	Mean	SD	Min	Max	Mean	SD	Min	Max
Total depth (m)	–	–	–	–	–	–	–	–	–	–	–	–
pH	7.39	0.36	6.70	7.73	7.68	0.67	7.20	8.15	8.41	0.30	8.20	8.62
Temp (°C)	18.31	0.80	17.00	19.20	18.60	2.26	17.00	20.20	21.00	0.00	21.00	21.00
EC (µS/cm)	498	24	475	540	1980	665	1510	2450	17,555	1648	16,390	18,720
TDS (mg/L)	324	15	309	351	1287	432	982	1593	11,411	1071	10,654	12,168
SiO_2 (mg/L)	1.03	0.59	0.45	2.07	17.49	5.85	13.36	21.63	13.54	18.92	0.17	26.92
K^+ (mg/L)	6.10	0.42	5.66	6.60	24.58	2.84	22.57	26.60	74.11	5.17	70.45	77.76
Na^+ (mg/L)	11.74	12.45	6.46	39.97	112.21	90.13	48.48	175.94	1650	141.42	1550	1750
Ca^{2+} (mg/L)	41.09	2.01	36.77	43.05	119.63	68.46	71.22	168.03	617	316	393	840
Mg^{2+} (mg/L)	16.12	1.39	13.31	17.58	59.05	34.92	34.36	83.74	269.7	4.52	266.5	272.9
HCO_3^- (mg/L)	186.74	11.98	173.17	203.33	254.66	108.36	178.04	331.28	550.07	352.30	300.95	799.18
F^- (mg/L)	0.49	0.08	0.39	0.65	0.24	0.06	0.19	0.28	–	–	–	–
Cl^- (mg/L)	31.02	2.60	27.33	35.24	263.39	143.39	161.99	364.78	4272.9	1074.7	3513	5032.8
NO_2^- (mg/L)	–	–	–	–	–	–	–	–	–	–	–	–
NO_3^- (mg/L)	1.86	1.33	0.80	4.59	35.58	13.80	25.82	45.33	0.68	0.03	0.66	0.70
SO_4^{2-} (mg/L)	34.3	1.4	32.4	36.3	308.5	92.4	243.2	373.8	3700	132.9	3606	3794
Mn (mg/L)	0.004	0.005	0.001	0.014	0.198	0.198	0.058	0.338	0.072	0.099	0.001	0.142
Fe (mg/L)	0.045	0.005	0.037	0.053	0.137	0.093	0.071	0.203	0.171	0.004	0.168	0.174
As (mg/L)	0.0007	0.0005	0.0002	0.002	0.002	0.001	0.001	0.003	0.021	0.001	0.021	0.022
B (mg/L)	0.056	0.039	0.026	0.134	0.072	0.006	0.068	0.076	2.438	0.477	2.100	2.775
Li (mg/L)	0.001	0.0002	0.0005	0.001	0.002	0.0004	0.001	0.002	0.033	0.009	0.027	0.039
Sr (mg/L)	0.324	0.023	0.298	0.362	0.859	0.342	0.617	1.100	7.845	0.148	7.740	7.950
Al (mg/L)	0.117	0.098	0.010	0.266	0.056	0.057	0.016	0.097	0.016	0.001	0.016	0.017
Cr (mg/L)	0.011	0.001	0.009	0.013	0.011	0.002	0.010	0.012	0.033	0.023	0.017	0.050
Co (mg/L)	0.000	0.000	0.000	0.000	0.001	0.000	0.001	0.001	0.002	0.000	0.001	0.002
Cu (mg/L)	0.002	0.000	0.001	0.002	0.003	0.001	0.003	0.003	0.010	0.002	0.008	0.011
Zn (mg/L)	0.006	0.002	0.003	0.008	0.008	0.001	0.008	0.009	0.011	0.000	0.011	0.011
Ba (mg/L)	0.018	0.001	0.017	0.019	0.032	0.013	0.023	0.041	0.019	0.002	0.017	0.020
Ni (mg/L)	0.006	0.000	0.005	0.007	0.009	0.002	0.008	0.011	0.027	0.002	0.025	0.029
$\delta^{18}O$-H_2O (‰)	2.00	0.13	1.74	2.13	1.29	0.65	0.83	1.75	6.31	0.59	5.89	6.73
$\delta^{2}H$-H_2O (‰)	18.49	1.19	15.94	19.20	14.02	4.72	10.69	17.36	31.16	4.36	28.08	34.25
$\delta^{15}N$-NO_3 (‰)	5.03	8.47	−12.76	11.52	11.43	4.60	8.18	14.68	6.45	2.51	4.68	8.23
$\delta^{18}O$-NO_3 (‰)	9.02	6.65	−4.69	14.24	5.51	1.67	4.33	6.70	3.28	1.52	2.20	4.35

Table 1. *Cont.*

Parameter	Wastewater Drain (n = 4)				Groundwater (n = 55)			
	Mean	SD	Min	Max	Mean	SD	Min	Max
Total depth (m)	–	–	–	–	32.11	20.74	9.00	156.00
pH	7.24	0.48	6.56	7.70	7.08	0.36	6.20	7.75
Temp (°C)	19.38	0.85	18.50	20.50	20.29	6.63	18.50	25.30
EC (µS/cm)	1410	276	1080	1680	1476	688	700	3650
TDS (mg/L)	917	179	702	1092	960	447	455	2373
SiO_2 (mg/L)	8.99	3.65	4.16	13.02	63.53	65.57	18.70	237.32
K^+ (mg/L)	17.97	6.79	7.79	21.50	11.64	6.50	4.64	31.61
Na^+ (mg/L)	132.45	88.85	7.87	200.78	147.27	108.88	7.47	579.77
Ca^{2+} (mg/L)	62.39	11.73	46.16	71.24	102.65	49.69	19.07	366.72
Mg^{2+} (mg/L)	21.67	3.10	17.02	23.47	35.50	16.05	9.63	85.81
HCO_3^- (mg/L)	306.25	66.54	246.29	367.02	390.90	120.78	165.16	753.35
F^- (mg/L)	0.28	0.01	0.27	0.29	0.25	0.19	0.14	1.38
Cl^- (mg/L)	208.16	88.72	129.54	312.22	170.34	110.86	33.31	603.98
NO_2^- (mg/L)	107.24	63.74	50.13	176.01	–	–	–	–
NO_3^- (mg/L)	18.24	34.58	0.67	70.11	49.90	109.22	0.42	651.79
SO_4^{2-} (mg/L)	114.88	57.77	58.31	165.67	200.66	175.32	2.47	827.15
Mn (mg/L)	0.227	0.150	0.102	0.429	0.831	0.720	0.0019	3.380
Fe (mg/L)	0.088	0.020	0.069	0.108	0.089	0.060	0.040	0.430
As (mg/L)	0.001	0.001	0.001	0.002	0.002	0.001	0.000	0.006
B (mg/L)	0.137	0.038	0.090	0.177	0.094	0.112	0.016	0.577
Li (mg/L)	0.005	0.002	0.004	0.007	0.003	0.004	0.000	0.023
Sr (mg/L)	0.928	0.535	0.425	1.410	0.986	0.947	0.128	6.590
Al (mg/L)	0.030	0.005	0.024	0.036	0.012	0.006	0.007	0.046
Cr (mg/L)	0.013	0.001	0.011	0.013	0.011	0.003	0.007	0.023
Co (mg/L)	0.001	0.000	0.001	0.001	0.001	0.000	0.000	0.003
Cu (mg/L)	0.002	0.000	0.002	0.003	0.003	0.004	0.001	0.025
Zn (mg/L)	0.020	0.014	0.010	0.041	0.022	0.056	0.003	0.404
Ba (mg/L)	0.023	0.005	0.019	0.030	0.122	0.087	0.011	0.341
Ni (mg/L)	0.011	0.001	0.009	0.012	0.009	0.003	0.005	0.023
$\delta^{18}O$-H_2O (‰)	2.08	0.15	1.93	2.29	1.23	1.46	−1.86	3.32
$\delta^{2}H$-H_2O (‰)	19.10	0.87	18.19	20.20	13.99	8.83	−4.46	26.61
$\delta^{15}N$-NO_3 (‰)	8.13	6.10	−0.16	13.86	8.92	16.87	−23.40	75.42
$\delta^{18}O$-NO_3 (‰)	−1.69	13.54	−13.25	13.20	10.53	12.85	−14.32	39.79

–: not detected.

3.3. Groundwater Recharge

The stable isotopes of water ($\delta^2H/\delta^{18}O$-H_2O) provide valuable insights into the origin and movement of groundwater, rendering them indispensable tracers in hydrological investigations. The Global Meteoric Water Line (GMWL) is a crucial tool to interpret the isotopic tracers as it is a reference for understanding fractionation and mixing processes in natural water circulation [63,64]. The Nile River, irrigation canals, and drains were the potentially important shallow groundwater sources. However, before they seeped into the groundwater, the water can undergo isotopic fractionation associated with evapotranspiration during the infiltration from the ground surface, leading to enriched values. The Quaternary aquifer can be isotopically depleted because of the hydraulic connection with the depleted underlain Tertiary aquifer.

The stable isotope composition of water samples gathered from the study area is depicted on the conventional $\delta^{18}O$-H_2O and δ^2H-H_2O diagram (Figure 3). The groundwater values for $\delta^{18}O$-H_2O and δ^2H-H_2O exhibit considerable diversity, spanning from −1.86‰ to 3.32‰ and from −4.46‰ to 26.61‰, respectively (Table 1). The water samples deviate below the GMWL (Figure 3), suggesting that they had been influenced by evaporation, in contrast with the old Nile water sample before Aswan High Dam (AHD) compiled by [65], which was distributed close to the GMWL. Most of the samples were distributed near the Nile water, the irrigation return flow, and wastewater endmembers (Figure 3), suggesting that these sources primarily recharged the shallow groundwater.

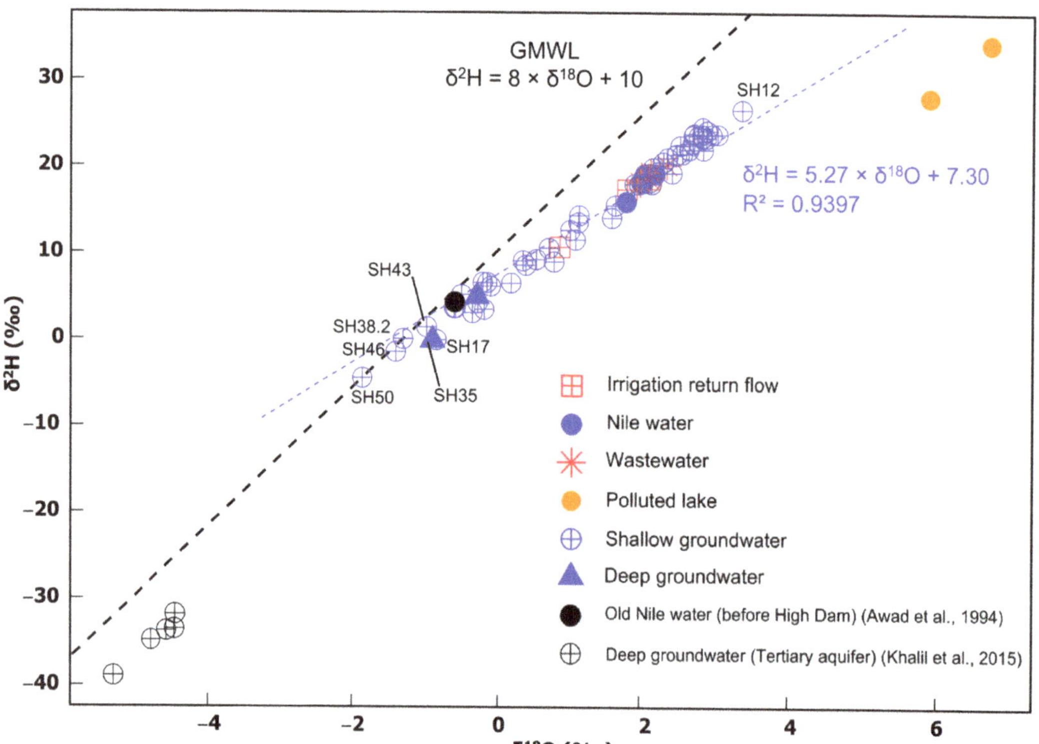

Figure 3. Scatterplot of water staple isotopes (δ^2H-H_2O and $\delta^{18}O$-H_2O) in groundwater and end-member samples in the study area [55,65].

In particular, the groundwater in the middle and western regions of the study area has undergone modern recharge, as indicated by its isotopic composition closely resembling that of the Nile water. In the west region, the isotopic values of groundwater experienced noticeable enrichment, likely due to an increase in evaporation (Figure 4). Several factors can contribute to this phenomenon. Firstly, the availability of Nile water and the prevalent traditional irrigation method, such as flood irrigation, enhance evaporation due to prolonged time on the surface before seeping downward. Secondly, the shallow depth of the water table contributed to the enrichment of the isotopic signature. Lastly, the presence of a thick upper Holocene unit exacerbated the evaporation process. Conversely, in the middle region, where this layer was absent, and there was a shortage of Nile water coupled with increased groundwater abstraction to compensate for this shortage, the isotopic composition of the groundwater remained closer to that of the Nile water and drainage systems. In the east region, where the upper Holocene layer is thick and there is shortage of the Nile water, the abstraction of the groundwater increased, which resulted in well nos. SH43, SH17, SH35, SH50, SH46, and SH38.2 being affected by mixing with the Tertiary aquifer. These findings are consistent with previous studies conducted by [51,66]. These results show that, although the NO_3^- groundwater pollution in the west/east regions was low, it can result from irrigation water enriched with CF and SON. This low concentration was due to the dilution of excess irrigation water and the Nile water recharging in the west.

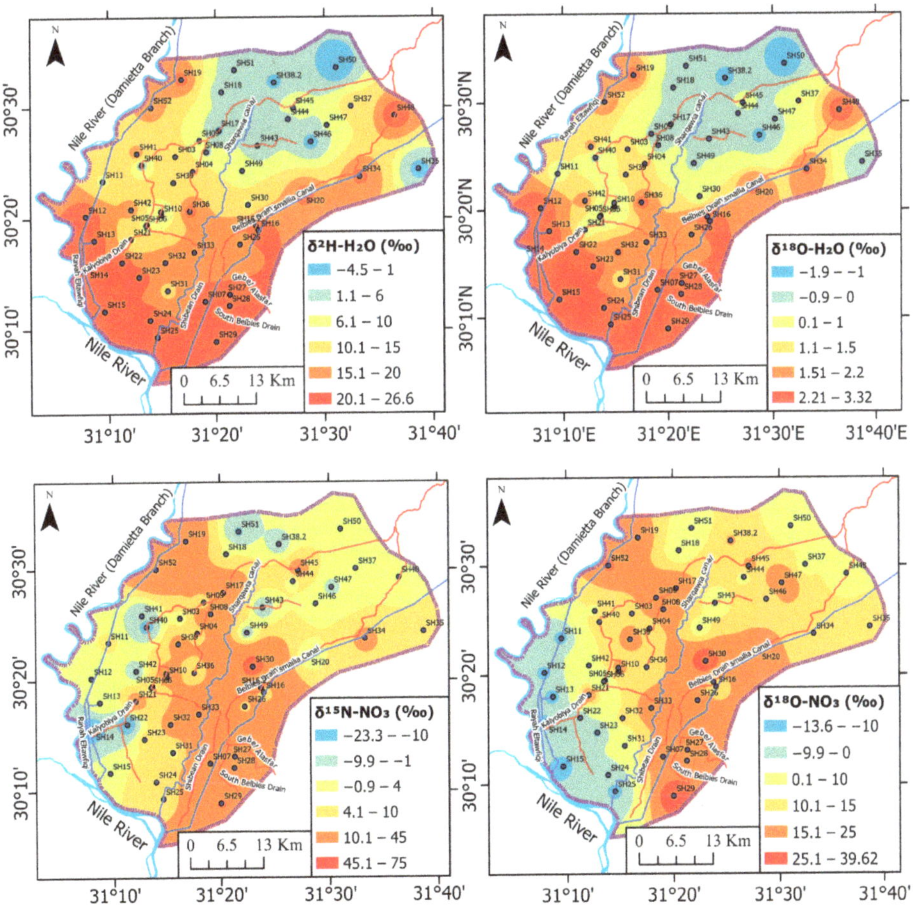

Figure 4. Spatial distribution maps of the δ^2H-H_2O, $\delta^{18}O$-H_2O, $\delta^{15}N$-NO_3, and $\delta^{18}O$-NO_3 of the groundwater in the study area.

In contrast, in the east, the dilution that resulted from the mixing with the deep groundwater contributed to the low NO_3^- concentrations. However, the relatively high NO_3^- (28.8 mg/L in well SH35) southeast of the study area could have originated from the lateral groundwater flow with high NO_3^- concentration due to relatively intense abstraction (250 m3/day). In contrast, in the middle region, more wastewater is loaded into the groundwater due to unconfined conditions.

3.4. Sources and Transformations of NO_3^-

Cl^- serves as a reliable indicator of wastewater and fertilizer impacts, which can contribute to elevated Cl^- levels and is unaffected by chemical, physical, and biological processes [45,46,67]. The correlation between Cl^- and NO_3^-/Cl^- is widely utilized for source identification of NO_3^- sources in water, specifically for distinguishing agricultural or/and wastewater sources [45,47] and for distinguishing between the influences of dilution and denitrification on NO_3^- levels [68]. A high NO_3^-/Cl^- ratio coupled with low Cl^- levels indicates that agricultural sources (e.g., chemical fertilizer) and soil organic nitrogen (SON) are the main contributors of NO_3^- in the groundwater. Conversely, a low NO_3^-/Cl^- ratio and high Cl^- content suggest wastewater sources as the primary origin of NO_3^- contamination [27,69].

Most of the groundwater samples had NO_3^-/Cl^- ratios < 1 and relatively high Cl^- concentrations (Figure 5a), suggesting a probable dominance of wastewater as the NO_3^- source [68,70]. However, a few samples in the middle region (SH26 and SH36; Figure 5a) exhibited high NO_3^-/Cl^- ratios > 1, which can be linked to a combination of SON and CF with wastewater inputs [70] (Figure 5a). The significant correlation between SO_4^{2-} and NO_3^- of the groundwaters in the middle region supported the dominant role of wastewater (Figures 5b and 6b) [71–73]. A few points were scattered from the correlation, which can be ascribed to the evaporate (e.g., gypsum) dissolution in the arid region [74,75] or as a result of using CF enriched with SO_4^{2-} [13,76].

In addition, most of the groundwater samples in the study area exhibited TDS concentrations below 1000 mg/L associated with a low $(NO_3^- + Cl^-)/HCO_3^-$ ratio. The correlation is strongly positive ($R^2 = 0.71$; $p < 0.01$) in the middle region, which can be attributed to the substantial reduction condition and then denitrification, which attenuates the NO_3^- and/or produce more HCO_3^- [70,77] (Figure 5c). This speculation can then be consistent with the theoretical denitrification defined in Equation (1) [78,79]. In the west and east regions, this relationship is a weak positive correlation as a result of the impacts of SON and CF [73,80] associated with biodegradation of organic matter, which is rich in the upper Holocene layer in these regions and consequently increase HCO_3^- concentration [81].

$$NO_3^- + 1.08\,CH_3OH + 0.24\,H_2CO_3 \rightarrow 0.56\,C_5H_7O_2N + 0.47\,N_2 + HCO_3^- + 1.68\,H_2O \quad (1)$$

The dual NO_3^- isotope approach ($\delta^{15}N$-NO_3 and $\delta^{18}O$-NO_3) is widely recognized as a powerful technique for constraining NO_3^- sources and behaviors [9,38,45,50,82]. Different NO_3^- sources exhibit characteristic isotopic signatures of $\delta^{15}N$-NO_3 and $\delta^{18}O$-NO_3, which can be used to trace their origins [38,83,84]. Figure 7a shows the $\delta^{15}N$-NO_3 and $\delta^{18}O$-NO_3 values of the groundwaters, Nile water, and potential endmembers in the study area. The isotopic compositions of $\delta^{15}N$-NO_3 and $\delta^{18}O$-NO_3 in the groundwaters showed wide ranges, ranging from −23.4‰ to 75.4‰ (mean: 8.92‰) and from −14.3‰ to 39.8‰ (mean: 10.53‰), respectively (Table 1). The groundwaters in the middle region have $\delta^{15}N$-NO_3 and $\delta^{18}O$-NO_3 ranging from 4‰ to 24.4‰ and from 10‰ to 23.9‰, respectively (Figures 4 and 7). As expected, the data are distributed near the endmembers for SON and CF and wastewater, suggesting that these anthropogenic sources should be the major sources of the groundwaters (Figure 7a). Due to their analogous isotopic compositions and transport routes, we merged SON and CF as one source. The direct input of atmospheric precipitation should be minor because the study area experiences minimal precipitation (Figure 7a).

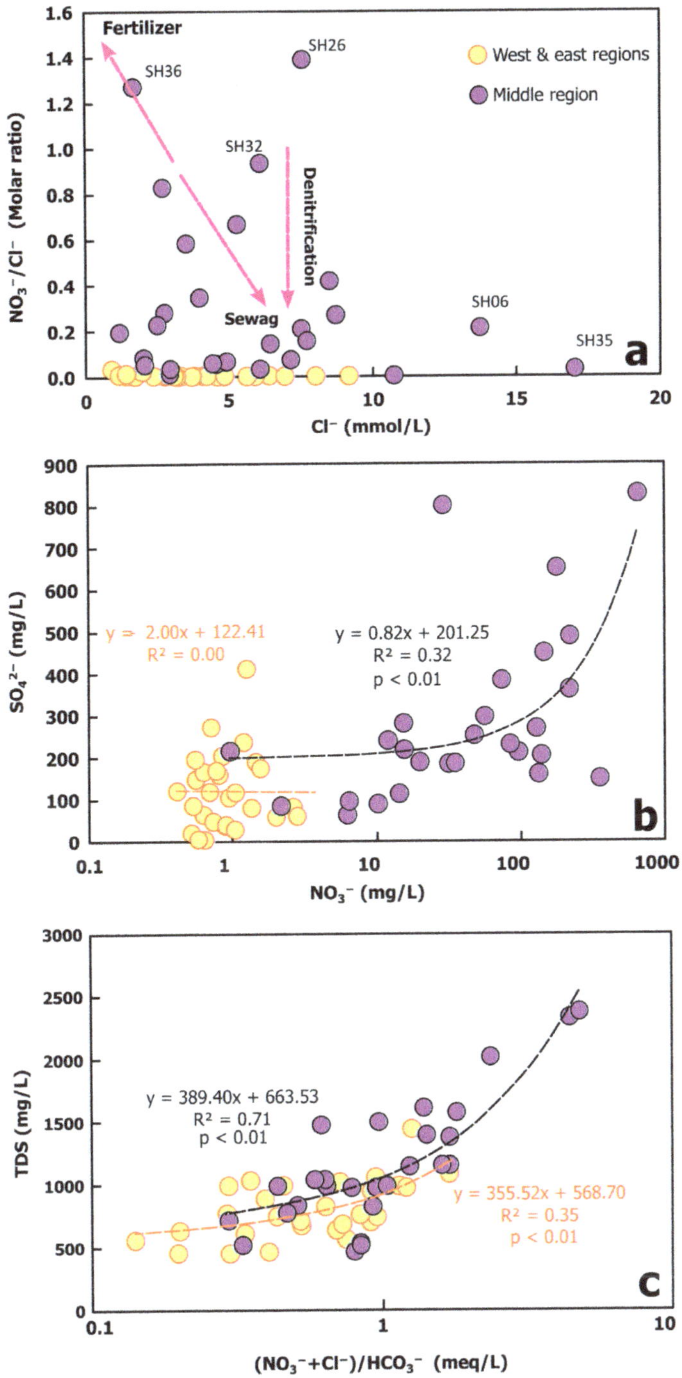

Figure 5. Plot of (**a**) NO_3^- + Cl^- versus Cl^-; (**b**) NO_3^- versus SO_4^{2-}; and (**c**) TDS versus (NO_3^- + Cl^-)/HCO_3^- of the groundwater samples in the study area.

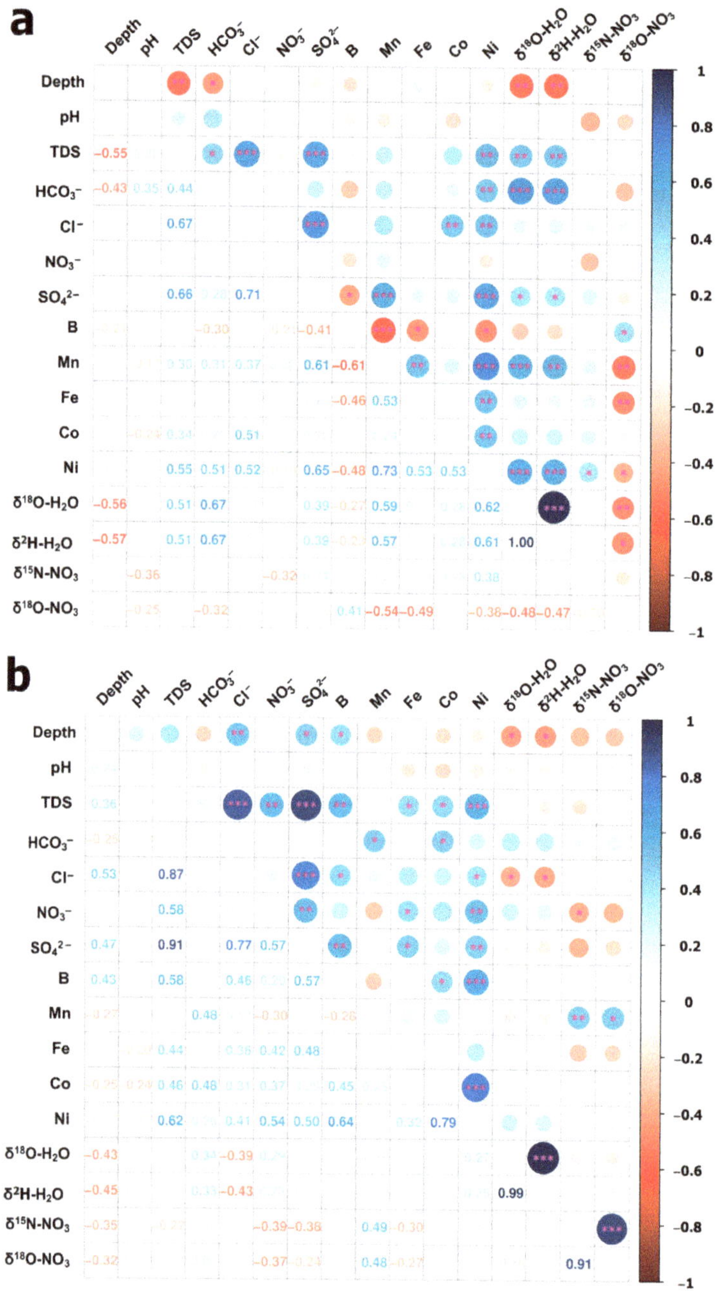

Figure 6. Plot of the Pearson correlation matrix of hydrogeochemical and isotope data of the shallow groundwater in the (**a**) west/east regions and (**b**) middle region. * $p < 0.05$; ** $p < 0.01$; *** $p < 0.001$.

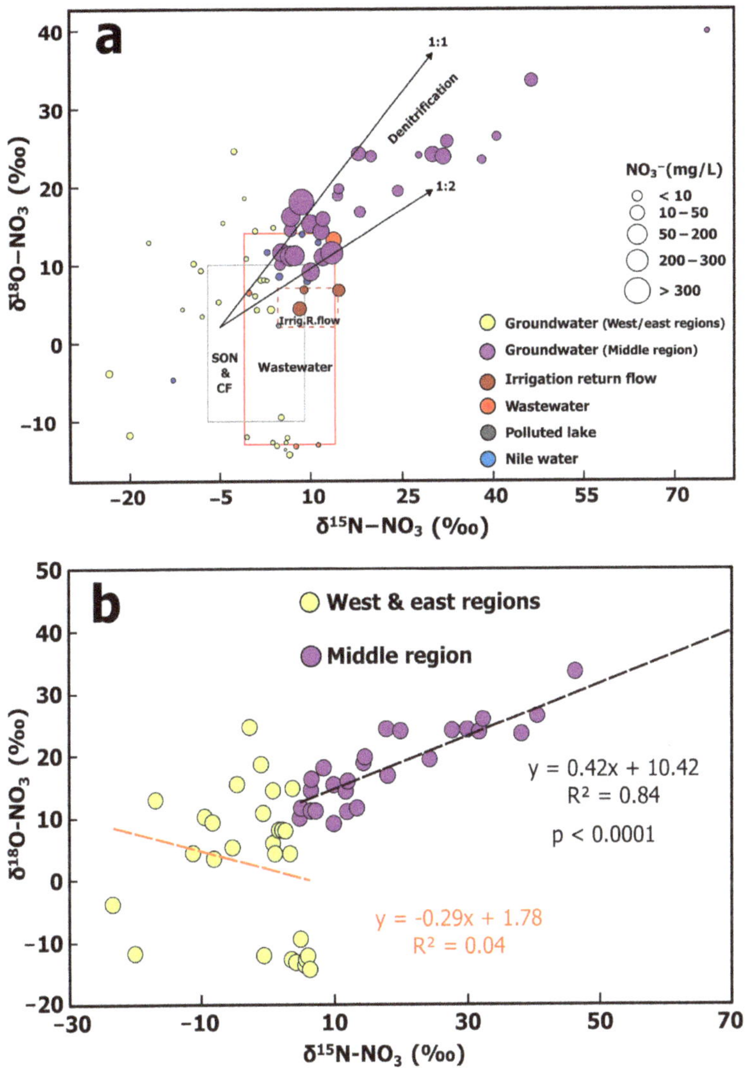

Figure 7. (**a**) Relationship between δ^{15}N-NO_3 and δ^{18}O-NO_3 values of the groundwater and endmember samples, with the ranges of potential nitrate sources. (**b**) δ^{15}N-NO_3 versus δ^{18}O-NO_3 of the groundwater samples in the study area.

Notably, a number of the isotopic values, especially those in the middle region, are higher than all the endmembers, and there is a significant positive correlation between the dual isotopes with a slope of 0.48 ($R^2 = 0.39$, $p < 0.0001$) (Figures 4, 6b and 7), which is the typical isotopic signal for biological NO_3^- removal [38]. Denitrification in groundwaters is well documented, leading to an exponential co-increase in δ^{15}N-NO_3 and δ^{18}O-NO_3, with values exceeding 100‰ recorded [38]. The significant denitrification is also supported by the simultaneous decreasing NO_3^- levels and increasing dual isotopic values (Figures 6b and 7a). The observation gives us the confidence that the prevalent denitrification process counterbalanced the excess NO_3^- loading to the groundwater to some extent (Figure 7a). The presence of clay lenses and clayey sediments in the aquifer can enhance the denitrification [38,85] by providing soluble organic carbon and anaerobic conditions [77]. However, more excess NO_3^- loading into the groundwater can inhibit this process and result in high NO_3^- levels (e.g., SH26 and SH32).

By contrast, no relationships existed between the dual isotopes in the samples collected in the west/east regions (Figures 6a and 7b), indicating minor denitrification. The groundwater samples in the west and east regions exhibited relatively depleted and constrained $\delta^{15}N\text{-}NO_3$ values (Figures 4 and 7a). Their $\delta^{15}N\text{-}NO_3$ and $\delta^{18}O\text{-}NO_3$ compositions ranged from −23.3‰ to 4‰ and from −13.6‰ to 10‰, respectively. Most of these samples had $\delta^{15}N\text{-}NO_3$ values lower than most ^{15}N-depleted SON and CF endmember, which can be ascribed to nitrification processes. Nitrification has a substantial fractionation effect, with ^{14}N being preferentially nitrified in NH_4^+ plentiful environments. The anthropogenic activities should discharge a large amount of NH_4^+ containing wastewater and irrigation discharge to the groundwaters, resulting in NH_4^+ plentiful conditions. Yet, the overall anaerobic environments in the groundwaters would inhibit nitrification. The speculation is supported by the low NO_3^- concentrations (Figures 4 and 7a). Collectively, the low isotopic values and concentrations of NO_3^- in the west/east regions should be attributed to the nitrification process, which also can be enhanced in the west due to the long residence time of water on the surface before it seeps into the groundwater [86]. Notably, although sample SH35 is collected from a deep well (depth = 156 m), it showed a NO_3^- value of 28.8 mg/L; this observation can result from NO_3^- recharging to the deep zone through the lateral groundwater flow.

Overall, according to the water chemistry and isotopic tracers, we found that wastewater and SON and CF were the major sources of shallow groundwaters. Denitrification is the primary process that occurs in the middle region accompanied by excess NO_3^- loadings, whereas the west/east regions can be affected by zonal nitrification.

3.5. Identification of NO_3^- Transformation Process Coupled with Mn Oxides Reduction

The presence of trace elements in groundwater provides valuable insights into NO_3^- dynamics [12,87]. Among these trace elements, Mn^{2+}, Fe^{2+}, and Cu^{2+} play crucial roles in denitrification by enhancing the rate of this process [44]. Notably, the relationship between oxidation and reduction of Mn and Fe in natural waters extends beyond the typical redox conditions necessary for their mobilization [88]. Mn and Fe in minerals serve as electron acceptors for denitrifying bacteria [77], thereby facilitating denitrification and ultimately enhancing denitrification rates [43,44,89].

The concentrations of Mn^{2+} in the groundwaters ranged from 0.002 mg/L to 3.38 mg/L, with a mean value of 0.83 mg/L. Furthermore, it is noteworthy that 65.5% of the groundwater samples examined exceeded the permissible limit of 0.4 mg/L set by the WHO (Figure S6). There were significant correlations between the Mn^{2+} concentrations and the $\delta^{15}N\text{-}NO_3$ and $\delta^{18}O\text{-}NO_3$ values in the middle region ($p < 0.01$; Figures 6b and 8a,b), implying the role of Mn in regulating the prevalent denitrification. The speculation was also supported by the negative correlation between Mn^{2+} and NO_3^- concentrations (Figures 6b and 8c). In addition, in the west/east region, the above correlations were absent (Figures 6a and 8a,c), further giving us the confidence that the biogeochemistry of Mn and denitrification were tightly coupled. This observed relationship was in line with the results of an experimental study that showed that MnO_2 addition enhanced the denitrifying bacteria's metabolism and removed about 99% of the NO_3^- and released Mn^{2+} in weak acidic and neutral conditions [43].

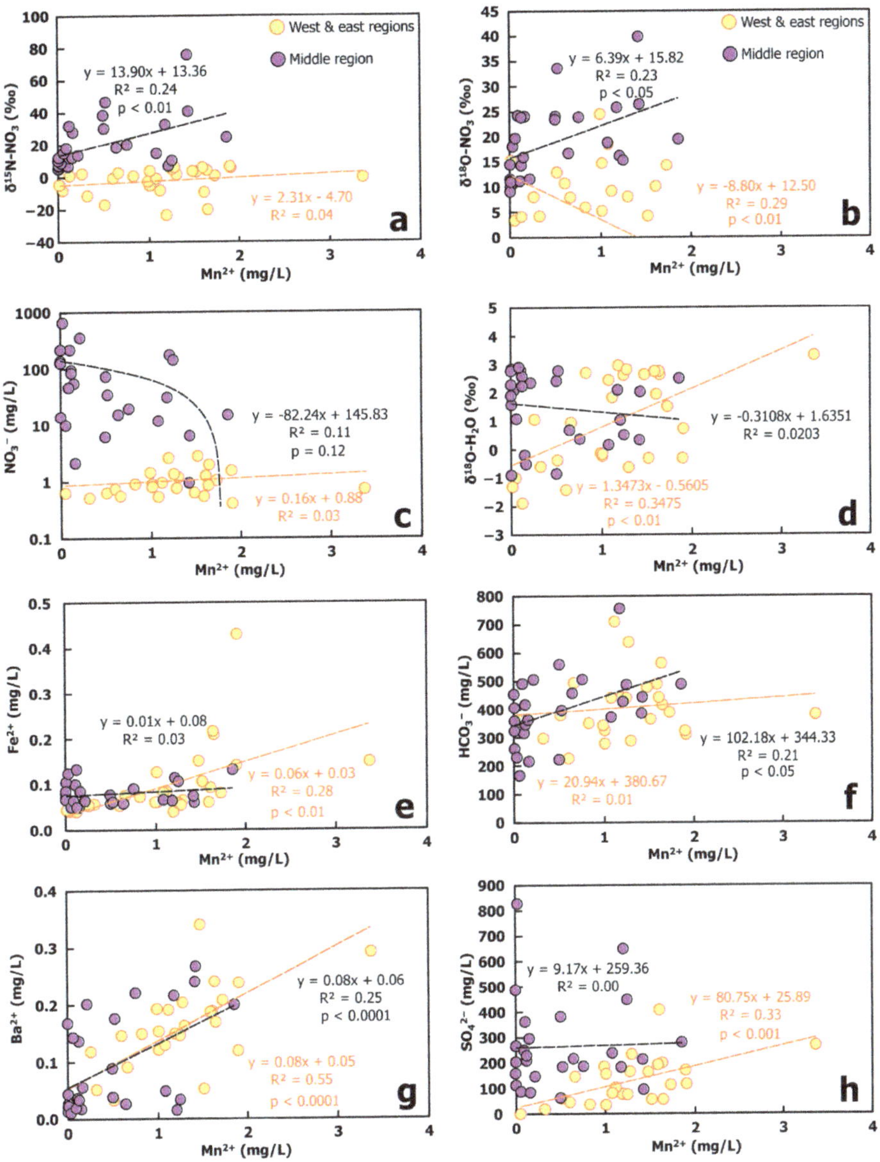

Figure 8. Scatterplot of (**a**) Mn^{2+} versus $\delta^{15}N$-NO_3; (**b**) Mn^{2+} versus $\delta^{18}O$-NO_3; (**c**) Mn^{2+} versus NO_3^-; (**d**) Mn^{2+} versus $\delta^{18}O$-H_2O; (**e**) Mn^{2+} versus Fe^{2+}; (**f**) Mn^{2+} versus HCO_3^-; (**g**) Mn^{2+} versus Ba^{2+}; and (**h**) Mn^{2+} versus SO_4^{2-} of the groundwater in this study.

The potential source of Mn in the study area is the upper Holocene layer [90], which is also enriched in organic matter and Mn minerals, leading to reduced environments [91]. Mn is commonly present in insoluble Mn(IV) and Mn(III), and their reduction will generate soluble Mn^{2+} [88,90]. In addition, Mn can result from CF utilized in the fruit farms in the western region [92]. We observed co-increasing of Mn^{2+} with HCO_3^-, Fe^{2+}, and Ba^{2+} (Figures 6 and 8e–g). As HCO_3^-, Fe^{2+}, and Ba^{2+} concentrations are closely related to chemical weathering, high Mn^{2+} in the groundwaters should be associated with chemical weathering processes. The mafic rocks are enriched with Mn [88] and the origin of the Nile silts and the Quaternary deposits as a result of weathering of these rocks [93], which lead to sediment enriched with Mn oxides. In contrast, there was no significant relationship between Mn^{2+} and SO_4^{2-} in the middle region (Figures 6b and 8h), indicating that the anthropogenic contribution of Mn^{2+} was less significant. However, Mn^{2+} and SO_4^{2-} showed a significant positive correlation in the west/east regions (Figures 6a and 8h), indicating the direct input of Mn^{2+} from anthropogenic sources [81,91]. That is, Mn was largely in an oxidized form in this layer before entering the water body, and it cannot participate in the reduction of NO_3^-. As the predominant factors influencing the concentrations of Mn^{2+} in the shallow groundwaters in the west/east and middle regions of the study area are quite different, the increase in Mn^{2+} concentrations and the associated redox process in the middle region enhanced denitrification, while in the west/east regions, denitrification and Mn^{2+} concentrations were decoupled.

However, reducing the Fe/Mn oxides coupled with nitrification–denitrification under anaerobic conditions and replacing the need for O_2 can be prevalent in these areas, where NH_4^+ oxidized and turned to NO_3^- or N_2 (i.e., anammox) based on Equation (2) [78].

$$NH_4^+ + 1.31\ NO_2^- + 0.066\ HCO_3^- + 0.13\ H^+ \rightarrow 1.02\ N_2 + 0.26\ NO_3^- + 0.066\ CH_2O_{0.5}N_{0.15} + 2.03\ H_2O \quad (2)$$

These findings highlight the importance of considering the biogeochemistry of trace elements, particularly Mn^{2+}, in assessing and managing NO_3^- contamination in shallow groundwater. However, it is important to note that while denitrification mitigates NO_3^- excess, it cannot fully offset the increasing NO_3^- loading into the groundwater. Due to this process, there is a higher release of Mn^{2+}, which, when combined with an excess of NO_3^-, poses a potential health risk to the residents.

4. Conclusions

In the present investigation, the hydrogeochemical parameters and multi-isotope tracers were used to identify the sources and transformation of NO_3^- in the shallow groundwater in the southeast of the Nile Delta region for the first time. Our findings reveal that the NO_3^- values ranged from 0.42 mg/L to 650 mg/L and the groundwater in the middle region was highly impacted by NO_3^- pollution. The NO_3^- pollution in the west/east regions was relatively weak due to a thick upper Holocene layer or as a result of the dilution process by the Nile River recharge in the west or the mixing with deep groundwater in the east. According to the WHO guidelines for drinking water, the criterion for NO_3^- and Mn^{2+} is exceeded in 23.6% and 65.5% of shallow groundwater samples, respectively. The values of δ18O-NO_3 and δ15N-NO_3 between −14.32‰ and −39.79‰ and between −23.4‰ and –75.42‰, respectively, implied that wastewater and SON and CF were the main NO_3^- sources. NO_3^- in the west/east regions can be derived from SON and CF associated with zonal nitrification. While in the middle region, NO_3^- was more derived from the wastewater, and the denitrification was prevalent. Remarkably, the denitrification might be closely coupled with the biogeochemical cycling of Mn. Finally, the data obtained from this study bear significant implications across several dimensions. Firstly, it establishes the first reference for NO_3^- isotope investigations of the groundwater in Egypt. Secondly, it is poised to furnish valuable insights to coupling Mn^{+2} and NO_3^- biogeochemical dynamics in the arid region, delineating pivotal denitrification and Mn reduction relationships.

Supplementary Materials: The following supporting information can be downloaded at: https://www.mdpi.com/article/10.3390/w16010022/s1. Figure S1: Geology of the southeastern region of the Nile Delta; Figure S2: Land use/Land cover (LU/LC) 10 m resolution (year 2022) of the study area; Figure S3: Hydrogeological map of the study area; Figure S4: Hydrogeological cross-section A-A′; Figure S5: Spatial distribution maps of the groundwater in the study area: (a) EC; (b) pH; (c) Na^+; (d) HCO_3^-; Figure S6: Spatial distribution maps of the groundwater in the study area: (a) Ca^{2+}; (b) Mg^{2+}; (c) Mn^{2+}; (d) Fe^{2+}; Figure S7: Scatterplot of Mn^{2+} versus SO_4^{2-}/Cl^- of the groundwater in the study area; Figure S8: Scatterplot of (a) ln [NO_3^-] versus δ^{18}O-NO_3, (b) ln [NO_3^-] versus δ^{15}N-NO_3 of the groundwater in the middle region of the study area; Figure S9: Scatterplot of (a) ln [NO_3^-] versus δ^{18}O-NO_3, (b) ln [NO_3^-] versus δ^{15}N-NO_3 of the groundwater in the west/east regions of the study area; Table S1: Quality parameters of the groundwater samples in the study area; Table S2: Limit of Detection of the Inductively Coupled Plasma Optical Emission Spectrometer (ICP-OES, IRIS Intrepid II XSP, USA) and ion chromatography (IC, Dionex 120, USA). References [23,53,54,94–96] are cited in the Supplementary Materials.

Author Contributions: Conceptualization, A.M.K., Z.X. and W.L.; methodology, A.M.K., Z.X. and W.L.; formal analysis, A.M.K., H.J., W.L. and J.Z.; investigation, A.M.K. and A.M.N.; writing—original draft preparation, A.M.K.; writing—review and editing, A.M.K., Z.X., H.J., W.L., J.Z. and A.M.N.; supervision, Z.X.; funding acquisition, Z.X. All authors have read and agreed to the published version of the manuscript.

Funding: This research was financially supported by the National Key Research and Development Program of China (grant no. 2020YFA0607700), the National Natural Science Foundation of China (grant nos. 41730857 and 42273050), and the Key Research Program of the Institute of Geology & Geophysics, CAS (grant no. IGGCAS-202204). Wenjing Liu acknowledges support from the Youth Innovation Promotion Association CAS (2019067).

Data Availability Statement: Data are available on request.

Acknowledgments: Alaa M. Kasem would like to thank the CAS-TWAS President's Fellowship Programme.

Conflicts of Interest: The authors declare that they have no known competing financial interest or personal relationship that could have appeared to influence the work reported in this paper.

References

1. Li, P.; Karunanidhi, D.; Subramani, T.; Srinivasamoorthy, K. Sources and Consequences of Groundwater Contamination. *Arch. Environ. Contam. Toxicol.* **2021**, *80*, 1–10. [CrossRef] [PubMed]
2. Han, G.; Tang, Y.; Liu, M.; Van Zwieten, L.; Yang, X.; Yu, C.; Wang, H.; Song, Z. Carbon-Nitrogen Isotope Coupling of Soil Organic Matter in a Karst Region under Land Use Change, Southwest China. *Agric. Ecosyst. Environ.* **2020**, *301*, 107027. [CrossRef]
3. Breida, M.; Alami Younssi, S.; Ouammou, M.; Bouhria, M.; Hafsi, M. Pollution of Water Sources from Agricultural and Industrial Effluents: Special Attention to NO_3^-, Cr(VI), and Cu(II). In *Water Chemistry*; IntechOpen: Rijeka, Croatia, 2020.
4. Zendehbad, M.; Cepuder, P.; Loiskandl, W.; Stumpp, C. Source Identification of Nitrate Contamination in the Urban Aquifer of Mashhad, Iran. *J. Hydrol. Reg. Stud.* **2019**, *25*, 100618. [CrossRef]
5. Zhang, Q.; Miao, L.; Wang, H.; Hou, J.; Li, Y. How Rapid Urbanization Drives Deteriorating Groundwater Quality in a Provincial Capital of China. *Pol. J. Environ. Stud.* **2019**, *29*, 441–450. [CrossRef] [PubMed]
6. Han, G.; Liu, C.Q. Water Geochemistry Controlled by Carbonate Dissolution: A Study of the River Waters Draining Karst-Dominated Terrain, Guizhou Province, China. *Chem. Geol.* **2004**, *204*, 1–21. [CrossRef]
7. El Shinawi, A.; Zeleňáková, M.; Nosair, A.M.; Abd-Elaty, I. Geo-Spatial Mapping and Simulation of the Sea Level Rise Influence on Groundwater Head and Upward Land Subsidence at the Rosetta Coastal Zone, Nile Delta, Egypt. *J. King Saud. Univ. Sci.* **2022**, *34*, 102145. [CrossRef]
8. Elewa, H.H.; Shohaib, R.E.; Qaddah, A.A.; Nousir, A.M. Determining Groundwater Protection Zones for the Quaternary Aquifer of Northeastern Nile Delta Using GIS-Based Vulnerability Mapping. *Environ. Earth Sci.* **2013**, *68*, 313–331. [CrossRef]
9. Gutiérrez, M.; Biagioni, R.N.; Alarcón-Herrera, M.T.; Rivas-Lucero, B.A. An Overview of Nitrate Sources and Operating Processes in Arid and Semiarid Aquifer Systems. *Sci. Total Environ.* **2018**, *624*, 1513–1522. [CrossRef]
10. Huang, T.; Pang, Z.; Yuan, L. Nitrate in Groundwater and the Unsaturated Zone in (Semi)Arid Northern China: Baseline and Factors Controlling Its Transport and Fate. *Environ. Earth Sci.* **2013**, *70*, 145–156. [CrossRef]
11. Li, P.; Tian, R.; Xue, C.; Wu, J. Progress, Opportunities, and Key Fields for Groundwater Quality Research under the Impacts of Human Activities in China with a Special Focus on Western China. *Environ. Sci. Pollut. Res.* **2017**, *24*, 13224–13234. [CrossRef]
12. Arcega-Cabrera, F.; Sickman, J.O.; Fargher, L.; Herrera-Silveira, J.; Lucero, D.; Oceguera-Vargas, I.; Lamas-Cosío, E.; Robledo-Ardila, P.A. Groundwater Quality in the Yucatan Peninsula: Insights from Stable Isotope and Metals Analysis. *Groundwater* **2021**, *59*, 878–891. [CrossRef] [PubMed]
13. Biddau, R.; Dore, E.; Da Pelo, S.; Lorrai, M.; Botti, P.; Testa, M.; Cidu, R. Geochemistry, Stable Isotopes and Statistic Tools to Estimate Threshold and Source of Nitrate in Groundwater (Sardinia, Italy). *Water Res.* **2023**, *232*, 119663. [CrossRef] [PubMed]
14. Ibrahim, K.O.; Gomo, M.; Oke, S.A. Groundwater Quality Assessment of Shallow Aquifer Hand Dug Wells in Rural Localities of Ilorin Northcentral Nigeria: Implications for Domestic and Irrigation Uses. *Groundw. Sustain. Dev.* **2019**, *9*, 100226. [CrossRef]

15. Jakeman, A.J.; Barreteau, O.; Hunt, R.J.; Rinaudo, J.-D.; Ross, A.; Arshad, M.; Hamilton, S. *Integrated Groundwater Management: An Overview of Concepts and Challenges in Integrated Groundwater Management: Concepts, Approaches and Challenges*; Jakeman, A.J., Barreteau, O., Hunt, R.J., Rinaudo, J.-D., Ross, A., Eds.; Springer International Publishing: Cham, Switzerland, 2016; pp. 3–20, ISBN 978-3-319-23576-9.
16. Puckett, L.J. *Nonpoint and Point Sources of Nitrogen in Major Watersheds of the United States*; Water-Resources Investigations Report 94-4001; US Geological Survey: Menlo Park, CA, USA, 1994; Volume 94.
17. Wakida, F.T.; Lerner, D.N. Non-Agricultural Sources of Groundwater Nitrate: A Review and Case Study. *Water Res.* **2005**, *39*, 3–16. [CrossRef] [PubMed]
18. Zhang, Q.; Sun, J.; Liu, J.; Huang, G.; Lu, C.; Zhang, Y. Driving Mechanism and Sources of Groundwater Nitrate Contamination in the Rapidly Urbanized Region of South China. *J. Contam. Hydrol.* **2015**, *182*, 221–230. [CrossRef] [PubMed]
19. Davidson, E.A.; David, M.B.; Galloway, J.N.; Goodale, C.L.; Haeuber, R.; Harrison, J.A.; Howarth, R.W.; Jaynes, D.B.; Lowrance, R.R.; Thomas, N.B.; et al. Excess Nitrogen in the U.S. Environment: Trends, Risks, and Solutions. *Issues Ecol.* **2011**, *15*, 1–16.
20. Dan-Hassan, M.A.; Olasehinde, P.I.; Amadi, A.N.; Yisa, J.; Jacob, J.O. Spatial and Temporal Distribution of Nitrate Pollution in Groundwater of Abuja, Nigeria. *Int. J. Chem.* **2012**, *4*, 104–112. [CrossRef]
21. Ward, M.H.; de Kok, T.M.; Levallois, P.; Brender, J.; Gulis, G.; Nolan, B.T.; VanDerslice, J. Workgroup Report: Drinking-Water Nitrate and Health—Recent Findings and Research Needs. *Environ. Health Perspect.* **2018**, *113*, 1607–1614. [CrossRef]
22. Jia, H.; Howard, K.; Qian, H. Use of Multiple Isotopic and Chemical Tracers to Identify Sources of Nitrate in Shallow Groundwaters along the Northern Slope of the Qinling Mountains, China. *Appl. Geochem.* **2020**, *113*, 104512. [CrossRef]
23. WHO. *Guidelines for Drinking Water Quality*; Fourth Edition Incorporating The First Addendum; World Health Organization: Geneva, Switzerland, 2017.
24. Rahman, A.; Mondal, N.C.; Tiwari, K.K. Anthropogenic Nitrate in Groundwater and Its Health Risks in the View of Background Concentration in a Semi Arid Area of Rajasthan, India. *Sci. Rep.* **2021**, *11*, 9279. [CrossRef]
25. Shakerkhatibi, M.; Mosaferi, M.; Pourakbar, M.; Ahmadnejad, M.; Safavi, N.; Banitorab, F. Comprehensive Investigation of Groundwater Quality in the North-West of Iran: Physicochemical and Heavy Metal Analysis. *Groundw. Sustain. Dev.* **2019**, *8*, 156–168. [CrossRef]
26. Šrajbek, M.; Kranjčević, L.; Kovač, I.; Biondić, R. Groundwater Nitrate Pollution Sources Assessment for Contaminated Wellfield. *Water* **2022**, *14*, 255. [CrossRef]
27. Romanelli, A.; Soto, D.X.; Matiatos, I.; Martínez, D.E.; Esquius, S. A Biological and Nitrate Isotopic Assessment Framework to Understand Eutrophication in Aquatic Ecosystems. *Sci. Total Environ.* **2020**, *715*, 136909. [CrossRef]
28. Pastén-Zapata, E.; Ledesma-Ruiz, R.; Harter, T.; Ramírez, A.I.; Mahlknecht, J. Assessment of Sources and Fate of Nitrate in Shallow Groundwater of an Agricultural Area by Using a Multi-Tracer Approach. *Sci. Total Environ.* **2014**, *470–471*, 855–864. [CrossRef] [PubMed]
29. Lee, C.M.; Choi, H.; Kim, Y.; Kim, M.S.; Kim, H.K.; Hamm, S.Y. Characterizing Land Use Effect on Shallow Groundwater Contamination by Using Self-Organizing Map and Buffer Zone. *Sci. Total Environ.* **2021**, *800*, 149632. [CrossRef]
30. Chen, R.; Hu, Q.; Shen, W.; Guo, J.; Yang, L.; Yuan, Q.; Lu, X.; Wang, L. Identification of Nitrate Sources of Groundwater and Rivers in Complex Urban Environments Based on Isotopic and Hydro-Chemical Evidence. *Sci. Total Environ.* **2023**, *871*, 162026. [CrossRef]
31. Hosono, T.; Tokunaga, T.; Kagabu, M.; Nakata, H.; Orishikida, T.; Lin, I.-T.; Shimada, J. The Use of Δ15N and Δ18O Tracers with an Understanding of Groundwater Flow Dynamics for Evaluating the Origins and Attenuation Mechanisms of Nitrate Pollution. *Water Res.* **2013**, *47*, 2661–2675. [CrossRef]
32. Jin, Z.; Qin, X.; Chen, L.; Jin, M.; Li, F. Using Dual Isotopes to Evaluate Sources and Transformations of Nitrate in the West Lake Watershed, Eastern China. *J. Contam. Hydrol.* **2015**, *177–178*, 64–75. [CrossRef]
33. Lorette, G.; Sebilo, M.; Buquet, D.; Lastennet, R.; Denis, A.; Peyraube, N.; Charriere, V.; Studer, J.-C. Tracing Sources and Fate of Nitrate in Multilayered Karstic Hydrogeological Catchments Using Natural Stable Isotopic Composition (Δ15N-NO_3^- and Δ18O-NO_3^-). Application to the Toulon Karst System (Dordogne, France). *J. Hydrol.* **2022**, *610*, 127972. [CrossRef]
34. Xing, M.; Liu, W. Using Dual Isotopes to Identify Sources and Transformations of Nitrogen in Water Catchments with Different Land Uses, Loess Plateau of China. *Environ. Sci. Pollut. Res.* **2016**, *23*, 388–401. [CrossRef]
35. Yue, F.-J.; Li, S.-L.; Liu, C.-Q.; Zhao, Z.-Q.; Ding, H. Tracing Nitrate Sources with Dual Isotopes and Long Term Monitoring of Nitrogen Species in the Yellow River, China. *Sci. Rep.* **2017**, *7*, 8537. [CrossRef] [PubMed]
36. Zhang, Q.; Wang, H.; Wang, L. Tracing Nitrate Pollution Sources and Transformations in the Over-Exploited Groundwater Region of North China Using Stable Isotopes. *J. Contam. Hydrol.* **2018**, *218*, 1–9. [CrossRef] [PubMed]
37. Zhang, Y.; Li, F.; Zhang, Q.; Li, J.; Liu, Q. Tracing Nitrate Pollution Sources and Transformation in Surface- and Ground-Waters Using Environmental Isotopes. *Sci. Total Environ.* **2014**, *490*, 213–222. [CrossRef] [PubMed]
38. Kendall, C.; Elliott, E.M.; Wankel, S.D. Tracing Anthropogenic Inputs of Nitrogen to Ecosystems. In *Stable Isotopes in Ecology and Environmental Science*, 2nd ed.; Michener, R., Lajtha, K., Eds.; Blackwell Publishing Ltd.: Oxford, UK, 2007; pp. 375–449.
39. Spalding, R.F.; Hirsh, A.J.; Exner, M.E.; Little, N.A.; Kloppenborg, K.L. Applicability of the Dual Isotopes Δ15N and Δ18O to Identify Nitrate in Groundwater beneath Irrigated Cropland. *J. Contam. Hydrol.* **2019**, *220*, 128–135. [CrossRef] [PubMed]
40. Cao, M.; Hu, A.; Gad, M.; Adyari, B.; Qin, D.; Zhang, L.; Sun, Q.; Yu, C.P. Domestic Wastewater Causes Nitrate Pollution in an Agricultural Watershed, China. *Sci. Total Environ.* **2022**, *823*, 153680. [CrossRef]

41. Minet, E.P.; Goodhue, R.; Meier-Augenstein, W.; Kalin, R.M.; Fenton, O.; Richards, K.G.; Coxon, C.E. Combining Stable Isotopes with Contamination Indicators: A Method for Improved Investigation of Nitrate Sources and Dynamics in Aquifers with Mixed Nitrogen Inputs. *Water Res.* **2017**, *124*, 85–96. [CrossRef]
42. Li, S.; Jiang, H.; Guo, W.; Zhang, W.; Zhang, Q. From Soil to River: Revealing the Mechanisms Underlying the High Riverine Nitrate Levels in a Forest Dominated Catchment. *Water Res.* **2023**, *241*, 120155. [CrossRef]
43. Gao, Z.; Su, J.; Ali, A.; Wang, X.; Bai, Y.; Wang, Y.; Wang, Z. Denitrification Strategy of Pantoea Sp. MFG10 Coupled with Microbial Dissimilatory Manganese Reduction: Deciphering the Physiological Response Based on Extracellular Secretion. *Bioresour. Technol.* **2022**, *355*, 127278. [CrossRef]
44. Labbé, N.; Parent, S.; Villemur, R. Addition of Trace Metals Increases Denitrification Rate in Closed Marine Systems. *Water Res.* **2003**, *37*, 914–920. [CrossRef]
45. Liu, C.-Q.; Li, S.-L.; Lang, Y.-C.; Xiao, H.-Y. Using Δ15N- and Δ18O-Values To Identify Nitrate Sources in Karst Ground Water, Guiyang, Southwest China. *Environ. Sci. Technol.* **2006**, *40*, 6928–6933. [CrossRef]
46. Panno, S.V.; Hackley, K.C.; Hwang, H.H.; Greenberg, S.E.; Krapac, I.G.; Landsberger, S.; O'Kelly, D.J. Characterization and Identification of Na-Cl Sources in Ground Water. *Groundwater* **2006**, *44*, 176–187. [CrossRef] [PubMed]
47. Wang, W.; Song, X.; Ma, Y. Identification of Nitrate Source Using Isotopic and Geochemical Data in the Lower Reaches of the Yellow River Irrigation District (China). *Environ. Earth Sci.* **2016**, *75*, 936. [CrossRef]
48. Gómez-Alday, J.J.; Hussein, S.; Arman, H.; Alshamsi, D.; Murad, A.; Elhaj, K.; Aldahan, A. A Multi-Isotopic Evaluation of Groundwater in a Rapidly Developing Area and Implications for Water Management in Hyper-Arid Regions. *Sci. Total Environ.* **2022**, *805*, 150245. [CrossRef] [PubMed]
49. Elagouz, M.H.; Abou-Shleel, S.M.; Belal, A.A.; El-Mohandes, M.A.O. Detection of Land Use/Cover Change in Egyptian Nile Delta Using Remote Sensing. *Egypt. J. Remote Sens. Space Sci.* **2020**, *23*, 57–62. [CrossRef]
50. Eissa, M.; Ali, M.; Zaghlool, E.; Stash, O.S. Hydrochemical and Stable Isotopes Indicators for Detecting Sources of Groundwater Contamination Close to Bahr El-Baqar Drain, Eastern Nile Delta, Egypt. *Water Sci.* **2019**, *33*, 54–64. [CrossRef]
51. Hussien, R.; Ahmed, M.; Aly, A.I. Tracking Anthropogenic Nitrogen-Compound Sources of Surface and Groundwater in Southwestern Nile Delta: Hydrochemical, Environmental Isotopes, and Modeling Approach. *Environ. Sci. Pollut. Res.* **2023**, *30*, 22115–22136. [CrossRef]
52. CAPMAS. Arab. Republic of Egypt—General. Census for Population, Housing and Establishments. 2017. Available online: https://censusinfo.capmas.gov.eg/metadata-en-v4.2/index.php/catalog/621 (accessed on 17 December 2023).
53. *RIGW-IWACO Hydrogeological Map of Egypt, 1:2000,000*; Research Institute of Groundwater, Water Research Center, Ministry of Public Works and Water Resources, Arab Republic of Egypt: Cairo, Egypt, 1992.
54. Hefny, K.; Khalil, J.B. General Hydrogeological Condition of Greater Cairo Area. *Water Sci. J Oct.* **1989**, *6*, 13–19.
55. Khalil, M.M.; Tokunaga, T.; Yousef, A.F. Insights from Stable Isotopes and Hydrochemistry to the Quaternary Groundwater System, South of the Ismailia Canal, Egypt. *J. Hydrol.* **2015**, *527*, 555–564. [CrossRef]
56. Awad, M.A.; El Arabi, N.E.; Hamza, M.S. Use of Solute Chemistry and Isotopes to Identify Sources of Ground-Water Recharge in the Nile Aquifer System, Upper Egypt. *Groundwater* **1997**, *35*, 223–228. [CrossRef]
57. Abu Salem, H.; Gemail, K.S.; Nosair, A.M. A Multidisciplinary Approach for Delineating Wastewater Flow Paths in Shallow Groundwater Aquifers: A Case Study in the Southeastern Part of the Nile Delta, Egypt. *J. Contam. Hydrol.* **2021**, *236*, 103701. [CrossRef]
58. American Public Health Association. *Standard Methods for the Examination of Water and Wastewater*, 23rd ed.; Rice, E.W., Bridgewater, L., American Public Health Association, Eds.; American Public Health Association: Washington, DC, USA, 2017.
59. Sigman, D.M.; Casciotti, K.L.; Andreani, M.; Barford, C.; Galanter, M.; Böhlke, J.K. A Bacterial Method for the Nitrogen Isotopic Analysis of Nitrate in Seawater and Freshwater. *Anal. Chem.* **2001**, *73*, 4145–4153. [CrossRef] [PubMed]
60. Weigand, M.A.; Foriel, J.; Barnett, B.; Oleynik, S.; Sigman, D.M. Updates to Instrumentation and Protocols for Isotopic Analysis of Nitrate by the Denitrifier Method. *Rapid Commun. Mass. Spectrom.* **2016**, *30*, 1365–1383. [CrossRef] [PubMed]
61. El-Fakharany, M.A.; Mansour, N.M.; Yehia, M.M.; Monem, M. Evaluation of Groundwater Quality of the Quaternary Aquifer through Multivariate Statistical Techniques at the Southeastern Part of the Nile Delta, Egypt. *Sustain. Water Resour. Manag.* **2017**, *3*, 71–81. [CrossRef]
62. Hegazy, D.; Abotalib, A.Z.; El-Bastaweesy, M.; El-Said, M.A.; Melegy, A.; Garamoon, H. Geo-Environmental Impacts of Hydrogeological Setting and Anthropogenic Activities on Water Quality in the Quaternary Aquifer Southeast of the Nile Delta, Egypt. *J. Afr. Earth Sci.* **2020**, *172*, 103947. [CrossRef]
63. Craig, H. Isotopic Variations in Meteoric Waters. *Science* **1961**, *133*, 1702–1703. [CrossRef] [PubMed]
64. Han, R.; Liu, W.; Zhang, J.; Zhao, T.; Sun, H.; Xu, Z. Hydrogeochemical Characteristics and Recharge Sources Identification Based on Isotopic Tracing of Alpine Rivers in the Tibetan Plateau. *Environ. Res.* **2023**, *229*, 115981. [CrossRef]
65. Awad, M.A.; Hamza, M.S.; Farid, M.S. Studies on the Recharge of the Aquifer Systems in the Southern Portion of the Nile Delta Using Radioisotopes and Hydro Chemistry. *Isot. Radiat. Res.* **1994**, *26*, 17.
66. Hamza, M.S.; Aly, A.I.M.; Swailem, F.M.; Nada, A. Environmentally Stable Isotopes and Groundwater Recharge in the Eastern Nile Delta. *Int. J. Water Resour. Dev.* **1987**, *3*, 228–232. [CrossRef]
67. Zhang, W.; Jiang, H.; Guo, W.; Li, S.; Zhang, Q. Unexpectedly High Nitrate Levels in a Pristine Forest River on the Southeastern Qinghai-Tibet Plateau. *J. Hazard. Mater.* **2023**, *458*, 132047. [CrossRef]

68. Widory, D.; Petelet-Giraud, E.; Négrel, P.; Ladouche, B. Tracking the Sources of Nitrate in Groundwater Using Coupled Nitrogen and Boron Isotopes: A Synthesis. *Environ. Sci. Technol.* **2005**, *39*, 539–548. [CrossRef]
69. Jin, J.; Wang, Z.; Zhao, Y.; Ding, H.; Zhang, J. Delineation of Hydrochemical Characteristics and Tracing Nitrate Contamination of Groundwater Based on Hydrochemical Methods and Isotope Techniques in the Northern Huangqihai Basin, China. *Water* **2022**, *14*, 3168. [CrossRef]
70. Torres-Martínez, J.A.; Mora, A.; Mahlknecht, J.; Daesslé, L.W.; Cervantes-Avilés, P.A.; Ledesma-Ruiz, R. Estimation of Nitrate Pollution Sources and Transformations in Groundwater of an Intensive Livestock-Agricultural Area (Comarca Lagunera), Combining Major Ions, Stable Isotopes and MixSIAR Model. *Environ. Pollut.* **2021**, *269*, 115445. [CrossRef] [PubMed]
71. Spoelstra, J.; Leal, K.A.; Senger, N.D.; Schiff, S.L.; Post, R. Isotopic Characterization of Sulfate in a Shallow Aquifer Impacted by Agricultural Fertilizer. *Groundwater* **2021**, *59*, 658–670. [CrossRef] [PubMed]
72. Torres-Martínez, J.A.; Mora, A.; Knappett, P.S.K.; Ornelas-Soto, N.; Mahlknecht, J. Tracking Nitrate and Sulfate Sources in Groundwater of an Urbanized Valley Using a Multi-Tracer Approach Combined with a Bayesian Isotope Mixing Model. *Water Res.* **2020**, 115962. [CrossRef] [PubMed]
73. Papazotos, P.; Vasileiou, E.; Perraki, M. The Synergistic Role of Agricultural Activities in Groundwater Quality in Ultramafic Environments: The Case of the Psachna Basin, Central Euboea, Greece. *Environ. Monit. Assess.* **2019**, *191*, 317. [CrossRef] [PubMed]
74. Ahmed, M.A.; Abdel Samie, S.G.; Badawy, H.A. Factors Controlling Mechanisms of Groundwater Salinization and Hydrogeochemical Processes in the Quaternary Aquifer of the Eastern Nile Delta, Egypt. *Environ. Earth Sci.* **2013**, *68*, 369–394. [CrossRef]
75. Rashed, M.; Awad, S.R.; Salam, M.A.; Smidt, E. Monitoring of Groundwater in Gabal El Asfar Wastewater Irrigated Area (Greater Cairo). *Water Sci. Technol.* **1995**, *32*, 163–169. [CrossRef]
76. Al-Gamal, S.; El-Sayed, S.; Atta, E. Exploring Surface-and Groundwater Interactions in East Delta Aquifer Using Conventional and Nonconventional Techniques. *J. Appl. Geol. Geophys.* **2018**, *6*, 48–58.
77. Jørgensen, P.R.; Urup, J.; Helstrup, T.; Jensen, M.B.; Eiland, F.; Vinther, F.P. Transport and Reduction of Nitrate in Clayey till underneath Forest and Arable Land. *J. Contam. Hydrol.* **2004**, *73*, 207–226. [CrossRef]
78. Desireddy, S.; Pothanamkandathil Chacko, S. A Review on Metal Oxide (FeOx/MnOx) Mediated Nitrogen Removal Processes and Its Application in Wastewater Treatment. *Rev. Environ. Sci. Biotechnol.* **2021**, *20*, 697–728. [CrossRef]
79. Gerardi, M.H. Nitrogen: Environmental and Wastewater Concerns. In *Nitrification and Denitrification in the Activated Sludge Process*; John Wiley & Sons, Inc.: Hoboken, NJ, USA, 2002; pp. 1–9.
80. Barzegar, R.; Moghaddam, A.A.; Tziritis, E.; Fakhri, M.S.; Soltani, S. Identification of Hydrogeochemical Processes and Pollution Sources of Groundwater Resources in the Marand Plain, Northwest of Iran. *Environ. Earth Sci.* **2017**, *76*, 297. [CrossRef]
81. Zhao, M.; Jiang, Y.; Jia, Y.; Lian, X.; Feng, F.; Shang, C.; Zang, Y.; Xi, B. Anthropogenic Perturbation Enhances the Release of Geogenic Mn to Groundwater: Evidence from Hydrogeochemical Characteristics. *Sci. Total Environ.* **2023**, *891*, 164450. [CrossRef]
82. Ji, X.; Shu, L.; Chen, W.; Chen, Z.; Shang, X.; Yang, Y.; Dahlgren, R.A.; Zhang, M. Nitrate Pollution Source Apportionment, Uncertainty and Sensitivity Analysis across a Rural-Urban River Network Based on Δ15N/Δ18O-NO_3^- Isotopes and SIAR Modeling. *J. Hazard. Mater.* **2022**, *438*, 129480. [CrossRef]
83. Kendall, C.; McDonnell, J.J. *Isotope Tracers in Catchment Hydrology*; Elsevier: Amsterdam, The Netherlands, 1998; ISBN 008092915X.
84. Wankel, S.D.; Kendall, C.; Francis, C.A.; Paytan, A. Nitrogen Sources and Cycling in the San Francisco Bay Estuary: A Nitrate Dual Isotopic Composition Approach. *Limnol. Oceanogr.* **2006**, *51*, 1654–1664. [CrossRef]
85. Rivett, M.O.; Buss, S.R.; Morgan, P.; Smith, J.W.N.; Bemment, C.D. Nitrate Attenuation in Groundwater: A Review of Biogeochemical Controlling Processes. *Water Res.* **2008**, *42*, 4215–4232. [CrossRef] [PubMed]
86. Jiang, H.; Liu, W.; Li, Y.; Zhang, J.; Xu, Z. Multiple Isotopes Reveal a Hydrology Dominated Control on the Nitrogen Cycling in the Nujiang River Basin, the Last Undammed Large River Basin on the Tibetan Plateau. *Environ. Sci. Technol.* **2022**, *56*, 4610–4619. [CrossRef] [PubMed]
87. Michener, R.; Lajtha, K. (Eds.) *Stable Isotopes in Ecology and Environmental Science*; Blackwell Publishing Ltd.: Oxford, UK, 2007; ISBN 9780470691854.
88. Homoncik, S.C.; MacDonald, A.M.; Heal, K.V.; Ó Dochartaigh, B.É.; Ngwenya, B.T. Manganese Concentrations in Scottish Groundwater. *Sci. Total Environ.* **2010**, *408*, 2467–2473. [CrossRef] [PubMed]
89. Yoch, D.C. Manganese, an Essential Trace Element for N_2 Fixation by *Rhodospirillum rubrum* and *Rhodopseudomonas capsulata*: Role in Nitrogenase Regulation. *J. Bacteriol.* **1979**, *140*, 987–995. [CrossRef]
90. Bennett, P.C.; El Shishtawy, A.M.; Sharp, J.M.; Atwia, M.G. Source and Migration of Dissolved Manganese in the Central Nile Delta Aquifer, Egypt. *J. Afr. Earth Sci.* **2014**, *96*, 8–20. [CrossRef]
91. Hou, Q.; Zhang, Q.; Huang, G.; Liu, C.; Zhang, Y. Elevated Manganese Concentrations in Shallow Groundwater of Various Aquifers in a Rapidly Urbanized Delta, South China. *Sci. Total Environ.* **2020**, *701*, 134777. [CrossRef]
92. Salem, M.G.; El-Awady, M.H.; Amin, E. Enhanced Removal of Dissolved Iron and Manganese from Nonconventional Water Resources in Delta District, Egypt. *Energy Procedia* **2012**, *18*, 983–993. [CrossRef]
93. Hamimi, Z.; El-Barkooky, A.; Martínez Frías, J.; Fritz, H.; Abd El-Rahman, Y. (Eds.) *The Geology of Egypt*, 1st ed.; Springer International Publishing: Cham, Switzerland, 2020; ISBN 978-3-030-15264-2.

94. EEBS. Geological, Hydrological and Geomorphological Properties of Soil for Egyptian Governorates. In *Soil Atlas of Egypt*; Part Egyptian Education Building Society Publication: Cairo, Egypt, 2004; pp. 1–290.
95. Egyptian Geological Survey. *Geologic Map of Egypt, Scale 1:2,000,000*; Egyptian Geological Survey and Mining Authority: Cairo, Egypt, 1981.
96. Karra, K.; Kontgis, C.; Statman-Weil, Z.; Mazzariello, J.C.; Mathis, M.; Brumby, S.P. Global Land Use/Land Cover with Sentinel 2 and Deep Learning. In Proceedings of the 2021 IEEE International Geoscience and Remote Sensing Symposium IGARSS, Brussels, Belgium, 11–16 July 2021; pp. 4704–4707.

water

MDPI

Review

Rice Fields and Aquatic Insect Biodiversity in Italy: State of Knowledge and Perspectives in the Context of Global Change

Tiziano Bo [1,2], Anna Marino [1,2,*], Simone Guareschi [1,2], Alex Laini [1,2] and Stefano Fenoglio [1,2]

1 Department of Life Sciences and Systems Biology, University of Turin, 10123 Turin, Italy; tiziano.bo@unito.it (T.B.); simone.guareschi@unito.it (S.G.); alex.laini@unito.it (A.L.); stefano.fenoglio@unito.it (S.F.)
2 ALPSTREAM—Alpine Stream Research Center, Parco del Monviso, 12030 Ostana, Italy
* Correspondence: anna.marino@unito.it

Abstract: Rice fields are one of the most important and extensive agro-ecosystems in the world. Italy is a major non-Asian rice producer, with a significant proportion of its yield originating from a vast area within the Po Valley, a region nourished by the waters of the Alps. While the biodiversity of these rice fields has been extensively documented for certain faunal groups, such as birds, there remains a paucity of research on the biodiversity of aquatic insects. A further challenge is the limited dissemination of findings, which have been primarily published in "gray" literature (local journals, newsletters and similar). Moreover, rice fields are of particular significance in the field of invasion biology, given their role in the arrival and spread of alien species. While the efficacy of rice fields as a substitute for the now-disappeared lowland natural environments is well documented, it is equally evident that traditional rice-growing techniques can require an unsustainable use of water resources, which threatens the biodiversity of the surrounding lotic systems. Here, we summarize and review multiple sources of entomological information from Italian rice fields, analyzing both publications in ISI journals and papers published in local journals (gray literature). In the near future, strategies that reduce the demand for irrigation, promote the cultivation of drought-tolerant crops, and utilize precision farming techniques will be implemented. The challenge will be balancing the need to reduce water withdrawal from rivers with the maintenance of wetlands where possible to support this pivotal component of regional biodiversity.

Keywords: rice paddies; freshwater insects; animal biodiversity; global warming; agro-ecosystems

1. Rice Fields as Ecosystem and Their Biodiversity

Rice is the second most widely grown cereal crop and it is the staple food for over half the world's human population. Consequently, rice paddies are one of the most important and extensive agro-ecosystems in several regions of the world [1,2]. Traditionally, the largest rice production areas are located in Asia, where ten countries are responsible for 85% of global rice production [3]. In particular, China and India are the top rice producers, followed by Bangladesh, Indonesia, Vietnam, Thailand, Burma, Philippines, Japan, and Pakistan. This means that, since ancient times, huge portions of these territories have been turned into rice fields, with enormous social, environmental, and ecological consequences. Outside Asia, rice production is prevalently concentrated in some parts of South America and Africa. In Europe, Italy is the main producer of rice, and 93% of Italian production comes from what is called the 'golden triangle of rice,' in the northwestern part of the country, an area of around 250,000 hectares between the provinces of Vercelli, Novara and Pavia (Figure 1).

Figure 1. Distribution (blue color) of the main rice-producing areas in Italy.

In Italy, rice cultivation has expanded since the second half of the 19th century, due to the availability of water from the Alps and the construction of a vast network of artificial canals. These canals distribute water in the lowland areas and are primarily fed by the Cavour Canal, a significant hydraulic work realized in 1866 and fed by the Po and other Alpine lotic systems. Such a concentration of agricultural activity has led to a profound transformation of the landscape and territory of northwestern Italy, with effects on local biodiversity that are not yet well known, because these often depend on the type and evolution of cultivation techniques. Additionally, though rice fields are a man-made environment, their biodiversity represents an interesting subject of study. In fact, rice fields are generally recognized as important substitute habitats for many aquatic species in many agro-ecosystems throughout the world [4]. Rice fields, along with their connected aquatic (e.g., canals, rivers) and terrestrial (e.g., adjacent crop and grass cultivations) habitats, constitute a mosaic of rapidly changing environmental situations, extremely rich in ecotones, that can potentially harbor a rich and unique biological diversity.

Moreover, these agro-ecosystems represent an intriguing area of study in the context of entomological biodiversity, given their potential for the inadvertent introduction of alien species. In fact, despite the dominance of aquatic insects in most inland waters, their unparalleled taxonomic diversity, and their occupation of nearly all trophic niches, there is a notable absence of invasive insects in freshwater, with the exception of a few examples, some of which are specifically observed in rice fields [5]. The importance of rice fields in insect biodiversity studies is evidenced by the number of related publications, with a specific bibliographic review using the terms *rice fields + insects + biodiversity* in Scopus (II/04/2025) yielding a total of 121 scientific papers (Figure 2). In particular, studies on the biodiversity of paddy fields have been carried out in different parts of the world, such as Sri Lanka [6], California [7], Sumatra [8], Egypt [9], Kenya [10], Brazil [11], and France [12];

however, a similar contribution is very scarce for the Italian rice fields, that constitute, as mentioned before, the largest rice cultivation area in Europe.

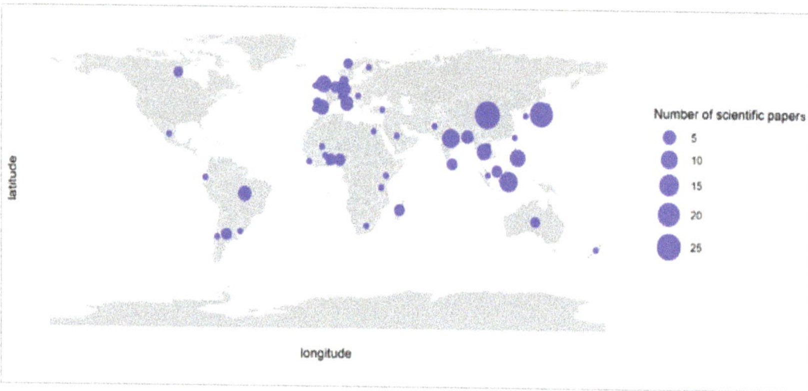

Figure 2. World distribution of papers related to *rice fields + insects + biodiversity*.

The objective of this review is to collate the published information on the entomological biodiversity of Italian rice fields, which is available in both grey and scientific literature. This paper represents the inaugural attempt to collate and disseminate this information to an international scientific audience. Furthermore, we aim to examine potential evolutionary scenarios in light of climate change, which is rapidly altering the environmental conditions of this area and will consequently have profound repercussions on both agricultural practices and biodiversity.

2. Insect Biodiversity of Italian Rice Fields

The biodiversity of Italian rice paddies has been widely explored with regard to certain systematic groups, such as birds, e.g., [13,14], while information regarding aquatic insects is scarcer and, above all, very scattered. These artificial environments host a peculiar aquatic invertebrate fauna that is potentially rich and diverse. In Table 1 we report a list of the most common freshwater insects that inhabit rice fields.

Table 1. Most common freshwater family insects that inhabit rice fields.

Order	Families
Ephemeroptera	Baetidae, Caenidae, Ephemerellidae
Odonata-Zygoptera	Calopterygidae, Lestidae, Platycnemididae, Coenagrionidae
Odonata-Anisoptera	Gomphidae, Aeshnidae, Cordulegasteridae, Libellulidae, Corduliidae
Heteroptera	Gerridae, Nepidae, Corixidae, Notonectidae, Naucoridae, Pleidae, Veliidae, Hydrometridae, Ochteridae
Trichoptera	Leptoceridae, Hydropsychidae, Phryganeidae, Lepidostomatidae, Limnephilidae
Lepidoptera	Crambidae, Noctuidae
Diptera	Chironomidae, Ceratopogonidae, Culicidae, Chaoboridae, Psychodidae, Stratiomyidae, Limoniidae, Tipulidae, Tabanidae, Ephydridae, Syrphidae, Sciomyzidae, Empididae, Muscidae, Cordyluridae
Coleoptera	Gyrinidae, Dytiscidae, Haliplidae, Elmidae, Dryopidae, Helophoridae, Hydrophilidae, Limnebiidae

Among the earliest studies that analyzed the entomofauna of Italian rice fields, we can report the publications of [15,16] which focused on aquatic Diptera and Trichoptera. A significant portion of these early works focused on species and groups of agricultural interest, setting a course that continued in the following decades [17–19]. Evidently, the stimulus for these studies derived from a clear interest in pest control, with the larval stages of certain Trichoptera species (primarily those of the family Phryganeidae) and Diptera (mainly Chironomidae and Tipulidae) being the focus of particular investigation. Apart from these groups, Stratyiomidae [20] and Ephydridae [21] were also reported because they are harmful to rice cultivation.

However, it is only in recent times that the interest of entomologists has shifted towards groups of no agricultural interest. This shift is largely due to the rapid changes in the agricultural landscape of the Po Valley since the Second World War, including habitat simplification, the mechanization of cultivation practices, and, most notably, the progressive disappearance of natural wetlands. In this scenario, rice fields have assumed an increasingly important role as surrogate environments, essentially the only habitat in which lowland aquatic insect populations can survive. For instance, the Odonata, a very ancient order of Palaeoptera, are mostly associated with lentic or semi-lentic environments and thus are among the most studied and best-known groups in rice fields, partly due to their aesthetic appeal, apical trophic role, and ease of observation. Noteworthy studies in this field include [22–24].

The other Paleopteran group, Ephemeroptera, is known from only very few species in Italian rice fields. These species have a high environmental tolerance, eurythermic and euryoxybiont habits, and rapid life cycles, such as *Baetis rhodani*, *Caenis horaria*, and *Serratella ignita* [25,26]. The other representatives of this ancient insect's order prefer lotic habitats characterized by running water and high oxygen levels. Among Hemimetabolous orders, in Italian rice fields, both Orthoptera and Hemiptera are reported. The first order is not strictly aquatic, inhabits rice field banks and is riparian [27], thus it is of minor importance in this paper. Aquatic and semiaquatic bugs (Heteroptera) are well represented in these anthropogenic habitats, mainly with the families Corixidae, Notonectidae, Gerridae, Pleidae, Nepidae [28]. Many of these families are predators, while others have a phytophagous diet consisting of plant remains and straw resulting from rice cutting. Homoptera are also reported for Italian rice fields, such as *Sipha gliceriae* and *Rhopalosiphum padi* [29].

Among holometabolous groups, Diptera is one of the most studied, partially because an important line of entomological research in rice fields has focused on the role that these environments play as a habitat for numerous species of hematophagous flies and their possible predators, as reported by [30], especially those belonging to the Culicidae family. The importance of these studies is heightened by the fact that rising temperatures and the globalization of transport have increased the risk of spreading new diseases such as Chikungunya, Dengue and other infections [31–37]. Among the most common Diptera found in rice fields, we report the following families: Chironomidae, Ceratopogonidae, Culicidae, Psychodidae, Stratiomyidae, Limoniidae, Tipulidae, Ephydridae, Syrphidae, Sciomyzidae, Empididae, and Muscidae.

Water beetles (order: Coleoptera) are a characteristic group inhabiting lentic and semi-lentic environments (ponds, oxbow lakes, marshes, abandoned meanders) and have found in rice fields a perfect surrogate for what had been their natural habitats in the Po floodplains. Among these, the Dytiscidae family stands out for its richness and diversity. These beetles colonize rice fields in both their larval and adult stages and serve as apex predators in these often-fishless environments, e.g., [30,38–40]. Among Dytiscidae, *Hydrogliphagus geminus* has been documented as an early colonizer of rice fields, and has been shown to be an effective controller of mosquito larvae [30]. Interestingly, a faunistic note on the distribution of Noteridae in southern Piedmont reports that rice fields are among the few habitats where these organisms are common and abundant [41]. Species from the family Hydrophilidae are also listed in these habitats, with the recurrent presence of *Hydrous piceus*, the largest beetle in Europe [17,42]. Apart from these predaceous groups, other strictly aquatic families can be found in paddy fields: Elmidae, Dryopidae, Helophoridae, Limnebiidae. Furthermore, Chrysomelidae Donacinae are occasionally found in northern Italian rice fields [43–45]. From another perspective, Curculionidae and Erirhinidae are groups of growing importance in Italian rice fields, because of the arrival of the invasive *Sitophilus oryzae* and *Lissorhoptrus oryzophilus* [46,47], which represents a serious agricultural problem and requires dedicated and continuous management.

A few Lepidoptera are reported as potential pests in rice cultivation. In particular *Ostrinia nubilalis*, *Paraponyx* spp. (Crambidae) and *Mythimna unipuncta* (Noctuidae) represent a threat because of their stem boring habits [48,49].

Apart from more faunistic oriented studies (Figure 3), there are also several studies with a broader focus, ranging from older research (e.g., [50]) to those that are more recently published (e.g., [51–53]).

Figure 3. Some common aquatic insects in Italian rice fields. Examples, from left to right: *Hydroporus* sp. (Dytiscidae), *Nepa cinerea* (Nepidae), *Laccobius* sp. (Hydrophilidae).

However, despite the large territorial extension, studies on the rice field's freshwater invertebrates in Italy are few, are often dated and are mostly spread out, with an important diffusion in grey literature.

In Table 2 we show the results of a Scopus query on selected keywords related to aspects of applied entomology, targeting scientific publications produced globally or focused on the Italian situation. In this context, the scarcity of studies on the Italian context is presumably attributable to two factors. Firstly, there appears to be an effective absence of research in this area. Secondly, and perhaps more pertinently, these studies are predominantly disseminated through non-international journals and grey literature (as evident in our References section). About 80% of the studies related to paddy-rice entomofauna in Italy are published in non-ISI journals.

Table 2. Number of scientific papers (Scopus) related to the main applied entomology keywords in rice fields (February 2025).

Keywords	N° Publications Worldwide	N° Publications Related to Italy
Rice + aquatic Insects	197	2
Rice fields + aquatic Insects	152	3
Rice fields + freshwater invertebrates	17	1
Rice fields + Diptera Culicidae	202	3
Rice fields + Coleoptera Curculionidae	94	2
Rice fields + *Bacillus thuringensis*	258	4
Rice fields + insecticides	1260	12
Rice fields + *Diflubenzoron*	11	1

3. Rice Fields and Insect Biodiversity in the Current Global Change Scenario: Issues, Problems and New Agricultural Approaches

As previously mentioned, Italian rice fields represent an important biodiversity hotspot for some groups of insects, e.g., [26,40]. In particular, they represent the only and last lentic or semi-lentic systems of the lowland areas of the Po Valley, once characterized by abandoned meanders, wetlands and oxbows, and gradually transformed over the centuries into one of the most important and productive agricultural landscapes in the world [54]. However, the management of rice fields is a contentious issue, particularly with regard to water supply. While the presence of water in the rice fields ensures the maintenance of the aforementioned biodiversity, in the context of the current climate change, there is an increasing need to decrease water consumption in agriculture, especially because the rivers in this area can be under multiple stressors (both natural and anthropogenic) for long periods of the year. For example, the decline in snowfall in the Alps and the marked increase in both air and water temperatures have indeed precipitated a disruption in the hydrological regime of most northern Italian rivers, which now exhibit diminished flow rates and frequently experience a lack of surface water for several months each year. This phenomenon, characterized by the adoption of an intermittent flow pattern, signifies a true "Mediterraneanisation" that has prompted numerous rivers to transition from a perennial to an abnormal intermittent regime, e.g., [55]. This phenomenon carries profound implications for the biodiversity of lotic environments [56] and the chemical, microbiological, and ecological quality of the affected water bodies [57]. Considering these challenges, there is an urgent need to explore novel sustainable agricultural techniques in rice cultivation.

For instance, it should be noticed that, in the rice-growing area of the Po valley, it has recently become a "good practice" to carry out winter flooding of the paddy fields after rice cutting (Figure 4).

(a) (b) (c)

Figure 4. Different phases of the agricultural cycle in rice fields, depicting (**a**) winter flooding, (**b**) spring low-level maintenance and (**c**) late summer pre-cutting drought.

In this way, the water demand in spring is greatly reduced and greenhouse gas emissions fall significantly [52,58,59]. The reduced need for water withdrawal from surrounding lotic systems like rivers and streams is another positive aspect of crucial importance for regional freshwater biodiversity. Considering data from recent decades on temperatures and patterns of snow and rain, this agricultural practice seems to be a promising way to follow in the coming years. Consequently, even during the winter months, rice paddies have the capacity to provide habitats for a variety of invertebrates, thereby enhancing the "metabolic power" of these human-made lentic ecosystems. Moreover, at these latitudes, e.g., [60], the flooding in the cold months rather than in the summer months mimics a more natural condition that can be beneficial to the indigenous fauna and detrimental to alien species (see details below).

4. Rice Fields and Their Role in the Biological Invasion Context

Rice fields also represent an important context of study regarding invasion biology. In fact, artificial and highly modified water bodies frequently experience the arrival of multiple introduced species, via intentional and unintentional releases, due to their landscape position (e.g., in an anthropogenic matrix) and level of connectivity [61].

In any geographical context, rice fields act as intermittent, human-regulated systems characterized by alternating dry and wet phases. This highly dynamic condition can be particularly stressful for exclusively aquatic invaders, yet in specific contexts, it may allow for multiple colonizations through different phases of the hydroperiod [62].

Aquatic insects represent a dominant component of the invertebrate fauna; however, they are rarely invasive in freshwater ecosystems [5], with few notable exceptions, such as the Asian tiger mosquito (*Aedes albopicutus*). Flooded rice fields are not an exception to this, with fish, crustaceans and mollusks identified as the main aquatic invasive taxa so far (see references below).

Rice fields in Mediterranean areas undergo summer inundation, a phenomenon rarely observed in natural temporary water bodies in the region, which may specifically facilitate invasion by warm and (sub-) tropical species. For instance, several non-native crustaceans (e.g., ostracods) have been passively dispersed through rice cultivation from Asia to European regions such as Spain and Italy, e.g., [63,64]. Similarly, seasonally flooded rice fields in the Ebro Delta (N Spain) have been invaded by the South American apple snail *Pomacea maculate*, which is quite resistant to high temperatures and dry conditions [65], as well as by numerous invasive fish and the red swamp crayfish *Procambarus clarkii* [66]. The latter example being particularly challenging due to the crayfish's digging behavior and the structural damage they can cause to draining structures [67], as well as their consumption of rice seed and plants [68].

In temperate northern Italy, spring inundations were the most common practice, creating temporal lowland ecosystems in an intensive agricultural landscape, often rich in plant and animal invaders. Semi-aquatic species, like those of the family Erirhinidae (Coleoptera), have been particularly highlighted in rice fields. The abovementioned Rice Water Weevil, *Lissorhoptrus oryzophilus* (Kuschel), native of North America, has been detected in most of the Italian paddy fields area due to both the active dispersal of adults and accidental movements caused by human transportation [69,70]. In this context, the use of chemical products like pyrethroid insecticides (e.g., alfacipermetrine) seems to control the species' population in the short term, despite affecting other aquatic life forms (e.g., insects and other invertebrates [12]). The occurrences of the aquatic fern *Azolla filiculoides* in Italy, which is native to the warm, temperate and subtropical Americas, suggest the necessity of further research specifically on the potentially co-occurring introduction of the species *Stenopelmus rufinasus* (Curculionidae), a semi-aquatic specialist herbivore known to feed on its leaves and which is already recorded in other Mediterranean countries [71]. This is of special concern considering the use of *Azolla* as a biofertilizer in rice fields outside its native range (e.g., Italy [72]). Recently, the presence of adults of *Halyomorpha halys* (Stål) (Hemiptera, Pentatomidae), native to East Asia, feeding on panicles has been highlighted in northern Italy and this marks the first evidence of an association between this species and rice, a crop not previously recorded as a host plant [73]. Overall, these ecosystems, both in Italy and other geographic contexts, can easily serve as gateway for the arrival and spread of species, facilitating the secondary dispersions of non-native species in other ecosystems (e.g., nearby wetlands [74]), highlighting the need for more interdisciplinary research on this topic.

5. Future Perspectives in Rice Field Entomological Research

Monitoring insects in rice fields with traditional methods is challenging because of sampling- and identification-related issues. Collecting insects from large areas requires a huge sampling effort, especially considering that sampling must be repeated over time to cope with species-specific life cycles [75]. This problem can lead to rare species, taxa with a localized distribution or pests at the initial stages of outbreak not being effectively detected. Moreover, most aquatic insects are present in water at the larval/nymphal stage (e.g., dipterans, mayflies, dragonflies) for which the identification at species level is limited by the lack of diagnostic characters and suitable identification keys. In recent years, DNA barcoding has emerged as a valuable and promising tool to overcome the above-mentioned obstacles, e.g., [76]. DNA metabarcoding is a technique for the genetic identification of organisms from a composite sample of organisms (bulk metabarcoding) or an environmental matrix (environmental DNA [77]). Metabarcoding usually archives a higher taxonomic resolution than morphological identification and differentiates cryptic species lacking morphological diagnostic characters [78,79]. Moreover, it can be used for studying intraspecific diversity for inferring biogeographical patterns [80,81] or fine-scale community assembly processes [76].

Although genetic identification of insects through DNA metabarcoding is appealing, only a limited number of studies have used this technique in rice fields. In rice fields, environmental DNA has been successfully used for targeting vertebrates, including snakes [82], anura [83] and birds [84]. It has also been used for the detection of insects such as the charismatic giant water bug *Kirkaldyia deyrolli* (Heteroptera: Belostomatidae [85]) and parasitoid wasps [86], though few works have targeted the entire insect community or groups of orders [87]. Interestingly, occurrences of insects in rice fields can also be obtained by targeting the stomach contents of fish (in case of presence) when performing diet analysis, e.g., [88]. DNA metabarcoding is a cost-effective technique that can be used in monitoring the biodiversity of rice fields. However, it suffers from some limitations that must be addressed when designing an eDNA monitoring campaign. Small or rare taxa can still be missed, especially when using bulk metabarcoding, where the biomass of abundant taxa can exceed those of small and rare taxa by orders of magnitude. Problems with reference datasets used for the taxonomic assignment of sequences can potentially affect the number of species found with DNA metabarcoding [89,90]. For example, sequences assigned to the wrong species and the lack of sequences for some species are major problems that will probably be mitigated in the near future due to the ongoing barcoding and metabarcoding initiatives [91]. Currently, DNA metabarcoding can be used as a complement to the traditional monitoring of rice fields, with the aim to create more comprehensive species inventories or as an early detection method of both agriculture pests and non-native species in general.

6. Conclusions

There is a great amount of evidence that climate change is profoundly altering the characteristics and dynamics of natural systems on a global scale [92]. Freshwater environments have been identified as being particularly vulnerable to climate change, due to the increase in water temperatures and the disruption of hydrologic cycles, with implications for their biodiversity [93–95]. This phenomenon can be potentially evident also in man-made aquatic agro-ecosystems, such as rice fields, where fluctuations in precipitation, temperature, and evaporation have been shown to exert a pivotal role, both directly (e.g., by altering environmental conditions) and indirectly (e.g., by leading to variations in agricultural practices). The aforementioned problem is of particular significance within the study area of this research. Indeed, Italy's rice-growing area, which is one of the largest in the world outside of Asia, coincides with one of the most anthropized regions on the planet and is fed by water from the Alps, one of the areas in which climate change is occurring at a significantly faster rate [96].

In the near future, strategies that reduce the demand for irrigation, promote the cultivation of drought-tolerant crops and utilize precision farming techniques will probably be implemented. Consequently, it is likely that the Italian rice-growing areas will experience a significant reduction in wetland areas and aquatic habitats. The main challenge, for multiple stakeholders with diverse interests, will inevitably be to, where possible, balance the need to reduce water diversion from rivers with the maintenance of permanent wetlands, in order to support this essential role of aquatic insects in regional biodiversity and in freshwater metabolic processes.

This review, in addition to underlining the importance of rice paddies from a strictly biological and conservation point of view, seeks to represent a small starting point to provide management guidance and tools for cultivating more sustainably agricultural environments essential for human livelihood. The implementation of monitoring, verification and control plans in sample rice growing areas can be a first step towards understanding, protecting and combining the conservation of species with more environmentally friendly agricultural practices, without affecting the final yield of the crop. Finally, greater knowledge and presence in the field can promptly signal the arrival of alien insect species that are potentially harmful and invasive.

Author Contributions: Conceptualization, investigation, writing, review and editing, all authors; funding acquisition, T.B. and S.F. All authors have read and agreed to the published version of the manuscript.

Funding: This research was supported by the Agritech National Research Centre related to Spoke 4 "*Multifunctional and resilient agriculture and forestry systems for the mitigation of climate change risks*" funded by the European Union Next-generation EU (PNRR)—Mission 4 Component 2, investment 1.4—D.D. 1032 17/06/2022, CN00000022.

Data Availability Statement: No new data were created or analyzed in this study.

Acknowledgments: The authors thank E. Guafa and A. Millán (University of Murcia) for their useful suggestions, M. Marcucci for his kind support, and A. Morisi for the photos of invertebrates and the valuable and continuous teachings.

Conflicts of Interest: The authors declare no conflict of interest.

References

1. FAO—Food and Agricultural Organization of the United Nations. *OECD-FAO Agricul Outlook 2011–2030*; FAO: Rome, Italy, 2009.
2. Kumar, N.; Chhokar, R.S.; Meena, R.P.; Kharub, A.S.; Gill, S.C.; Tripathi, S.C.; Gupta, O.P.; Mangrauthia, S.K.; Sundaram, R.M.; Sawant, C.P.; et al. Challenges and opportunities in productivity and sustainability of rice cultivation system: A critical review in Indian perspective. *Cereal Res. Commun.* **2021**, *50*, 573–601. [CrossRef] [PubMed]
3. Muthayya, S.; Sugimoto, J.D.; Montgomery, S.; Maberly, G.F. An overview of global rice production, supply, trade, and consumption. *Ann. N. Y. Acad. Sci.* **2014**, *1324*, 7–14. [CrossRef]
4. Chester, E.T.; Robson, B.J. Anthropogenic refuges for freshwater biodiversity: Their ecological characteristics and management. *Biol. Conserv.* **2013**, *166*, 64–75. [CrossRef]
5. Fenoglio, S.; Bonada, N.; Guareschi, S.; López-Rodríguez, M.J.; Millán, A.; de Figueroa, J.M.T. Freshwater ecosystems and aquatic insects: A paradox in biological invasions. *Biol. Lett.* **2016**, *12*, 20151075. [CrossRef] [PubMed]
6. Edirisinghe, J.P.; Bambaradeniya, C.N. Rice fields: An ecosystem rich in biodiversity. *J. Natl. Sci. Found. Sri Lanka* **2006**, *34*, 57. [CrossRef]
7. Hesler, L.S.; Grigarick, A.A.; Oraze, M.J.; Palrang, A.T. Arthropod fauna of conventional and organic rice fields in California. *J. Econ. Entomol.* **1993**, *86*, 149–158. [CrossRef]
8. Herlinda, S.; Karenina, T.; Irsan, C.; Pujiastuti, Y. Arthropods inhabiting flowering non-crop plants and adaptive vegetables planted around paddy fields of freshwater swamps of South Sumatra, Indonesia. *Biodiversitas* **2019**, *20*, 3328–3339. [CrossRef]

9. Hendawi, A.; Sherif, M.; Abada, A.; El-Abashi, M. Aquatic and semiaquatic insects occurring in Egyptian rice fields and hazardous effects of insecticides. *Egypt. J. Agric. Res.* **2005**, *83*, 493–501.
10. Service, M.W. Mortalities of the immature stages of species B of the *Anopheles gambiae* complex in Kenya: Comparison between rice fields and temporary pools, identification of predators, and effects of insecticidal spraying. *J. Med. Entomol.* **1997**, *13*, 535–545. [CrossRef]
11. Panizzon, J.P.; Macedo, V.R.M.; Machado, V.; Fiuza, L.M. Microbiological and physical–chemical water quality of the rice fields in Sinos River's basin, Southern Brazil. *Environ. Monit. Assess.* **2013**, *185*, 2767–2775. [CrossRef]
12. Suhling, F.; Befeld, S.; Häusler, M.; Katzur, K.; Lepkojus, S.; Mesleard, F. Effects of insecticide applications on macroinvertebrate density and biomass in rice-fields in the Rhône-delta, France. *Hydrobiologia* **2000**, *431*, 69–79. [CrossRef]
13. Fasola, M.; Ruiz, X. The Value of Rice Fields as Substitutes for Natural Wetlands for Waterbirds in the Mediterranean Region. *Col. Waterbirds* **1996**, *19*, 122–128. [CrossRef]
14. Longoni, V. Rice fields and waterbirds in the Mediterranean region and the Middle East. *Waterbirds* **2010**, *33* (Suppl. S1), 83–96. [CrossRef]
15. Del Guercio, G. I Friganeidi nuocciono al riso. I tafani del riso. Le larve delle tipule nocive al riso (In Italian). *Redia* **1911**, *7*, 466–467.
16. Cavazza, F. Ricerche intorno alle specie dannose alla coltivazione del riso (*Oryza sativa*) e specialmente al Chironomus cavazzai Kieff. *Boll. Lab. Zool. Gen. Agric.* **1914**, *6*, 320–331.
17. Moretti, G.P. Note sulla fauna entomologica delle risaie. *Atti Soc. Ital. Sci. Nat.* **1932**, *71*, 61–85.
18. Moretti, G.P. I tricotteri delle risaie. *Atti Soc. Ital. Sci. Nat.* **1934**, *73*, 104–116.
19. Cocchi, G. Ricerche sui Ditteri Chironomidi dannosi al riso nella bassa Bolognese. *Boll. Oss. Mal. Piante Bologna* **1966**, *1*, 39–64.
20. Goidanich, A. Contributi alla conoscenza dell'entomofauna di risaia. 1. Gli Straziomidi: Mancati nemici del riso. *Risicoltura* **1939**, *29*, 221–230.
21. Zangheri, S. Un dittero minatore del riso nel basso ferrarese (*Hydrellia griseola* Fallen, Dipt. *Ephydridae*). *Boll. Soc. Entomol. Ital.* **1956**, *86*, 12–16.
22. Giugliano, L.; Hardersen, S.; Santini, G. Odonata communities in retrodunal ponds: A comparison of sampling methods. *Int. J. Odonatol.* **2012**, *15*, 13–23. [CrossRef]
23. Golfieri, B.; Hardersen, S.; Maiolini, B.; Surian, N. Odonates as indicators of the ecological integrity of the river corridor: Development and application of the Odonate River Index (ORI) in northern Italy. *Ecol. Indic.* **2016**, *61*, 234–247. [CrossRef]
24. Giuliano, D.; Bogliani, G. Odonata in rice agroecosystems: Testing good practices for their conservation. *Agric. Ecosyst. Environ.* **2019**, *275*, 65–72. [CrossRef]
25. Lupi, D.; Savoldelli, S.; Rocco, A.; Rossaro, B. Italian rice agroecosystems: A threat to insect biodiversity? *Landsc. Manag. Funct. Biodivers. IOBC Bull.* **2012**, *75*, 127–131.
26. Lupi, D.; Rocco, A.; Rossaro, B. Benthic macroinvertebrates in Italian rice fields. *J. Limnol.* **2013**, *72*, 184–200. [CrossRef]
27. Giuliano, D.; Cardarelli, E.; Bogliani, G. Grass management intensity affects butterfly and orthopteran diversity on rice field banks. *Agric. Ecosyst. Environ.* **2018**, *267*, 147–155. [CrossRef]
28. Cianferoni, F.; Graziani, F.; Dioli, P.; Ceccolini, F. Review of the occurrence of *Halyomorpha halys* (Hemiptera: Heteroptera: Pentatomidae) in Italy, with an update of its European and World distribution. *Biologia* **2018**, *73*, 599–607. [CrossRef]
29. AA.VV. "Il riso", coordinamento scientifico di A. Ferrero. In *Collana Coltura & Cultura*; Bologna Script, Ed.; ideata e coordinata da R. Angelini; Bayer CropScience: Bologna, Italy, 2008.
30. Bellini, R.; Pederzani, F.; Pilani, R.; Veronesi, R.; Maini, S. *Hydroglyphus pusillus* (Fabricius) (Coleoptera Dytiscidae): Its role as a mosquito larvae predator in rice fields. *Boll. Dell'istituto Di Entomol. 'G. Grandi' Dell' Univ. Di Bologna* **2000**, *54*, 155–163.
31. Toma, L.; Cipriani, M.; Goffredo, M.; Romi, R.; Lelli, R. First report on entomological field activities for the surveillance of West Nile disease in Italy. *Vet. Ital* **2008**, *44*, 499–512.
32. Di Luca, M.; Boccolini, D.; Severini, F.; Toma, L.; Barbieri, F.M.; Massa, A.; Romi, R. A 2-year entomological study of potential malaria vectors in Central Italy. *Vector-Borne Zoonotic Dis.* **2009**, *9*, 703–711. [CrossRef]
33. Boccolini, D.; Toma, L.; Luca, M.D.; Severini, F.; Cocchi, M.; Bella, A.; Romi, R. Impact of environmental changes and human-related factors on the potential malaria vector, *Anopheles labranchiae* (Diptera: Culicidae), in Maremma, Central Italy. *J. Med. Entomol.* **2012**, *49*, 833–842. [CrossRef] [PubMed]
34. Rumpf, S.B.; Gravey, M.; Brönnimann, O.; Luoto, M.; Cianfrani, C.; Mariethoz, G.; Guisan, A. From white to green: Snow cover loss and increased vegetation productivity in the European Alps. *Science* **2022**, *376*, 1119–1122. [CrossRef] [PubMed]
35. Rosà, R.; Marini, G.; Bolzoni, L.; Neteler, M.; Metz, M.; Delucchi, L.; Rizzoli, A. Early warning of West Nile virus mosquito vector: Climate and land use models successfully explain phenology and abundance of *Culex pipiens* mosquitoes in north-western Italy. *Parasites Vectors* **2014**, *7*, 269. [CrossRef] [PubMed]
36. Calzolari, M.; Mosca, A.; Montarsi, F.; Grisendi, A.; Scremin, M.; Roberto, P.; Albieri, A. Distribution and abundance of *Aedes caspius* (Pallas, 1771) and *Aedes vexans* (Meigen, 1830) in the Po Plain (northern Italy). *Parasites Vectors* **2024**, *17*, 452. [CrossRef]

37. González, M.A.; Chaskopoulou, A.; Georgiou, L.; Frontera, E.; Cáceres, F.; Masia, M.; Figuerola, J. Mosquito management strategies in European rice fields: Environmental and public health perspectives. *J. Environ. Manag.* **2024**, *370*, 122534. [CrossRef]
38. Lupi, D.; Jucker, C.; Rocco, A. Rice fields as a hot spot of water beetles (Coleoptera Adephaga and Polyphaga). *Redia* **2014**, *97*, 95–112.
39. Zanella, F. Biocenosi delle risaie con particolare riferimento ai Culicidi. *Disinfest. Ig. Ambient.* **2000**, *17*, 12–20.
40. Bo, T.; Fenoglio, S. Biodiversità e gestione delle risaie: Un caso di studio inerente alle comunità di invertebrati acquatici della Lomellina (PV). *Pianura* **2024**, *44*, 112–121.
41. Bosi, G.; Bo, T.; Fenoglio, S. Alcune considerazioni sulla distribuzione di Noteridae e Dityscidae (Coleoptera) nella provincia di Alessandria. *Riv. Piemont. Stor. Nat.* **2009**, *30*, 79–93.
42. Campadelli, G. Due abitatori dell'acqua: *Hydrous piceus* L. e *Dytiscus marginalis* L. *Nat. Mont.* **1982**, *29*, 59–63.
43. Campaioli, S.; Ghetti, P.F.; Minelli, A.; Ruffo, S. *Manuale per il Riconoscimento dei Macroinvertebrati Delle Acque Dolci Italiane (Vol. I)*; Provincia Autonoma di Trento: Trento, Italy, 1994.
44. Campaioli, S.; Ghetti, P.F.; Minelli, A.; Ruffo, S. *Manuale per il Riconoscimento dei Macroinvertebrati Delle Acque Dolci Italiane (Vol. II)*; Provincia Autonoma di Trento: Trento, Italy, 1999.
45. Giudici, M.L.; Villa, B. The rice leaf bug, *Trigonotylus caelestialium* Kirkaldy, on rice in Italy. In Proceedings of the International Temperate Rice Conference, Novara, Italy, 25–28 June 2007; pp. 146–147.
46. Gelosi, A. Punteruolo del riso (*Sitophilus oryzae* Linneus) [Biologia e lotta]. *Inf. Fitopatol.* **1982**, *32*, 31–34.
47. Lupi, D.; Cenghialta, C.; Giudici, M.L.; Villa, B.; Tabacchi, M. Prime acquisizioni sulla biologia e sul contenimento di *Lissorhoptrus oryzophilus* (punteruolo acquatico del riso) in Lombardia. In *Atti Giornate Fitopatologiche*; CLUEB: Bologna, Italy, 2008; pp. 245–246.
48. Giudici, M.L.; Villa, B. The armyworm *Mythimna unipuncta* (Haworth) found on rice in Italy. In Proceedings of the Conference "Challenges and Opportunities for Sustainable Rice-Based Production Systems", Turin, Italy, 13–15 September 2004; pp. 13–15.
49. Magagnoli, S.; Lanzoni, A.; Masetti, A.; Depalo, L.; Albertini, M.; Ferrari, R.; Burgio, G. Sustainability of strategies for *Ostrinia nubilalis* management in Northern Italy: Potential impact on beneficial arthropods and aflatoxin contamination in years with different meteorological conditions. *Crop Prot.* **2021**, *142*, 105529. [CrossRef]
50. Supino, F. Note sulla fauna delle risaie. *R. Ist. Lomb. Sci. Lett.* **1932**, *54*, 1–12.
51. Savini, D.; Occhipinti-Ambrogi, A. Bad Moon Rising: Il gambero rosso della Louisiana, una minaccia per gli ecosistemi acquatici della Lombardia. *Mem. Della Soc. Ital. Sci. Nat. Mus. Civ. Stor. Nat. Milano* **2008**, *36*, 19–20.
52. Kraehmer, H.; Thomas, C.; Vidotto, F. Rice production in Europe. In *Rice Production Worldwide*; Springer: Cham, Switzerland, 2017; pp. 93–116. [CrossRef]
53. Rossaro, B.; Marziali, L.; Cortesi, P. The effects of tricyclazole treatment on aquatic invertebrates in a rice paddy field. *CLEAN–Soil Air Water* **2014**, *42*, 29–35. [CrossRef]
54. Guareschi, S.; Laini, A.; Viaroli, P.; Bolpagni, R. Integrating habitat-and species-based perspectives for wetland conservation in lowland agricultural landscapes. *Biodivers. Conserv.* **2020**, *29*, 153–171. [CrossRef]
55. Piano, E.; Doretto, A.; Falasco, E.; Fenoglio, S.; Gruppuso, L.; Nizzoli, D.; Viaroli, P.; Bona, F. If Alpine streams run dry: The drought memory of benthic communities. *Aquat. Sci.* **2019**, *81*, 32. [CrossRef]
56. Gruppuso, L.; Falasco, E.; Fenoglio, S.; Marino, A.; Nizzoli, D.; Piano, E.; Bona, F. Dataset of a flow intermittency study: Benthic communities of 13 alpine intermittent rivers. *Data Brief* **2024**, *54*, 110449. [CrossRef]
57. Marino, A.; Bertolotti, S.; Macrì, M.; Bona, F.; Bonetta, S.; Falasco, E.; Fenoglio, S. Impact of wastewater treatment and drought in an Alpine region: A multidisciplinary case study. *Heliyon* **2024**, *10*, e35290. [CrossRef]
58. Vitali, A.; Moretti, B.; Lerda, C.; Said-Pullicino, D.; Celi, L.; Romani, M.; Fogliatto, S.; Vidotto, F. Conservation tillage in temperate rice cropping systems: Crop production and soil fertility. *Field Crops Res.* **2024**, *308*, 109–276. [CrossRef]
59. Vitali, A.; Russo, F.; Moretti, B.; Romani, M.; Vidotto, F.; Fogliatto, S.; Celi, L.; Said-Pullicino, D. Interaction between water, crop residue and fertilization management on the source-differentiated nitrogen uptake by rice. *Biol. Fertil. Soils* **2024**, *60*, 757–772. [CrossRef]
60. Monteleone, B.; Borzí, I. Drought in the Po Valley: Identification, Impacts and Strategies to Manage the Events. *Water* **2024**, *16*, 1187. [CrossRef]
61. Francis, R.A.; Chadwick, M.A. Invasive alien species in freshwater ecosystems: A brief overview. In *A Handbook of Global Freshwater Invasive Species*; Francis, R.A., Ed.; Routledge: London, UK, 2012; pp. 3–24.
62. Guareschi, S.; South, J. Biological invasions in intermittent rivers and streams: Current knowledge, and future frontiers. *Ecosistemas* **2024**, *33*, 2600. [CrossRef]
63. Rossi, V.; Benassi, G.; Veneri, M.; Bellavere, C.; Menozzi, P.; Moroni, A.; Mckenzie, K.G. Ostracoda of the Italian ricefields thirty years on: New synthesis and hypothesis. *J. Limnol.* **2003**, *62*, 1–8. [CrossRef]
64. Valls, L.; Rueda, J.; Mesquita-Joanes, F. Rice fields as facilitators of freshwater invasions in protected wetlands: The case of Ostracoda (Crustacea) in the Albufera Natural Park (E Spain). *Zool. Stud.* **2014**, *53*, 68. [CrossRef]

65. Oosterom, M.V.L.-V.; Casas-Ruiz, J.P.; Gampe, D.; López-Robles, M.A.; Ludwig, R.; Núñez-Marcé, A.; Muñoz, I. Responses of a native and a recent invader snail to warming and dry conditions: The case of the lower Ebro River. *Aquat. Ecol.* **2019**, *53*, 497–508. [CrossRef]
66. Clavero, M.; López, V.; Franch, N.; Pou-Rovira, Q.; Queral, J.M. Use of seasonally flooded rice fields by fish and crayfish in a Mediterranean wetland. *Agric. Ecosyst. Environ.* **2015**, *213*, 39–46. [CrossRef]
67. Arce, J.A.; Diéguez-Uribeondo, J. Structural damage caused by the invasive crayfish *Procambarus clarkii* (Girard, 1852) in rice fields of the Iberian Peninsula: A study case. *Fundam. Appl. Limnol.* **2015**, *186*, 259–269. [CrossRef]
68. Souty-Grosset, C.; Anastácio, P.M.; Aquiloni, L.; Banha, F.; Choquer, J.; Chucholl, C.; Tricarico, E. The red swamp crayfish *Procambarus clarkii* in Europe: Impacts on aquatic ecosystems and human well-being. *Limnologica* **2016**, *58*, 78–93. [CrossRef]
69. Lupi, D.; Colombo, M.; Giudici, M.L.; Villa, B.; Cenghialta, C.; Passoni, D. On the spatial spread of the rice water weevil, *Lissorhoptrus oryzophilus* Kuschel (Coleoptera: Erirhinidae), in Italy. *J. Ent. Acar. Res.* **2010**, *42*, 81–90. [CrossRef]
70. Lupi, D.; Jucker, C.; Rocco, A.; Giudici, M.L.; Boattin, S.; Colombo, M. Current status of the rice water weevil *Lissorhoptrus oryzophilus* in Italy: Eleven-year invasion. *EPPO Bull.* **2015**, *45*, 123–127. [CrossRef]
71. Dana, D.; Viva, S. Stenopelmus rufinasus Gyllenhal 1836 (Coleoptera: Erirhinidae) naturalized in Spain. *Coleopt. Bull.* **2006**, *60*, 41–42. [CrossRef]
72. Bocchi, S.; Malgioglio, A. Azolla-Anabaena as a biofertilizer for rice paddy fields in the Po Valley, a temperate rice area in Northern Italy. *Int. J. Agron.* **2010**, *2010*, 152158. [CrossRef]
73. Lupi, D.; Dioli, P.; Limonta, L. First evidence of *Halyomorpha halys* (Stål)(Hemiptera Heteroptera, Pentatomidae) feeding on rice (*Oryza sativa* L.). *J. Entomol. Res.* **2017**, *49*, 67–71. [CrossRef]
74. Van Leeuwen, C.H.; Huig, N.; Van der Velde, G.; Van Alen, T.A.; Wagemaker, C.A.; Sherman, C.D.; Figuerola, J. How did this snail get here? Several dispersal vectors inferred for an aquatic invasive species. *Freshw. Biol.* **2013**, *58*, 88–99. [CrossRef]
75. Preti, M.; Verheggen, F.; Angeli, S. Insect pest monitoring with camera-equipped traps: Strengths and limitations. *J. Pest. Sci.* **2021**, *94*, 203–217. [CrossRef]
76. Laini, A.; Stubbington, R.; Beermann, A.J.; Burgazzi, G.; Datry, T.; Viaroli, P.; Wilkes, M.; Zizka VM, A.; Saccò, M.; Leese, F. Dissecting biodiversity: Assessing the taxonomic, functional and phylogenetic structure of an insect metacommunity in a river network using morphological and metabarcoding data. *Eur. J. Biol.* **2023**, *90*, 320–332. [CrossRef]
77. Taberlet, P.; Coissac, E.; Pompanon, F.; Brochmann, C.; Willerslev, E. Towards next-generation biodiversity assessment using DNA metabarcoding. *Mol. Ecol.* **2012**, *21*, 2045–2050. [CrossRef]
78. Gauthier, M.; Konecny-Dupré, L.; Nguyen, A.; Elbrecht, V.; Datry, T.; Douady, C.; Lefébure, T. Enhancing DNA metabarcoding performance and applicability with bait capture enrichment and DNA from conservative ethanol. *Mol. Ecol. Resour.* **2020**, *20*, 79–96. [CrossRef]
79. Jones, F.C. Taxonomic sufficiency: The influence of taxonomic resolution on freshwater bioassessments using benthic macroinvertebrates. *Environ. Rev.* **2008**, *16*, 45–69. [CrossRef]
80. Elbrecht, V.; Vamos, E.E.; Steinke, D.; Leese, F. Estimating intraspecific genetic diversity from community DNA metabarcoding data. *PeerJ* **2018**, *6*, e4644. [CrossRef] [PubMed]
81. Turon, X.; Antich, A.; Palacín, C.; Præbel, K.; Wangensteen, O.S. From metabarcoding to metaphylogeography: Separating the wheat from the chaff. *Ecol. Appl.* **2020**, *30*, e02036. [CrossRef]
82. Nishizawa, R.; Nakao, R.; Ushimaru, A.; Minamoto, T. Development of environmental DNA detection assays for snakes in paddy fields in Japan. *Landsc. Ecol. Eng.* **2023**, *19*, 3–10. [CrossRef]
83. Kim, K.; Kwon, S.; Jang, Y. Can eDNA Present in Aquatic Environments of Rural Areas Help Identify Species Diversity in the Order Anura? *Water* **2024**, *16*, 21. [CrossRef]
84. Katayama, N.; Yamamoto, S.; Baba, Y.G.; Ito, K.; Yamasako, J. Complementary role of environmental DNA for line-transect bird surveys: A field test in a Japanese rice landscape. *Ecol. Indic.* **2024**, *166*, 112442. [CrossRef]
85. Ogata, S.; Nishiwaki, A.; Yamazoe, K.; Sugai, K.; Takahara, T. Discovery of unknown new ponds occupied by the endangered giant water bug (Hemiptera: Heteroptera: Belostomatidae) by combining environmental DNA and capture surveys. *Entomol. Sci.* **2023**, *26*, e12540. [CrossRef]
86. Wang, L.; Wu, H.; He, W.; Lai, G.; Li, J., Liu, S.; Zhou, Q. Diversity of Parasitoid Wasps and Comparison of Sampling Strategies in Rice Fields Using Metabarcoding. *Insects* **2024**, *15*, 4. [CrossRef]
87. Kim, K.; Kwon, S.; Noh, A. Detection of frog and aquatic insects by environmental DNA in paddy water ecology. *Korean J. Agric. Sci.* **2023**, *50*, 299–312. [CrossRef]
88. Wu, M.; Lu, R.; Huang, W.; Liu, H.; Zou, Y.; Tao, L.; Sun, Y.; Wang, Q.; Tang, K. Major diet of common carp (*Cyprinus carpio* L.) over different developmental stages in rice-field: Agroecological interactions between fishes and rice in Sichuan, China, based on DNA metabarcoding approach. *Glob. Ecol. Conserv.* **2024**, *56*, e03298. [CrossRef]
89. Elbrecht, V.; Peinert, B.; Leese, F. Sorting things out: Assessing effects of unequal specimen biomass on DNA metabarcoding. *Ecol. Evol.* **2017**, *7*, 6918–6926. [CrossRef]

90. Keck, F.; Couton, M.; Altermatt, F. Navigating the seven challenges of taxonomic reference databases in metabarcoding analyses. *Mol. Ecol. Resour.* **2023**, *23*, 742–755. [CrossRef] [PubMed]
91. Weigand, H.; Beermann, A.J.; Čiampor, F.; Costa, F.O.; Csabai, Z.; Duarte, S.; Geigerg, M.F.; Grabowski, M.; Rimet, F.; Rulik, B.; et al. DNA barcode reference libraries for the monitoring of aquatic biota in Europe: Gap-analysis and recommendations for future work. *Sci. Total Environ.* **2019**, *678*, 499–524. [CrossRef] [PubMed]
92. IPCC. *Climate Change 2021: The Physical Science Basis. Contribution of Working Group I to the Sixth Assessment Report of the Intergovernmental Panel on Climate Change*; Cambridge University Press: Cambridge, UK; New York, NY, USA, 2021. [CrossRef]
93. Heino, J.; Virkkala, R.; Toivonen, H. Climate change and freshwater biodiversity: Detected patterns, future trends and adaptations in northern regions. *Biol. Rev.* **2009**, *84*, 39–54. [CrossRef] [PubMed]
94. Fenoglio, S.; Bo, T.; Cucco, M.; Mercalli, L.; Malacarne, G. Effects of global climate change on freshwater biota: A review with special emphasis on the Italian situation. *Ital. J. Zool.* **2010**, *77*, 374–383. [CrossRef]
95. Dudgeon, D.; Strayer, D.L. Bending the curve of global freshwater biodiversity loss: What are the prospects? *Biol. Rev.* **2025**, *100*, 205–226. [CrossRef]
96. Gobiet, A.; Kotlarski, S.; Beniston, M.; Heinrich, G.; Rajczak, J.; Stoffel, M. 21st century climate change in the European Alps—A review. *Sci. Total Environ.* **2014**, *493*, 1138–1151. [CrossRef]

water

MDPI

Review

Natural Factors of Microplastics Distribution and Migration in Water: A Review

Xianjin An [1,2,*], Yanling Wang [1,2], Muhammad Adnan [3,4], Wei Li [5] and Yaqin Zhang [1,2]

1 School of Karst Science, Guizhou Normal University, Guiyang 550001, China
2 State Engineering Technology Institute for Karst Desertification Control, Guiyang 550001, China
3 State Key Laboratory of Environmental Geochemistry, Institute of Geochemistry, Chinese Academy of Sciences, Guiyang 550081, China
4 University of Chinese Academy of Sciences, Beijing 100049, China
5 School of Science, Guiyang University, Guiyang 550005, China
* Correspondence: xianjin_an@163.com

Abstract: Microplastics are widely present worldwide and are of great concern to scientists and governments due to their toxicity and ability to serve as carriers of other environmental pollutants. The abundance of microplastics in different water bodies varied significantly, mainly attributed to the initial emission concentration of pollutants and the migration ability of pollutants. The migration process of microplastics determines the abundance, fate, and bioavailability of microplastics in water. Previous studies have proved that the physicochemical properties of water bodies and the properties of microplastics themselves are important factors affecting their migration, but the change in external environmental conditions is also one of the main factors controlling the migration of microplastics. In this paper, we focus on the effects of meteorological factors (rainfall, light, and wind) on the distribution and migration of microplastics and conclude that the influence of meteorological factors on microplastics mainly affects the inflow abundance of microplastics, the physical and chemical properties of water, and the dynamics of water. At the same time, we briefly summarized the effects of aquatic organisms, water substrates, and water topography on microplastics. It is believed that aquatic organisms can affect the physical and chemical properties of microplastics through the physical adsorption and in vivo transmission of aquatic plants, through the feeding behavior, swimming, and metabolism of animals, and through the extracellular polymers formed by microorganisms, and can change their original environmental processes in water bodies. A full understanding of the influence and mechanism of external environmental factors on the migration of microplastics is of great theoretical significance for understanding the migration law of microplastics in water and comprehensively assessing the pollution load and safety risk of microplastics in water.

Keywords: microplastics; natural factor; distribution; migration; water

1. Introduction

Microplastics (MPs) refer to plastic particles with a size between 1 μm and 5 mm and are also a class of macromolecular polymers with high heterogeneity. These microplastics are widely distributed in aquatic and terrestrial ecosystems and pose potential risks to ecosystems and humans through inhalation, ingestion, and skin contact, as well as through the food chain [1]. It has become a hot topic of concern in global political and academic circles. However, it elaborated the concept of MPs for the first time, which also triggered the rapid growth of microplastic research in environmental media [2]. Plastic debris in the environment is a growing pollution problem, and a large number of studies have shown that MPs are ubiquitous on the planet and in polar regions, seawater, rivers, lakes, urban water bodies, underground water bodies, soils, the Qinghai-Tibet Plateau [3–6], food [7], and organisms [8]. "Plastics—the Fast Facts 2023" reported that global rate production rose from 1.7 million tons in 1950 to 367 million tons in 2020, in addition to 403 million tons in

2022 due to the COVID-19 pandemic in 2019 [9]. As of 2015, humans produced at least 6.9 billion tons of plastic waste [10], and 11% of plastic waste enters aquatic ecosystems every year [11]. According to the previous study, by 2030, about 53 million metric tons of plastic waste will still be entering the water environment each year, even with global efforts to reduce and manage plastic waste [12]. Therefore, the migration of microplastics in the aquatic environment affects the security of the entire ecosystem.

Similar to plankton in size, aquatic organisms quickly ingest MPs; the bisphenol A, phthalates, flame retardants, and coloring metals carried by them are toxic and can accumulate in organisms, disrupt the stability of biological cells, cause cytotoxicity [13,14], affect the normal physiological function of cells [15], and destroy biological tissues and organs [16], which endanger the health of organisms. Recent studies have confirmed that microplastics have entered the human placenta, brain, and other tissues and organs, and, amazingly, nanoplastics can also break through the blood-brain barrier [17] and may also be an accomplice in causing Parkinson's disease [18]. At the same time, drinking water is also an important medium of microplastic pollution; every 1 L of plastic bottled water contains up to 240,000 microplastic particles, of which microplastics account for 10% and 90% may be nanoplastics [19]. Shockingly, there is direct evidence that microplastics and nanoplastics have penetrated human arteries and have helped increase the risk of serious diseases such as heart disease, stroke, and death [20]. It is worth noting that MPs are not only toxic in their constituents but also have a significant enrichment effect on environmental pollutants due to their small particle size and large specific surface area, and they are also good carriers of heavy metals [21], organic pollutants [22], and biotoxins [23] in the environment. There are a wide range of sources of MPs in natural water systems, including domestic sewage [24], atmospheric deposition [25], agricultural irrigation [26], seawater recharge [27], tire particles [28], etc.

Water bodies are important places for the convergence of primary and secondary microplastics in the environment, and, at the same time, water bodies are important functional elements in the environment, which play an extremely important role in global ecosystems and human health [17,20,25]. Primary microplastics are microplastics products mainly used for personal care products, while secondary microplastics are the products of mechanical stress, ultraviolet radiation, and biodegradation of large-sized plastics. They enter natural water bodies through atmospheric settlement, ground runoff, and domestic wastewater. The migration of microplastics in water is the key to the fate of microplastics. The migration behavior of microplastics in water mainly includes vertical sedimentation, suspension and floating, horizontal migration, aggregation, sedimentation, and undercurrent exchange [29–32]. Studies have pointed out that the properties of microplastics affect their migration process. The physical and chemical properties of water components can change the migration pathways of microplastics [31,33,34]. However, there is still a lack of review literature on the influence of external natural factors such as aquatic plants and rainfall on the migration of microplastics in water bodies [35–37]. Based on this, this paper briefly reviews the natural influencing factors of microplastic migration in water bodies, including meteorological factors, aquatic organisms, sediment, and topography of water bodies. Understanding the migration of microplastics in water is of great theoretical significance for comprehensively assessing the pollution load and safety risk of microplastics in water.

2. Meteorological Factors

Microplastics will be affected by various meteorological factors in the open environment of water bodies. Here, we review the effects of precipitation, sunlight radiation, and wind on microplastics' distribution, migration, and fate in water bodies (Figure 1).

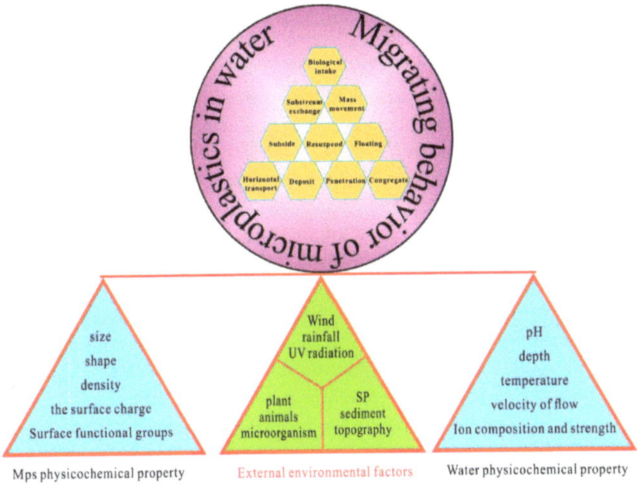

Figure 1. The transport of MPs is affected by external environmental factors.

2.1. Rainfall

Previous studies have shown that rainfall-induced surface runoff may be one of the important pathways for microplastics to enter water in the terrestrial environment. Rainfall can also flush suspended microplastics from the atmosphere into the water environment. At the same time, rainfall will also change the hydrodynamic conditions, which will affect the occurrence and migration of the original microplastics in the water body. We systematically summarized the existing literature, searched 126 relevant articles with TS = (microplastics AND (rainfall OR rain) AND water) using the WOS database, and finally sorted out 60 articles on the effects of rainfall on microplastics in water systems using the PRISMA [38] literature screening process (Table 1). Through systematic analysis, the paper concluded that rainfall affects the distribution and migration of microplastics in water in four aspects (Figure 2).

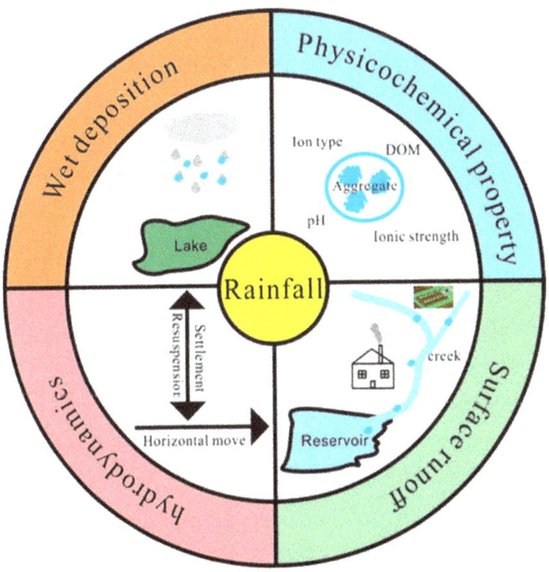

Figure 2. The mechanism of distribution and transport of MPs is affected by rainfall.

The intensity and process of rainfall significantly affect the abundance of microplastics in water, and heavy rainfall affects the dynamics of water, which in turn affects the migration of microplastics. Hydrodynamic conditions affect the sedimentation and resuspension of microplastics [39]. The effect of rainfall on the abundance of microplastics in water bodies mainly showed positive effects, but there were also different results (Table 1). However, it is pointed out that there is a negative correlation between the abundance of microplastics and the flow velocity in sediments, and strong hydrodynamic conditions increase the horizontal migration of microplastics, which is not conducive to the deposition of microplastics [40]. Studies have shown that the content of microplastics in the Vuachere and Venoge rivers in Switzerland increases significantly after rain [41]. Another study pointed out that rainfall intensity and drought periods are important factors affecting microplastics in freshwater environments [42]. It has been suggested that rainstorm erosion can bring terrestrial microplastics into water bodies and increase their abundance in water bodies [43,44]. The particulate matter brought into the water body during the heavy rainfall process interacts with the plastic particles, which converts the large-sized plastics into small-sized plastic particles, thereby changing the migration and occurrence state of microplastics. The microplastics deposited in the water body will also be resuspended with the disturbance of heavy rainfall, thereby increasing the horizontal movement of microplastics in the water body [45]. It was shown that sustained rainfall increases the abundance and diversity of microplastics in surface waters, while the opposite trend was observed for microplastics in sediments [46]. Rainfall intensity positively correlates with the abundance of microplastics in urban water bodies [47,48]. However, some studies have pointed out a negative or no correlation between the abundance of microplastics in water bodies and rainfall, as rainfall and flooding may dilute the abundance of microplastics in groundwater and affect the migration of microplastics [49,50]. However, that also suggested that rainfall can dilute microplastic pollution in flowing water bodies such as rivers, mainly because the abundance of microplastics in rainfall is lower than that in rivers. The microplastics in rivers are washed out to the downstream areas. Still, compared with closed and regional water bodies, rainfall will inevitably reduce the concentration of microplastics in the regional atmosphere and increase the abundance of microplastics in water [51]. Previous research has shown that flood events lead to the release of large amounts of microplastics in rivers [52]. The process of rainfall also has a certain effect on the distribution and migration of microplastics in water, and it concluded that the abundance of microplastics in rainwater is higher in the early stage of rainfall, and the concentration in water decreases as rainfall continues [53]. Drought, as opposed to rainfall, is also one of the factors affecting microplastics in water bodies, and studies have shown that the concentration of microplastics in rainwater is positively correlated with the number of dry days before rainfall [42,54]. During three consecutive days of rainfall, the concentration of microplastics in river surface water was twice that of coastal seawater, and the number of microplastics decreased by 90% at 2 h after rainfall [55], which also indicates that different stages of rainfall also change the distribution of microplastics in water bodies.

The type of rainfall may also change the migration capacity of microplastics in the water by affecting the physical and chemical properties of the water body; the difference in ionic composition in the rainwater will affect the ion content of the water body, which in turn will affect the migration of microplastics, and previou study has shown that there is a significant positive correlation between the abundance of microplastics in the sediment and the concentration of Ca^{2+}/Mg^{2+} in the overlying water [56]. The increase of ionic strength will enhance the agglomeration of microplastics, and at the same time, the electric double layer between the microplastics and the medium will be compressed, reducing the electrostatic repulsion, thus reducing the migration ability of microplastics [57]. The migration capacity of microplastics decreases with the increase in the concentration of high-valent cations. In the presence of high-valent cations, microplastic particles are more likely to agglomerate, so they tend to settle in water and have a weakened horizontal migration ability [58]. Prolonged acidic rainfall may lower the pH of water bodies. It has been shown that the migration capacity of microplastics in water usually increases with a pH increase. The increase in pH will reduce microplastic agglomeration, affect functional group deprotonation, increase the negative charge between microplastics and the medium, and enhance the electrostatic repulsion [59,60]. Small-scale water bodies are generally more susceptible to rainfall types than large-scale ones. Interestingly, it has been reported that coral reefs can trap microplastics, and acid rain can lead to bleaching and increased release of stored microplastics on coral reefs, allowing microplastics to redistribute [61]. As far as we know, the research on the direct impact of rainfall types on microplastics in water has not been reported, and it is necessary to systematically explain the effects of different rainfall types on the distribution and migration of microplastics in water from a combination of indoor quantitative experiments and field measurements.

It is easy to overlook that rainfall tends to change the content of dissolved organic matter in water bodies, thus affecting the migration of microplastics. When the dissolved organic matter in the water body increases, it will reduce the agglomeration capacity of microplastics and adsorb on the surface of microplastics, increase the negative charge, enhance electrostatic repulsion, and enhance the steric hindrance effect between microplastics and the medium, covering the deposition sites of microplastics on the surface of the medium, thereby increasing the migration ability of microplastics [62,63].

Table 1. Literature summary of microplastics in sewer systems during rainfall events.

Country	Event	Sample	Abundance *	Effect **	Ref.
China	Rain	Rainwater	146~8629 items/m^2	Positive	[50]
China	Rain	Pearl River	219.8 ± 160.5 n/L (before); 474 ± 259.7 n/L (after)	Positive	[46]
Nigeria	Rain	Oxbow Lake	3.70 items/L (dry season); 3.08 items/L (rainy season)	Negative	[64]
China	Rain	Rainwater	1.1 × 10^{13} particle/day (wet); 7.4 × 10^{12} particles/day (dry)	Positive	[65]
Canada	Rain	Catchments	33.5 pieces/L (rain); 19.1 pieces/L (baseflow)	Positive	[66]
Japan	Rain	Surface water	35 items/L (light rainfalls); 929 items/L (moderate); 331 items/L (heavy)	Positive	[67]
Australia	Rain, storm	Cooks River Estuary	0.4 particles/L (before storm and heavy rain); 17.38 particles/L (after)	Positive	[68]
China	Rain	Qing River	1.16 n/L (before); 1.04 n/L (after)	Negative	[51]
Brazil	Rain	Goiana Estuary	0.56 n/100 m^3 (rainy); 0.62 n/100 m^3 (dry)	Positive	[69]
China	Rain	Urban river	29.98 n/L (dry season); 90.99 n/L (wet season)	Positive	[70]
Sri Lanka	Rain	Beira Lake and Canal	lake: 0.011 (dry), 0.007 (wet); canal: 0.003 (wet), 0.002 (dry)	No	[71]
China	Rain	Runoff	6.0 items/L (beginning); 1.0~4.0 items/L (during); 0.7 items/L (end)	Positive	[72]
China	Rain, flood	Dafangying River	18.62 ± 7.12 items/m^3	Positive	[73]
China	Rain	Chaohu Lake	2133 ± 1534 n/m^3 (dry season); 1679 ± 1577 n/m^3 (wet season)	Negative	[74]
Brazil	Rain	Jurujuba Cove	14.4~202.8 n/L (rainy season); 91.2~137.4 n/L (dry season)	Negative	[75]
Brazil	Rain	Fish farms	81.12 items/L (dry); 236.96 items/L (rainy)	Positive	[76]
South Africa	Rain	Crocodile River	1058 (cool-dry season); 625 (hot-dry season); 625 particles/m^3 (hot-wet season)	Positive	[77]
Singapore	Rain	Sea	164.5 particles/mL	Positive	[78]
China	Rain	Lake	0.59 items/L	Positive	[79]
China	Rain	Rainwater	141 (spring); 140 (winter); 102 (summer); 78 particles/(m^2·d) (autumn)	Positive	[80]

Table 1. Cont.

Country	Event	Sample	Abundance *	Effect **	Ref.
China	Rain	Maowei Sea	2.8 particles/L (rainy season); 4.29 particles/L (dry season)	Negative	[81]
Mexico	Rain	Runoff	177.13 particles/L	Positive	[47]
USA	Rain	Estuarine rivers	90,007 pieces/km^2 (summer); 95425 pieces/km^2 (fall)	Positive	[82]
Indian	Rain, Storm	Manipal Lake	0.423 particles/L (monsoon); 0.117 particles/L (post-monsoon)	Positive	[83]
Malaysia	Rain, wind	Sepanggar Bay water	106.6 ± 23.0 (SWM); 63.0 ± 8.0 (NEM); 31.2 ± 6.7 particles/m^3 (INTER)	Positive	[84]
China	Rain	Rainwater	229 n/(m^2·d) (wet deposition); 125 n/(m^2·d) (dry deposition)	Positive	[85]
Colombian	Rain	Estuaries	0.33 items/m^3 (high rain); 0.085 items/L (low rain)	Positive	[86]
Canada	Rain	Urban runoff	186 particles/L	Positive	[87]
France	Rain	Liane River	35.5 (heavy rain); 5.1 (light rain); 12.4 particles/m^3 (no rain)	Positive	[88]
China	Rain	Karst groundwater	4.50 items/L	Positive	[89]
Brazil	Rain	Paraíba do Sul River	1~12 particles/m^3 (low water season); 1~18.3 particles/m^3 (high water season)	Positive	[90]
China	Rain	Xincun Lagoon Bay	60.9 ± 21.5 items/L (rainy season); 72.6 ± 23.7 items/L (dry season)	Negative	[91]
Turkey	Flood	Mediterranean Region river	539,189 MPs/km^2 (before flood); 7,699,716 MPs/km^2 (afterwards)	Positive	[92]
France	Stormwater	Catchment outlet	29 items/L	Positive	[93]
USA	Rain	Tampa Bay surface water	2.2 particles/L (rain in OTB site); 1.0 particles/L(average)	Positive	[94]
Indian	Monsoonal rainfall	Udyavara River	530.14 ± 352 particles/m^3	Positive	[95]
Vietnam	Rain	Saigon River	53 items/L (rainy season); 75 items/L (dry season)	No	[96]
India	Rain	Netravathi River	36.86 ± 23.12 (2020 monsoon); 70.5 ± 61.22 MP/m^3 (after)	Positive	[97]
Belgium	Rain	Flanders surface water	0.48 MPs/L	No	[98]
Argentina	Rain	Lake	100 (spring)~180 MPs/m^3 (summer)	Positive	[99]
China	Rain, flood	Yangtze Estuary	300 n/kg (1954 flood at ECS1); 1000 n/kg (1998 flood at CCYY1)	Positive	[100]
China	Rain, typhoons	Seawater	63.6 ± 37.4 items/L (before typhoon); 89.5 ± 20.6 items/L (after typhoon)	Positive	[101]
Indonesia	Rain	Jakarta River	9.80 ± 4.79 (rainy season); 8.01 ± 4.82 particles/m^3 (dry season)	Positive	[102]
Brazil	Rain	Acaraí Lagoon	1.4~3.4 n/L	Negative	[103]
Australia	Rain	Perth metropolitan waters	47,164 pieces/km^2 (heavy rain in May); 2461 pieces/km^2 (March)	Positive	[104]
India	Rain	Mandovi-Zuari estuarine	107 particles/L (wet season); 99 particles/L (dry season)	Positive	[105]
Australia	Rain	Storm drains	139.43 items/effort (before); 132.6 items/effort (during); 294.5 items/effort (after)	Positive	[106]
Finland	Rain	Surface flow wetland	104 MPs/m^3 (inflow); 200 (outflow addition deposition)	Positive	[107]
Thailand	Rain	Runoff	1.3 ± 1.3 particles/L (wet season); 2.8 ± 0.9 particles/L (dry season)	Positive	[108]
India	Rain	Sharavathi River sediment	2.5~57.5 pieces/kg (pre-monsoon); 0~15 pieces/kg (post-monsoon)	Positive	[109]
China	Rain	Donghu Lake	5.84 ± 2.95 items/L (equilibrium state); 8.27 ± 5.65 items/L (during rain); 7.60 ± 4.04 items/L (after rainfall)	Positive	[110]

Table 1. *Cont.*

Country	Event	Sample	Abundance *	Effect **	Ref.
China	Rain	WWTP	36.2~126.2 particles/L(rain); 38.9~75.3 particles/L (no rain)	Positive	[111]
China	River	Hanjiang rRiver	30.9 (base flow); 80.2~114.5 (flood)	Positive	[112]
China	Rain, typhoon	Surface seawater in Hong Kong	0.02 items/L (dry season); 0.10 items/L (wet season)	Positive	[113]
China	Rain	Harbor and coastal sediments	36.5 ± 52.5 items/kg (dry season); 22.6 ± 23.2 items/kg (wet season)	Positive	[114]
Italy	Rain, wind	Lake	0.82~1.24 particles/m^3 (before); 2.42~4.41 particles/m^3 (after)	Positive	[115]
Portugal	Rain	Estuary	263 items/kg (no rain); 205 items/kg (rain)	No	[116]
China	Rain	Jiaozhou Bay sea water	0.174 pieces/m^3 (heavy rain in May); 0.05 piece/m^3 (no rain in November)	Positive	[117]
USA	Rain	Outfalls	0.30 ± 0.10~0.80 ± 0.33 MP/L (rain)	Negative	[118]
Italy	Rain	Mugnone Creek	3.5×10^8 items/day (wet season in 2019); 5.2×10^6 (dry season 2020)	Positive	[119]
Lithuania	Rain	WWTP	2982 ± 54 MP/L (wet season); 1964 ± 50 MP/L (dry season)	Positive	[120]
Germany	Rain	Weser River	219.05 items/m^3 (no rain day); 14,536.1 items/m^3 (rain day)	Positive	[121]

* The average abundance of microplastics is mainly represented in the literature. Meanwhile, Get Data software v 2.2 (http://getdata-graph-digitizer.findmysoft.com/ (accessed on 29 April 2024)) was used to manually extract the measured data when the data information in the literature was presented graphically. ** represented the effect between the abundance of microplastics in water and the degree of rainfall.

2.2. *Ultraviolet Radiation*

The effect of sunlight radiation on the distribution and transport of microplastics in water bodies is twofold (Figure 3). On the one hand, sunlight radiation will change microplastics' physicochemical properties and surface morphology, making them more susceptible to aging and decomposition. In the natural environment, the microplastics in the water body easily decompose slowly into microplastics and nano-sized microplastics with smaller particle sizes under ultraviolet irradiation, especially the microplastics floating on the surface of the water body. The action of ultraviolet rays and oxygen will gradually age and decompose under long-term exposure to sunlight, and the surface morphology and functional groups will change accordingly; therefore, the light affects the distribution and migration law of microplastics through the changes in the particle size and surface characteristics of microplastics [122,123]. Photoaging will roughen the surface of microplastics, cracks will appear, and the density of polyethylene microplastics will increase, changing the water body's migration process. Previous research pointed out that the color of plastics may also be an influencing factor affecting the aging and degradation of microplastics in water, and sunlight exposure is the main reason for the aging of plastics, which can easily trigger the chain reaction and chain breakage of plastic polymers, resulting in the cracking of plastics into microplastics [124]. Photoaging often changes the color of the plastic polymer, and the color of the plastic itself can also affect the absorption of sunlight.

On the other hand, sunlight exposure will affect the change of atmospheric gradient and change the evaporation of water, so that sunlight radiation will bring about the linkage of wind and rain, and at the same time, sunlight radiation will affect the temperature of the water body and affect the change of dissolved oxygen and other water physical and chemical properties in the water. The change in the water's physical and chemical properties and the resulting change in biological activities will change the distribution and migration of microplastics. The study pointed out that the increase in solar radiation, the increase in temperature, and the melting of glaciers, ice, and snow will increase the total abundance of microplastics in the water body. The microplastic pollution in a remote lake on the Qinghai-Tibet Plateau is caused by the microplastics released by the melting of glaciers. Both light exposure and the resulting physical weathering affect the decomposition of plastics, change the size of microplastics, and then affect the migration and distribution of microplastics in water bodies. An increase in the temperature of a water body can lead to changes in the aquatic community. The study reported that predation by benthic organisms increases when water temperatures rise, so abundant biological activity may lead to the resuspension of deposited microplastics [125]. Sunlight exposure drives the circulation of global ocean currents, which is also an important factor affecting the migration of microplastics in water bodies. However, mathematical models were used to predict the transport pathways of plastic pollution in the wake of ocean circulation [126].

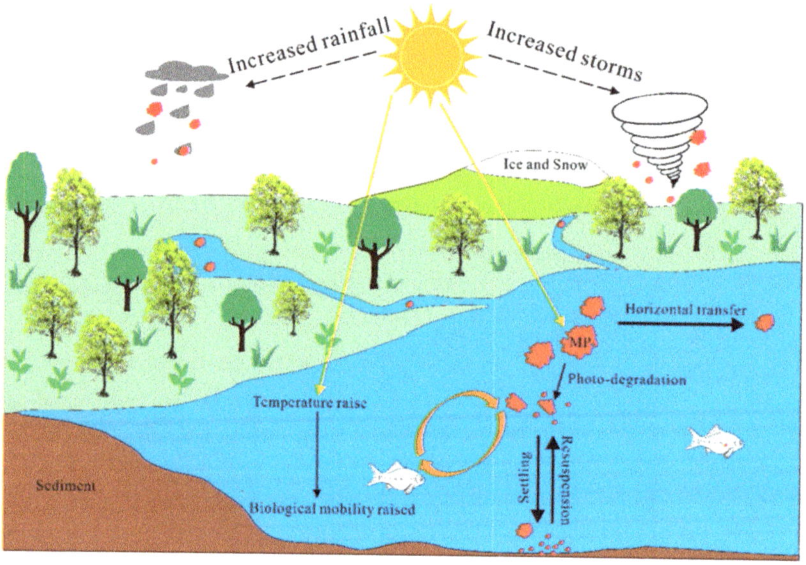

Figure 3. Conceptual diagram of the effects of ultraviolet radiation on microplastics in water.

2.3. Wind

Wind is an important path for microplastics to travel from land sources to water sinks, and also pulls microplastics to remote and alpine areas that are inaccessible to people, such as the Arctic [127] and Qinghai-Tibet Plateau [128]. Typhoons in China increased the concentration of microplastics in the aquatic environment of its immediate vicinity [101]. A previous study found that hurricane-induced turbulence redistributed microplastics in coastal waters [129]. It has been noted that during storm events, microplastics can be transported by rapid flow and deposited in deeper locations when turbulence slows down [130]. Microplastics in water of different depths before and after the storm in Santa Nika Bay were investigated, and it was found that the abundance of microplastics on the surface and mid-water increased after the storm, while the abundance of microplastics decreased on the seafloor [131]. Wind can significantly impact microplastics floating on the surface of the water, especially polystyrene foam. Floating microplastics increase their horizontal migration capacity under the influence of wind [132]. The sedimentation flux of microplastic polymers in water increases with the increase in wind speed, and high-density microplastic polymers are more affected by wind force than low-density microplastics [133]. So, it was found that in the northwestern Mediterranean, strong winds were followed by five times as many plastic particles floating on the ocean's surface, which facilitated the mixing of plastic particles in the surface water column and their vertical redistribution [134].

However, the abundance of microplastics in the waters of Lake Zurich and Lake Constance decreased after vertical mixing caused by strong winds, mainly because the winds before sampling in Lake Zurich and Lake Constance may have reduced the number of particles measured due to vertical mixing [41]. It is worth noting that the increased winds due to global warming strengthen the circulation of microplastics in the atmosphere, hydrosphere, and pedosphere. In the future, the impact of wind on microplastics should be viewed from a global perspective; however, large-scale field detection is very difficult, so it is necessary to develop corresponding models and locally measured data to explore the impact of wind on microplastics.

3. Aquatic Life

Aquatic organisms in the aquatic environment are crucial to the distribution and migration of microplastics, including the adsorption, interception, and internal absorption of aquatic plants, the carrying and ingestion of aquatic animals, and the interaction between microorganisms and microplastics in water (Figure 4).

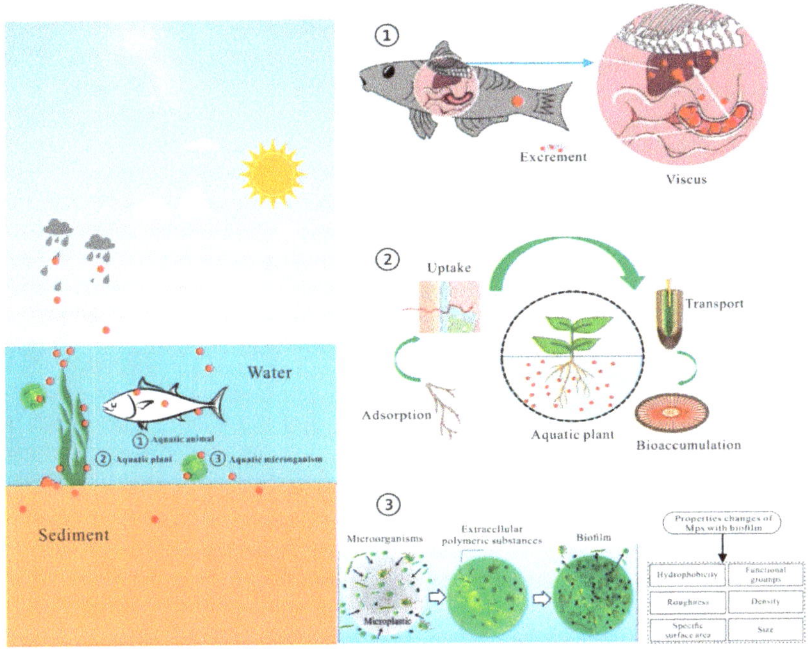

Figure 4. Schematic diagram of the interaction process between aquatic organisms and microplastics.

3.1. Aquatic Plants

The interaction between microplastics and aquatic plants affects the normal growth of aquatic plants and changes the occurrence state and migration ability of microplastics. Aquatic plants can affect the migration of microplastics in water through surface adsorption, absorption, transport, and accumulation, thereby changing the exposure concentration of free microplastics in water and reducing the bioavailability of other animals, plants, and humans [135–137] (Figure 3). Therefore, aquatic plants are considered to be an important way to slow down the migration of microplastics in water bodies. Another study reported that seaweed has the ability to trap microplastics [138]. The results showed that the three aquatic plants had the effect of capturing and removing polystyrene, and the roots of aquatic plants had the strongest ability to capture micro/nanoplastics in the water body, which could reach 6250 μg/g, and different root structures may affect the absorption and transport of micro/nanoplastics [135]. Polystyrene (PS) microplastics will be significantly aggregated on the root surface of plants, especially at the root tips; microplastics tightly adhere to the roots and can remain adhered to them after washing [139]. Smaller nanoscale microplastics move from adherent root hairs to columnar vascular bundles inside roots [140]. However, hydroponically grown rice seedlings were exposed to PS microplastics, and it was found that the microplastics were distributed in the rice root system and the intercellular space [141]. Studies have shown that small-sized microplastics can enter plants through the stomata on the leaf surface, binding microplastics to aquatic plants and changing their migration process [142].

3.2. *Aquatic Animals*

Aquatic animals are direct victims of exposure to microplastics in water bodies, and these animals alter the abundance and migration of microplastics in water bodies through ingestion and epidermal contact. Microplastics ingested by aquatic organisms accumulate in different organs in the body through the digestive system, and the microplastics accumulated in aquatic organisms of different nutritional levels can be transmitted within the food chain in the preying relationship between organisms, ultimately endangering human health. Currently, the research on microplastics in organisms mainly focuses on fish (Figure 5), followed by crustaceans, mollusks, and annelids [143]. Fish are also the main body, and the focus of attention is on the impact of microplastics in water bodies. Microplastics have been detected in more than 728 fish species [144,145]. Fish change the migration of microplastics in water through ingestion, transport, digestion, and excretion. Their ingestion behavior is the premise of controlling the transport, digestion, and excretion of microplastics in fish and the intake of microplastics by aquatic animals, including active ingestion (foraging or accidental ingestion) and passive ingestion (food chain transmission). In a survey on microplastic exposure in food fish off the east coast of Brazil, it was found that 62.5% of Atlantic mackerel were exposed to microplastics, while 33% of them were oblique megalodons. Similarly, aquatic organisms can also ingest microplastics in the terrestrial water environment, and the concentration of microplastics in carp, crucian carp, etc. is 1~6 [146]. Previous studies suggested that copepod aquatic organisms would ingest smaller microplastics, which would accumulate in the anterior midgut and eventually be excreted in dense feces, thus changing the migration path of microplastics in the water body [147]. Vroom et al. pointed out that PS in the intestinal tract of Calanus finmarchicus forms aggregates that can account for 30 to 90 percent of the intestinal volume, and these aggregates are excreted in the form of feces, which, combined with the diurnal vertical migration of aquatic animals, transports microplastics to deeper waters [148]. Different feeding patterns correlate with microplastic loading in aquatic animals [149]. The study showed that a certain concentration of microplastics was detected in the skin tissue, proving that the adhesion of the aquatic animal epidermis or the skin tissue is the carrier of microplastic migration in the water body [150]. Interestingly, recent studies have shown that microplastics have been detected in rotifers collected from all marine and freshwater sites and have shown that the grinding effect of rotifers chewing is considered to be an important producer of nanoplastics in the aquatic environment and that rotifers alter the migration of microplastics in the aquatic environment by affecting the size of the plastics [151].

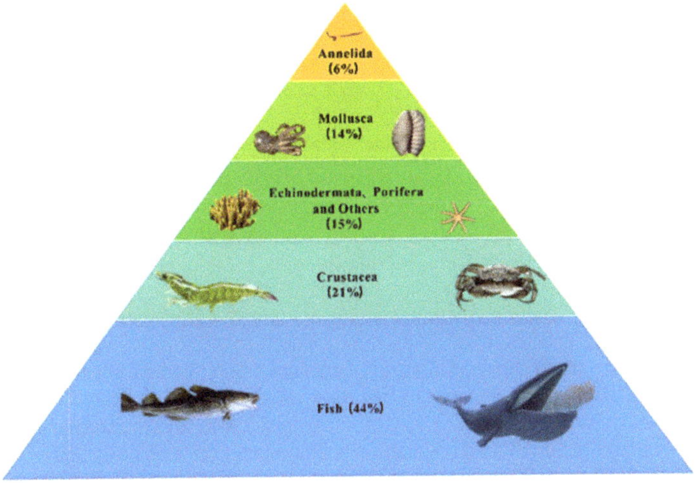

Figure 5. Species distribution of microplastics ecotoxicity studies in organisms.

3.3. Water Microorganisms

Biofilms are integral structures formed by the growth and development of protozoa, bacteria, algae, and fungi [152]. Bacterial colonies within biofilms secrete extracellular polymeric substances (EPSs). EPS is considered a viscous colloidal substance that plays a key role in microbial colonization and contaminant migration [153]. In the aquatic environment, microorganisms can attach to microplastics, forming biofilms, which affect the sedimentation and transport of microplastics in water [154–156]. The biofilm formed by microorganisms on the surface of microplastics will affect the surface morphology and physicochemical properties of microplastics and then affect the migration behavior of microplastics. Due to the presence of biofilms, the hydrophobicity, density, functional groups, size, surface, and roughness of microplastics change. As previously reported, the sedimentation rate of PET and PS attached to the biofilm was 1% and 4% faster than the original state [157]. Biofilm increases the roughness of the surface of microplastics. After biofilm formation, the surface roughness, number of pores, and surface area of microplastics increase [158]. Recent studies have shown that the biofilm on the surface of microplastics can be used as a kind of surface roughness, and the sedimentation rate can be changed by changing the resistance coefficient of microplastics during the sedimentation process. It has also been pointed out that the multilayer structure and anisotropy of microbial biofilms are not exactly equivalent to the roughness of traditional particles [159]. The spectral absorption summit of the functional groups of PE plastics was exposed to seawater changes, producing an additional peak at 1700, which is believed to be mainly due to the C==O produced by the decomposition of microplastics [156]. There are some differences in the study of biofilm alteration of the surface hydrophobicity of microplastics. Attachment to biofilm has been reported to reduce the hydrophobicity of PE [160], but it has also been reported that biocontaminated plastic surfaces will increase hydrophobicity [161]. At the same time, microorganisms are also considered an ideal way for microplastics to be degraded. The aerobic environment on the water body's surface will promote microorganisms to degrade microplastics into carbon dioxide and water. In the anoxic environment of deep-water bodies, microplastics will be decomposed into carbon dioxide, water, and methane under the action of anaerobic microplastics. In the process of degradation, the original structure of microplastics will change significantly between polymers. Between polymers and plastic additives, the characteristics of plastics will change and the specific surface area of microplastics exposed to the environment will increase, which will promote the attachment and reproduction of microorganisms and the occurrence of physicochemical reactions. After the interaction between microplastics and microorganisms, their particle size continues to decrease, and smaller microplastics are easily suspended on the surface of the water body again, changing the vertical migration path. These smaller microplastics are also more easily ingested by aquatic animals, changing their migration process.

4. Water Matrix: Suspended Particulate, Sediment, and Topography

Geographical location and land use were excluded from this study, considering that these two factors are closely related to human activities [162], so the effects of natural factors on microplastics in water bodies did not take into account the effects of geographical location and land use. The migration capacity of microplastics is related to the roughness of the medium.

4.1. Suspended Particulate

The rougher the surface of the sand grains, the easier it is for microplastics to be deposited and the lower their migration capacity. The suspended sediment in the water body can disrupt the stability of microplastics [163,164] and significantly affect the migration behaviors of microplastics, such as accumulation, suspension, and infiltration. Studies have pointed out that the aggregation behavior of microplastics and suspended sediment increases the density of microplastic aggregates, affecting the vertical distribution and long-term migration of microplastics [165,166]. The interaction between MPs and suspended sediment is influenced by water environmental conditions (e.g., hydrodynamic characteristics, concentration and type of ions, dissolved or granular organic/inorganic colloids, microorganisms, and phytoplankton) and physicochemical properties of MPs/sediment itself (e.g., particle Zeta potential size, charge distribution, and surface polar functional groups) [163,167]. Studies have reported that about 2–9% of all microplastic mass leaving Switzerland is transported by sediment [3]. In the water environment, microplastics combine with other suspended particles to change the density of microplastics so that they migrate from the surface water body to the deeper water body; once the microplastics change the suspension state, under the action of complex meteorological factors, the migration path will be more complex [168]. Studies have shown that microplastics form aggregates with natural mineral particles, such as kaolin and other clay particles, and negatively charged alginate and iron oxide in water form electrostatic interaction with microplastic particles, which makes the two form larger aggregates and affects the migration process of microplastics in water [167,169].

4.2. Sediments

Water sediments contain a large number of microplastics, which migrate due to changes in natural conditions such as wind, waves, and tides and can also move through human disturbances [170]. The roughness, particle size structure, and organic matter composition of the sediment all affect the migration and distribution of microplastics. It has been shown that microplastics deposited in coarse-grained sediments are more likely to be resuspended, while sediments with higher viscosity and rich organic matter are less likely to be resuspended [171]. The retention and accumulation of MPs and aggregates in sediments can gradually clog the pores between sediment particles and inhibit or even completely hinder the osmotic migration of MPs in sediments [172]. So, it has been pointed out that constructed wetlands can effectively remove microplastics in water bodies, with a removal efficiency of more than 90%, which significantly changes the migration process of microplastics in this water body, and the matrix particle size of wetlands and the design of wetlands are the key factors affecting the removal and migration of microplastics in water bodies [173]. In the treatment of rural domestic sewage, it was found that the smaller the particle size of the matrix, the higher the removal efficiency of microplastics, which may be caused by the greater friction of the microplastics per unit volume of sediment. Researchers reported that microplastics' removal and retention efficiency in the sand-based reactor was close to 100%, which was significantly higher than that of the gravel-based reactor [174]. The biofilm formed on the sedimentary matrix of the water body affects the migration of microplastics, and it was observed that the biofilm growing on the water matrix reduces the pore space and increases the viscosity of the matrix, thereby enhancing the substrate's ability to retain microplastics [175]. The nature of sediments is also a factor affecting the migration of microplastics in water, and it has been suggested that positively charged sediments are more likely to adsorb negatively charged microplastics [176].

4.3. Water Topography and Landform

The topography and appendages of the water body can affect the migration of microplastics in the water body. It has been shown that artificial structures such as dams, sand bars, and diversion walls can reduce a water body's flow and thus increase the microplastic sedimentation rate [177–179]. River morphology can affect the velocity of water bodies or the migration of microplastics through bend interception. Compared with curved channels, straight channels are more likely to cause the horizontal migration of microplastics [171]. The study showed that the abundance of microplastics in sediment samples from the straight channel of the Thames River in Canada was lower than that collected at the inner and outer bends [171]. River morphology can also affect the undercurrent exchange behavior of particles through advection pumping, scouring, and silting exchange. When the shape of the riverbed is uneven, the surface pressure of the riverbed changes, which drives the movement of pore water so that colloids and plastic particles enter the riverbed and migrate. In the riverbed scouring zone, water flow affects the release and accumulation of pore water, resulting in the exchange of riverbed and colloidal particles [180–182].

5. Conclusions and Outlook

The migration of microplastics in water determines the distribution, fate, and ecological risk of microplastics, and the current research mainly focuses on the influence of physical and chemical properties of water bodies on the migration of microplastics and the research on microplastics under natural conditions is not sufficient. This paper focuses on the distribution and migration of microplastics in water and focuses on the influence of meteorological factors on microplastics, concluding that the influence of meteorological factors on microplastics mainly affects the distribution and migration of microplastics by affecting the abundance of microplastics inflow, affecting the physical and chemical properties of water bodies and water dynamics. At the same time, we also briefly summarized the effects of aquatic organisms' water substrates and water topography on microplastics. It is believed that aquatic organisms can affect the physical and chemical properties of microplastics through the physical adsorption and in vivo transmission of aquatic plants, through the feeding behavior, moving, and metabolism of animals and the EPSs formed by microorganisms, changing their original environmental processes in the water body. Future research on microplastic migration should pay more attention to the following.

1. Carry out indoor quantitative simulation experiments and outdoor long-term observations on the migration of microplastics under changes in meteorological conditions and understand the migration mechanism and controlling factors of microplastics in the field at a large scale. Attention should be paid to the changes in water environmental conditions in different seasons to better evaluate microplastics' interannual variation characteristics. Global climate change has led to more frequent extreme weather events, and the transport of microplastics in water bodies and their sediments may be significantly affected by climate change, so it is necessary to strengthen the research on the global distribution and circulation of microplastics under climate change scenarios in the future.
2. In different types of water bodies, topography and landforms, as well as the types of particulate matter in water bodies, are very different, and more attention needs to be paid to the movement mechanism of microplastics in different types of lakes and rivers. It is important to clarify the agglomeration behavior and co-migration mechanism between particulate matter and microplastics in clear water to predict

microplastics' environmental behavior and risks accurately. The migration process of microplastics in seawater and surface water has been partially studied, but more attention needs to be paid to the migration of microplastics in groundwater and special ecological regions such as karst water [89].

3. The impact of water temperature change on the transport of microplastics and the impact of ocean current processes on the global migration of microplastics were investigated. Currently, most of the research on the migration mechanism of microplastics in water bodies is carried out in the laboratory, which is quite different from the actual environment in which microplastics are located, and the model microplastics used are also significantly different from those in the real environment.
4. The interactive migration model of external environmental factors, internal water characteristics, and the physical and chemical properties of microplastics can be constructed. Systematic research can be carried out on multiple factors and scales to have a more comprehensive understanding of the migration process of microplastics in the real environment and better supervise their ecological risks [126,169]. At the same time, large-scale data acquisition needs to be standardized before validating the migration model to increase the applicability of measurements. Strengthening the study of the movement mechanics of microplastics and the spatial mode of pollution, determining the risk location of pollutant accumulation, and better delineating the area with the greatest risk of potential negative effects of microplastic pollution will help to focus on water protection.
5. Microplastics in water bodies are important carriers of other pollutants, and the synergistic migration behavior and mechanism of bacteria and viruses of other pollutants, especially newly polluted antibiotics, and microplastics in the actual water environment are still unknown and need to be further studied and improved [183].
6. Microplastics are constantly changing in the actual environment and may be affected by various factors in the migration, decomposition, and transformation process. Their original physical and chemical properties will also change.
7. Combined with the characteristics of microplastic pollution and the law of migration, an efficient governance system with scientific classification, source treatment, and migration control as the main contents and a sub-regional, sub-type, and sub-time supervision mechanism should be established.

In general, external environmental factors affect the distribution and migration of microplastics, Therefore, understanding and mastering the change of external environmental conditions can effectively control the pollution distribution and migration behavior of microplastics in water bodies. For example, after strong weather events, the exposure abundance of microplastics in water bodies may increase significantly, so the intake of microplastics can be reduced by reducing the use of water bodies. By mastering the settlement law of microplastics in different rivers, we can selectively use water resources in different locations, and then control the exposure of microplastics in water bodies. However, for aquatic organisms, effective source control is the main way to reduce their environmental risks.

Author Contributions: Conceptualization, X.A. and Y.W.; writing—original draft preparation, X.A., W.L. and M.A.; supervision, X.A.; writing—review and editing, X.A., W.L. and M.A.; visualization, X.A., Y.Z. and W.L. All authors have read and agreed to the published version of the manuscript.

Funding: This research was funded by The Guizhou Education Department Youth Science and Technology Talents Growth Project, China (KY [2022]001); Fundamental Scientific Research Funds of Guiyang University, China (GYU-KY-[2022]); The Joint Foundation of Guizhou Province (LH [2017]7348); The Science and Technology Plan Project in 2023 of Guiyang city, Guizhou Province.

Data Availability Statement: Not applicable.

Acknowledgments: The authors thank the anonymous reviewers for their comments on quality improvement.

Conflicts of Interest: The authors declare no conflicts of interest.

References

1. Rillig, M.C.; Lehmann, A. Microplastic in terrestrial ecosystems. *Science* **2020**, *368*, 1430–1431. [CrossRef] [PubMed]
2. Thompson, R.C.; Olsen, Y.; Mitchell, R.P.; Davis, A.; Rowland, S.J.; John, A.W.; McGonigle, D.; Russell, A.E. Lost at sea: Where is all the plastic? *Science* **2004**, *304*, 838. [CrossRef] [PubMed]
3. Mennekes, D.; Nowack, B. Predicting microplastic masses in river networks with high spatial resolution at country level. *Nat. Water* **2023**, *1*, 523–533. [CrossRef]
4. Zhang, Y.; Kang, S.; Luo, X.; Shukla, T.; Gao, T.; Allen, D.; Allen, S.; Bergmann, M. Microplastics and nanoplastics pose risks on the Tibetan Plateau environment. *Sci. Bull.* **2024**, *69*, 589–592. [CrossRef] [PubMed]
5. Singh, S.; Bhagwat, A. Microplastics: A potential threat to groundwater resources. *Groundw. Sustain. Dev.* **2022**, *19*, 100852. [CrossRef]
6. de Souza Machado, A.A.; Lau, C.W.; Till, J.; Kloas, W.; Lehmann, A.; Becker, R.; Rillig, M.C. Impacts of microplastics on the soil biophysical environment. *Environ. Sci. Technol.* **2018**, *52*, 9656–9665. [CrossRef] [PubMed]
7. Janani, R.; Bhuvana, S.; Geethalakshmi, V.; Jeyachitra, R.; Sathishkumar, K.; Balu, R.; Ayyamperumal, R. Micro and nano plastics in food: A review on the strategies for identification, isolation, and mitigation through photocatalysis, and health risk assessment. *Environ. Res.* **2023**, *241*, 117666. [CrossRef] [PubMed]
8. Rebelein, A.; Int-Veen, I.; Kammann, U.; Scharsack, J.P. Microplastic fibers—Underestimated threat to aquatic organisms? *Sci. Total Environ.* **2021**, *777*, 146045. [CrossRef] [PubMed]
9. Plastics Europe. *Plastics—The Fast Facts 2023*; Plastics Europe: Brussels, Belgium, 2023.
10. Organisation for Economic Co-Operation and Development. *Global Plastics Outlook: Economic Drivers, Environmental Impacts and Policy Options*; OECD Publishing: Paris, France, 2022.
11. Jansen, M.; Barnes, P.; Bornman, J.; Rose, K.; Madronich, S.; White, C.; Zepp, R.; Andrady, A. The Montreal Protocol and the fate of environmental plastic debris. *Photochem. Photobiol. Sci.* **2023**, *22*, 1203–1211. [CrossRef]
12. Borrelle, S.B.; Ringma, J.; Law, K.L.; Monnahan, C.C.; Lebreton, L.; McGivern, A.; Murphy, E.; Jambeck, J.; Leonard, G.H.; Hilleary, M.A. Predicted growth in plastic waste exceeds efforts to mitigate plastic pollution. *Science* **2020**, *369*, 1515–1518. [CrossRef]
13. Wu, X.; Pan, J.; Li, M.; Li, Y.; Bartlam, M.; Wang, Y. Selective enrichment of bacterial pathogens by microplastic biofilm. *Water Res.* **2019**, *165*, 114979. [CrossRef]
14. Hollóczki, O.; Gehrke, S. Can nanoplastics alter cell membranes? *ChemPhysChem* **2020**, *21*, 9–12. [CrossRef]
15. Kim, H.-Y.; Ashim, J.; Park, S.; Kim, W.; Ji, S.; Lee, S.-W.; Jung, Y.-R.; Jeong, S.W.; Lee, S.-G.; Kim, H.-C. A preliminary study about the potential risks of the UV-weathered microplastic: The proteome-level changes in the brain in response to polystyrene derived weathered microplastics. *Environ. Res.* **2023**, *233*, 116411. [CrossRef] [PubMed]
16. Von Moos, N.; Burkhardt-Holm, P.; Köhler, A. Uptake and effects of microplastics on cells and tissue of the blue mussel *Mytilus edulis* L. after an experimental exposure. *Environ. Sci. Technol.* **2012**, *46*, 11327–11335. [CrossRef]
17. Kopatz, V.; Wen, K.; Kovács, T.; Keimowitz, A.S.; Pichler, V.; Widder, J.; Vethaak, A.D.; Hollóczki, O.; Kenner, L. Micro-and nanoplastics breach the blood–brain barrier (BBB): Biomolecular corona's role revealed. *Nanomaterials* **2023**, *13*, 1404. [CrossRef]
18. Liu, Z.; Sokratian, A.; Duda, A.M.; Xu, E.; Stanhope, C.; Fu, A.; Strader, S.; Li, H.; Yuan, Y.; Bobay, B.G. Anionic nanoplastic contaminants promote Parkinson's disease–associated α-synuclein aggregation. *Sci. Adv.* **2023**, *9*, eadi8716. [CrossRef] [PubMed]
19. Qian, N.; Gao, X.; Lang, X.; Deng, H.; Bratu, T.M.; Chen, Q.; Stapleton, P.; Yan, B.; Min, W. Rapid single-particle chemical imaging of nanoplastics by SRS microscopy. *Proc. Natl. Acad. Sci. USA* **2024**, *121*, e2300582121. [CrossRef]
20. Marfella, R.; Prattichizzo, F.; Sardu, C.; Fulgenzi, G.; Graciotti, L.; Spadoni, T.; D'Onofrio, N.; Scisciola, L.; La Grotta, R.; Frigé, C. Microplastics and nanoplastics in atheromas and cardiovascular events. *N. Engl. J. Med.* **2024**, *390*, 900–910. [CrossRef] [PubMed]
21. Lin, Z.; Hu, Y.; Yuan, Y.; Hu, B.; Wang, B. Comparative analysis of kinetics and mechanisms for Pb (II) sorption onto three kinds of microplastics. *Ecotoxicol. Environ. Saf.* **2021**, *208*, 111451. [CrossRef]
22. Fajardo, C.; Martín, C.; Costa, G.; Sánchez-Fortún, S.; Rodríguez, C.; de Lucas Burneo, J.J.; Nande, M.; Mengs, G.; Martín, M. Assessing the role of polyethylene microplastics as a vector for organic pollutants in soil: Ecotoxicological and molecular approaches. *Chemosphere* **2022**, *288*, 132460. [CrossRef]
23. Pestana, C.J.; Moura, D.S.; Capelo-Neto, J.; Edwards, C.; Dreisbach, D.; Spengler, B.; Lawton, L.A. Potentially poisonous plastic particles: Microplastics as a vector for cyanobacterial toxins microcystin-LR and microcystin-LF. *Environ. Sci. Technol.* **2021**, *55*, 15940–15949. [CrossRef] [PubMed]
24. Li, C.; Busquets, R.; Campos, L.C. Assessment of microplastics in freshwater systems: A review. *Sci. Total Environ.* **2020**, *707*, 135578. [CrossRef] [PubMed]
25. Chen, Q.; Shi, G.; Revell, L.E.; Zhang, J.; Zuo, C.; Wang, D.; Le Ru, E.C.; Wu, G.; Mitrano, D.M. Long-range atmospheric transport of microplastics across the southern hemisphere. *Nat. Commun.* **2023**, *14*, 7898. [CrossRef] [PubMed]
26. Yang, L.; Zhang, Y.; Kang, S.; Wang, Z.; Wu, C. Microplastics in soil: A review on methods, occurrence, sources, and potential risk. *Sci. Total Environ.* **2021**, *780*, 146546. [CrossRef] [PubMed]
27. Viaroli, S.; Lancia, M.; Re, V. Microplastics contamination of groundwater: Current evidence and future perspectives. A review. *Sci. Total Environ.* **2022**, *824*, 153851. [CrossRef] [PubMed]
28. Luo, Z.; Zhou, X.; Su, Y.; Wang, H.; Yu, R.; Zhou, S.; Xu, E.G.; Xing, B. Environmental occurrence, fate, impact, and potential solution of tire microplastics: Similarities and differences with tire wear particles. *Sci. Total Environ.* **2021**, *795*, 148902. [CrossRef] [PubMed]

29. Hasegawa, T.; Nakaoka, M. Trophic transfer of microplastics from mysids to fish greatly exceeds direct ingestion from the water column. *Environ. Pollut.* **2021**, *273*, 116468. [CrossRef] [PubMed]
30. Shiu, R.-F.; Chen, L.-Y.; Lee, H.-J.; Gong, G.-C.; Lee, C. New insights into the role of marine plastic-gels in microplastic transfer from water to the atmosphere via bubble bursting. *Water Res.* **2022**, *222*, 118856. [CrossRef]
31. Cai, Y.; Li, C.; Zhao, Y. A review of the migration and transformation of microplastics in inland water systems. *Int. J. Environ. Res. Public Health* **2021**, *19*, 148. [CrossRef]
32. Gao, S.; Li, Z.; Zhang, S. Trophic transfer and biomagnification of microplastics through food webs in coastal waters: A new perspective from a mass balance model. *Mar. Pollut. Bull.* **2024**, *200*, 116082. [CrossRef]
33. Arienzo, M.; Ferrara, L.; Trifuoggi, M. Research progress in transfer, accumulation and effects of microplastics in the oceans. *J. Mar. Sci. Eng.* **2021**, *9*, 433. [CrossRef]
34. Issac, M.N.; Kandasubramanian, B. Effect of microplastics in water and aquatic systems. *Environ. Sci. Pollut. Res.* **2021**, *28*, 19544–19562. [CrossRef]
35. Sang, W.; Chen, Z.; Mei, L.; Hao, S.; Zhan, C.; bin Zhang, W.; Li, M.; Liu, J. The abundance and characteristics of microplastics in rainwater pipelines in Wuhan, China. *Sci. Total Environ.* **2021**, *755*, 142606. [CrossRef]
36. Zhang, X.; Chen, Y.; Li, X.; Zhang, Y.; Gao, W.; Jiang, J.; Mo, A.; He, D. Size/shape-dependent migration of microplastics in agricultural soil under simulative and natural rainfall. *Sci. Total Environ.* **2022**, *815*, 152507. [CrossRef]
37. Yu, H.; Liu, M.; Gang, D.; Peng, J.; Hu, C.; Qu, J. Polyethylene microplastics interfere with the nutrient cycle in water-plant-sediment systems. *Water Res.* **2022**, *214*, 118191. [CrossRef]
38. Page, M.J.; McKenzie, J.E.; Bossuyt, P.M.; Boutron, I.; Hoffmann, T.C.; Mulrow, C.D.; Shamseer, L.; Tetzlaff, J.M.; Akl, E.A.; Brennan, S.E. The PRISMA 2020 statement: An updated guideline for reporting systematic reviews. *BMJ* **2021**, *372*, n71. [CrossRef]
39. Rao, Z.; Niu, S.; Zhan, N.; Wang, X.; Song, X. Microplastics in Sediments of River Yongfeng from Maanshan City, Anhui Province, China. *Bull. Environ. Contam. Toxicol.* **2020**, *104*, 166–172. [CrossRef]
40. Zhang, K.; Xiong, X.; Hu, H.; Wu, C.; Bi, Y.; Wu, Y.; Zhou, B.; Lam, P.K.S.; Liu, J. Occurrence and Characteristics of Microplastic Pollution in Xiangxi Bay of Three Gorges Reservoir, China. *Environ. Sci. Technol.* **2017**, *51*, 3794–3801. [CrossRef] [PubMed]
41. Faure, F.; Demars, C.; Wieser, O.; Kunz, M.; De Alencastro, L.F. Plastic pollution in Swiss surface waters: Nature and concentrations, interaction with pollutants. *Environ. Chem.* **2015**, *12*, 582–591. [CrossRef]
42. Schmidt, N.; Thibault, D.; Galgani, F.; Paluselli, A.; Sempéré, R. Occurrence of microplastics in surface waters of the Gulf of Lion (NW Mediterranean Sea). *Prog. Oceanogr.* **2018**, *163*, 214–220. [CrossRef]
43. Liu, K.; Wang, X.; Fang, T.; Xu, P.; Zhu, L.; Li, D. Source and potential risk assessment of suspended atmospheric microplastics in Shanghai. *Sci. Total Environ.* **2019**, *675*, 462–471. [CrossRef]
44. Schmidt, N.; Castro-Jiménez, J.; Oursel, B.; Sempéré, R. Phthalates and organophosphate esters in surface water, sediments and zooplankton of the NW Mediterranean Sea: Exploring links with microplastic abundance and accumulation in the marine food web. *Environ. Pollut.* **2021**, *272*, 115970. [CrossRef] [PubMed]
45. Flynn, K.F.; Chudyk, W.; Watson, V.; Chapra, S.C.; Suplee, M.W. Influence of biomass and water velocity on light attenuation of *Cladophora glomerata* L.(Kuetzing) in rivers. *Aquat. Bot.* **2018**, *151*, 62–70. [CrossRef]
46. Wu, J.; Ye, Q.; Sun, L.; Liu, J.; Huang, M.; Wang, T.; Wu, P.; Zhu, N. Impact of persistent rain on microplastics distribution and plastisphere community: A field study in the Pearl River, China. *Sci. Total Environ.* **2023**, *879*, 163066. [CrossRef] [PubMed]
47. de Jesus Piñon-Colin, T.; Rodriguez-Jimenez, R.; Rogel-Hernandez, E.; Alvarez-Andrade, A.; Wakida, F.T. Microplastics in stormwater runoff in a semiarid region, Tijuana, Mexico. *Sci. Total Environ.* **2020**, *704*, 135411. [CrossRef] [PubMed]
48. Rauert, C.; Charlton, N.; Okoffo, E.D.; Stanton, R.S.; Agua, A.R.; Pirrung, M.C.; Thomas, K.V. Concentrations of tire additive chemicals and tire road wear particles in an Australian urban tributary. *Environ. Sci. Technol.* **2022**, *56*, 2421–2431. [CrossRef] [PubMed]
49. Stanton, T.; Johnson, M.; Nathanail, P.; MacNaughtan, W.; Gomes, R.L. Freshwater microplastic concentrations vary through both space and time. *Environ. Pollut.* **2020**, *263*, 114481. [CrossRef] [PubMed]
50. Chen, Y.; Niu, J.; Xu, D.; Zhang, M.; Sun, K.; Gao, B. Wet deposition of globally transportable microplastics (<25 μm) hovering over the megacity of Beijing. *Environ. Sci. Technol.* **2023**, *57*, 11152–11162.
51. Wei, Y.; Dou, P.; Xu, D.; Zhang, Y.; Gao, B. Microplastic reorganization in urban river before and after rainfall. *Environ. Pollut.* **2022**, *314*, 120326. [CrossRef]
52. Ockelford, A.; Cundy, A.; Ebdon, J.E. Storm response of fluvial sedimentary microplastics. *Sci. Rep.* **2020**, *10*, 1865. [CrossRef]
53. Do, T.; Park, Y.; Lim, B.; Kim, S.; Chae, M.-Y.; Chun, C.-H. Effect of the first-flush phenomenon on the quantification of microplastics in rainwater. *Mar. Pollut. Bull.* **2023**, *187*, 114559. [CrossRef] [PubMed]
54. Hajiouni, S.; Mohammadi, A.; Ramavandi, B.; Arfaeinia, H.; De-la-Torre, G.E.; Tekle-Röttering, A.; Dobaradaran, S. Occurrence of microplastics and phthalate esters in urban runoff: A focus on the Persian Gulf coastline. *Sci. Total Environ.* **2022**, *806*, 150559. [CrossRef] [PubMed]
55. Cheung, P.K.; Hung, P.L.; Fok, L. River microplastic contamination and dynamics upon a rainfall event in Hong Kong, China. *Environ. Process.* **2019**, *6*, 253–264. [CrossRef]
56. Li, Y.; Ke, S.; Xu, D.; Zhuo, H.; Liu, X.; Gao, B. Preferential deposition of buoyant small microplastics in surface sediments of the Three Gorges Reservoir, China: Insights from biomineralization. *J. Hazard. Mater.* **2024**, *468*, 133693. [CrossRef] [PubMed]

57. Quevedo, I.R.; Tufenkji, N. Mobility of functionalized quantum dots and a model polystyrene nanoparticle in saturated quartz sand and loamy sand. *Environ. Sci. Technol.* **2012**, *46*, 4449–4457. [CrossRef]
58. Franchi, A.; O'Melia, C.R. Effects of natural organic matter and solution chemistry on the deposition and reentrainment of colloids in porous media. *Environ. Sci. Technol.* **2003**, *37*, 1122–1129. [CrossRef] [PubMed]
59. Sadri, B.; Pernitsky, D.; Sadrzadeh, M. Aggregation and deposition of colloidal particles: Effect of surface properties of collector beads. *Colloids Surf. A Physicochem. Eng. Asp.* **2017**, *530*, 46–52. [CrossRef]
60. Wang, M.; Gao, B.; Tang, D. Review of key factors controlling engineered nanoparticle transport in porous media. *J. Hazard. Mater.* **2016**, *318*, 233–246. [CrossRef] [PubMed]
61. Haapkylä, J.; Unsworth, R.K.; Flavell, M.; Bourne, D.G.; Schaffelke, B.; Willis, B.L. Seasonal rainfall and runoff promote coral disease on an inshore reef. *PLoS ONE* **2011**, *6*, e16893. [CrossRef]
62. Cai, L.; Hu, L.; Shi, H.; Ye, J.; Zhang, Y.; Kim, H. Effects of inorganic ions and natural organic matter on the aggregation of nanoplastics. *Chemosphere* **2018**, *197*, 142–151. [CrossRef]
63. Liu, J.; Ma, Y.; Zhu, D.; Xia, T.; Qi, Y.; Yao, Y.; Guo, X.; Ji, R.; Chen, W. Polystyrene Nanoplastics-enhanced Contaminant Transport: Role of Irreversible Adsorption in Glassy Polymeric Domain. *Environ. Sci. Technol.* **2018**, *52*, 2677–2685. [CrossRef]
64. Oni, B.A.; Ayeni, A.O.; Agboola, O.; Oguntade, T.; Obanla, O. Comparing microplastics contaminants in (dry and raining) seasons for Ox-Bow Lake in Yenagoa, Nigeria. *Ecotoxicol. Environ. Saf.* **2020**, *198*, 110656. [CrossRef] [PubMed]
65. Sun, J.; Peng, Z.; Zhu, Z.-R.; Fu, W.; Dai, X.; Ni, B.-J. The atmospheric microplastics deposition contributes to microplastic pollution in urban waters. *Water Res.* **2022**, *225*, 119116. [CrossRef]
66. Ross, M.S.; Loutan, A.; Groeneveld, T.; Molenaar, D.; Kroetch, K.; Bujaczek, T.; Kolter, S.; Moon, S.; Huynh, A.; Khayam, R. Estimated discharge of microplastics via urban stormwater during individual rain events. *Front. Environ. Sci.* **2023**, *11*, 1090267. [CrossRef]
67. Imbulana, S.; Tanaka, S.; Moriya, A.; Oluwoye, I. Inter-event and intra-event dynamics of microplastic emissions in an urban river during rainfall episodes. *Environ. Res.* **2024**, *243*, 117882. [CrossRef]
68. Hitchcock, J.N. Storm events as key moments of microplastic contamination in aquatic ecosystems. *Sci. Total Environ.* **2020**, *734*, 139436. [CrossRef]
69. Lima, A.R.A.; Barletta, M.; Costa, M.F. Seasonal distribution and interactions between plankton and microplastics in a tropical estuary. *Estuar. Coast. Shelf Sci.* **2015**, *165*, 213–225. [CrossRef]
70. Liu, S.; Li, Y.; Wang, F.; Gu, X.; Li, Y.; Liu, Q.; Li, L.; Bai, F. Temporal and spatial variation of microplastics in the urban rivers of Harbin. *Sci. Total Environ.* **2024**, *910*, 168373. [CrossRef]
71. Bandara, R.; Perera, M.; Gomes, P.I.; Yan, X.-F. Profiling Microplastic Pollution in Surface Water Bodies in the Most Urbanized City of Sri Lanka and Its Suburbs to Understand the Underlying Factors. *Water Air Soil Pollut.* **2023**, *234*, 157. [CrossRef]
72. Zhang, J.; Ding, W.; Zou, G.; Wang, X.; Zhao, M.; Guo, S.; Chen, Y. Urban pipeline rainwater runoff is an important pathway for land-based microplastics transport to inland surface water: A case study in Beijing. *Sci. Total Environ.* **2023**, *861*, 160619. [CrossRef]
73. Wu, J.; Jiang, Z.; Liu, Y.; Zhao, X.; Liang, Y.; Lu, W.; Song, J. Microplastic contamination assessment in water and economic fishes in different trophic guilds from an urban water supply reservoir after flooding. *J. Environ. Manag.* **2021**, *299*, 113667. [CrossRef] [PubMed]
74. Liu, H.; Sun, K.; Liu, X.; Yao, R.; Cao, W.; Zhang, L.; Wang, X. Spatial and temporal distributions of microplastics and their macroscopic relationship with algal blooms in Chaohu Lake, China. *J. Contam. Hydrol.* **2022**, *248*, 104028. [CrossRef]
75. Castro, R.O.; Silva, M.L.; Marques, M.R.C.; de Araújo, F.V. Evaluation of microplastics in Jurujuba Cove, Niterói, RJ, Brazil, an area of mussels farming. *Mar. Pollut. Bull.* **2016**, *110*, 555–558. [CrossRef]
76. Dantas Filho, J.V.; Pedroti, V.P.; Santos, B.L.T.; de Lima Pinheiro, M.M.; de Mira, Á.B.; da Silva, F.C.; e Silva, E.C.S.; Cavali, J.; Guedes, E.A.C.; de Vargas Schons, S. First evidence of microplastics in freshwater from fish farms in Rondônia state, Brazil. *Heliyon* **2023**, *9*, e15066. [CrossRef]
77. Nkosi, M.S.; Cuthbert, R.N.; Wu, N.; Shikwambana, P.; Dalu, T. Microplastic abundance, distribution, and diversity in water and sediments along a subtropical river system. *Environ. Sci. Pollut. Res.* **2023**, *30*, 91440–91452. [CrossRef]
78. Curren, E.; Leong, S.C.Y. Spatiotemporal characterisation of microplastics in the coastal regions of Singapore. *Heliyon* **2023**, *9*, e12961. [CrossRef]
79. Dong, H.; Wang, L.; Wang, X.; Xu, L.; Chen, M.; Gong, P.; Wang, C. Microplastics in a remote lake basin of the Tibetan Plateau: Impacts of atmospheric transport and glacial melting. *Environ. Sci. Technol.* **2021**, *55*, 12951–12960. [CrossRef] [PubMed]
80. Huang, Y.; He, T.; Yan, M.; Yang, L.; Gong, H.; Wang, W.; Qing, X.; Wang, J. Atmospheric transport and deposition of microplastics in a subtropical urban environment. *J. Hazard. Mater.* **2021**, *416*, 126168. [CrossRef]
81. Zhu, J.; Zhang, Q.; Huang, Y.; Jiang, Y.; Li, J.; Michal, J.J.; Jiang, Z.; Xu, Y.; Lan, W. Long-term trends of microplastics in seawater and farmed oysters in the Maowei Sea, China. *Environ. Pollut.* **2021**, *273*, 116450. [CrossRef] [PubMed]
82. Yonkos, L.T.; Friedel, E.A.; Perez-Reyes, A.C.; Ghosal, S.; Arthur, C.D. Microplastics in four estuarine rivers in the Chesapeake Bay, USA. *Environ. Sci. Technol.* **2014**, *48*, 14195–14202. [CrossRef]
83. Warrier, A.K.; Kulkarni, B.; Amrutha, K.; Jayaram, D.; Valsan, G.; Agarwal, P. Seasonal variations in the abundance and distribution of microplastic particles in the surface waters of a Southern Indian Lake. *Chemosphere* **2022**, *300*, 134556. [CrossRef]

84. Tang, C.N.; Kuwahara, V.S.; Leong, S.C.Y.; Moh, P.Y.; Yoshida, T. Effect of monsoon on microplastic bioavailability and ingestion by zooplankton in tropical coastal waters of Sabah. *Mar. Pollut. Bull.* **2023**, *193*, 115182. [CrossRef] [PubMed]
85. Liu, Z.; Bai, Y.; Ma, T.; Liu, X.; Wei, H.; Meng, H.; Fu, Y.; Ma, Z.; Zhang, L.; Zhao, J. Distribution and possible sources of atmospheric microplastic deposition in a valley basin city (Lanzhou, China). *Ecotoxicol. Environ. Saf.* **2022**, *233*, 113353. [CrossRef] [PubMed]
86. Garcés-Ordóñez, O.; Castillo-Olaya, V.; Espinosa-Díaz, L.F.; Canals, M. Seasonal variation in plastic litter pollution in mangroves from two remote tropical estuaries of the Colombian Pacific. *Mar. Pollut. Bull.* **2023**, *193*, 115210. [CrossRef] [PubMed]
87. Smyth, K.; Drake, J.; Li, Y.; Rochman, C.; Van Seters, T.; Passeport, E. Bioretention cells remove microplastics from urban stormwater. *Water Res.* **2021**, *191*, 116785. [CrossRef] [PubMed]
88. Pasquier, G.; Doyen, P.; Dehaut, A.; Veillet, G.; Duflos, G.; Amara, R. Vertical distribution of microplastics in a river water column using an innovative sampling method. *Environ. Monit. Assess.* **2023**, *195*, 1302. [CrossRef] [PubMed]
89. An, X.; Li, W.; Lan, J.; Adnan, M. Preliminary study on the distribution, source, and ecological risk of typical microplastics in karst groundwater in Guizhou Province, China. *Int. J. Environ. Res. Public Health* **2022**, *19*, 14751. [CrossRef] [PubMed]
90. da Costa, I.D.; Nunes, N.N.; Costa, L.L.; Zalmon, I.R. Is the Paraíba do Sul River colourful? Prevalence of microplastics in freshwater, south-eastern Brazil. *Mar. Freshw. Res.* **2022**, *73*, 1439–1449. [CrossRef]
91. Wei, Y.; Ma, W.; Xu, Q.; Sun, C.; Wang, X.; Gao, F. Microplastic distribution and influence factor analysis of seawater and surface sediments in a typical bay with diverse functional areas: A case study in Xincun lagoon, China. *Front. Environ. Sci.* **2022**, *10*, 829942. [CrossRef]
92. Gündoğdu, S.; Çevik, C.; Ayat, B.; Aydoğan, B.; Karaca, S. How microplastics quantities increase with flood events? An example from Mersin Bay NE Levantine coast of Turkey. *Environ. Pollut.* **2018**, *239*, 342–350. [CrossRef]
93. Treilles, R.; Gasperi, J.; Gallard, A.; Saad, M.; Dris, R.; Partibane, C.; Breton, J.; Tassin, B. Microplastics and microfibers in urban runoff from a suburban catchment of Greater Paris. *Environ. Pollut.* **2021**, *287*, 117352. [CrossRef] [PubMed]
94. McEachern, K.; Alegria, H.; Kalagher, A.L.; Hansen, C.; Morrison, S.; Hastings, D. Microplastics in Tampa Bay, Florida: Abundance and variability in estuarine waters and sediments. *Mar. Pollut. Bull.* **2019**, *148*, 97–106. [CrossRef] [PubMed]
95. Unnikrishnan, V.; Valsan, G.; Amrutha, K.; Sebastian, J.G.; Rangel-Buitrago, N.; Khaleel, R.; Chandran, T.; Reshma, S.; Warrier, A.K. A baseline study of microplastic pollution in a Southern Indian Estuary. *Mar. Pollut. Bull.* **2023**, *186*, 114468. [CrossRef] [PubMed]
96. Strady, E.; Kieu-Le, T.-C.; Gasperi, J.; Tassin, B. Temporal dynamic of anthropogenic fibers in a tropical river-estuarine system. *Environ. Pollut.* **2020**, *259*, 113897. [CrossRef] [PubMed]
97. Amrutha, K.; Warrier, A.K.; Rangel-Buitrago, N. Did the COVID-19 pandemic play a role in the spatial and temporal variations of microplastics? Evidence from a tropical river in southern India. *Mar. Pollut. Bull.* **2023**, *192*, 115088. [CrossRef] [PubMed]
98. Semmouri, I.; Vercauteren, M.; Van Acker, E.; Pequeur, E.; Asselman, J.; Janssen, C. Distribution of microplastics in freshwater systems in an urbanized region: A case study in Flanders (Belgium). *Sci. Total Environ.* **2023**, *872*, 162192. [CrossRef] [PubMed]
99. Alfonso, M.B.; Arias, A.H.; Piccolo, M.C. Microplastics integrating the zooplanktonic fraction in a saline lake of Argentina: Influence of water management. *Environ. Monit. Assess.* **2020**, *192*, 117. [CrossRef] [PubMed]
100. Chen, H.; Cheng, Y.; Wang, Y.; Ding, Y.; Wang, C.; Feng, X.; Fan, Q.; Yuan, F.; Fu, G.; Gao, B. Microplastics: A potential proxy for tracing extreme flood events in estuarine environments. *Sci. Total Environ.* **2024**, *918*, 170554. [CrossRef]
101. Wang, J.; Lu, L.; Wang, M.; Jiang, T.; Liu, X.; Ru, S. Typhoons increase the abundance of microplastics in the marine environment and cultured organisms: A case study in Sanggou Bay, China. *Sci. Total Environ.* **2019**, *667*, 1–8. [CrossRef]
102. Cordova, M.R.; Ulumuddin, Y.I.; Purbonegoro, T.; Puspitasari, R.; Afianti, N.F.; Rositasari, R.; Yogaswara, D.; Hafizt, M.; Iswari, M.Y.; Fitriya, N. Seasonal heterogeneity and a link to precipitation in the release of microplastic during COVID-19 outbreak from the Greater Jakarta area to Jakarta Bay, Indonesia. *Mar. Pollut. Bull.* **2022**, *181*, 113926. [CrossRef]
103. Lorenzi, L.; Reginato, B.C.; Mayer, D.G.; Dantas, D.V. Plastic floating debris along a summer-winter estuarine environmental gradient in a coastal lagoon: How does plastic debris arrive in a conservation unit? *Environ. Sci. Pollut. Res.* **2020**, *27*, 8797–8806. [CrossRef] [PubMed]
104. Hajbane, S.; Pattiaratchi, C. Plastic Pollution Patterns in Offshore, Nearshore and Estuarine Waters: A Case Study from Perth, Western Australia. *Front. Mar. Sci.* **2017**, *4*, 63. [CrossRef]
105. Gupta, P.; Saha, M.; Rathore, C.; Suneel, V.; Ray, D.; Naik, A.; Unnikrishnan, K.; Dhivya, M.; Daga, K. Spatial and seasonal variation of microplastics and possible sources in the estuarine system from central west coast of India. *Environ. Pollut.* **2021**, *288*, 117665. [CrossRef] [PubMed]
106. Bauer-Civiello, A.; Critchell, K.; Hoogenboom, M.; Hamann, M. Input of plastic debris in an urban tropical river system. *Mar. Pollut. Bull.* **2019**, *144*, 235–242. [CrossRef] [PubMed]
107. Büngener, L.; Postila, H.; Löder, M.G.; Laforsch, C.; Ronkanen, A.-K.; Heiderscheidt, E. The fate of microplastics from municipal wastewater in a surface flow treatment wetland. *Sci. Total Environ.* **2023**, *903*, 166334. [CrossRef]
108. Xue, W.; Maung, G.Y.T.; Otiti, J.; Tabucanon, A.S. Land use-based characterization and source apportionment of microplastics in urban storm runoffs in a tropical region. *Environ. Pollut.* **2023**, *329*, 121698. [CrossRef]
109. Amrutha, K.; Shajikumar, S.; Warrier, A.K.; Sebastian, J.G.; Sali, Y.A.; Chandran, T.; Sivadas, S.; Naik, R.; Amrish, V.N.; Kumar, A. Assessment of pollution and risks associated with microplastics in the riverine sediments of the Western Ghats: A heritage site in southern India. *Environ. Sci. Pollut. Res.* **2023**, *30*, 32301–32319. [CrossRef] [PubMed]

110. Jiang, Y.; Yang, Y.; Zhan, C.; Cheng, B. Impacts of rainfall and lakeshore soil properties on microplastics in inland freshwater: A case study in Donghu Lake, China. *Environ. Sci. Process. Impacts* **2024**, *26*, 891–901. [CrossRef]
111. Luo, Y.; Xie, H.; Xu, H.; Zhou, C.; Wang, P.; Liu, Z.; Yang, Y.; Huang, J.; Wang, C.; Zhao, X. Wastewater treatment plant serves as a potentially controllable source of microplastic: Association of microplastic removal and operational parameters and water quality data. *J. Hazard. Mater.* **2023**, *441*, 129974. [CrossRef]
112. Lu, X.; Wang, X.; Liu, X.; Singh, V.P. Dispersal and transport of microplastic particles under different flow conditions in riverine ecosystem. *J. Hazard. Mater.* **2023**, *442*, 130033. [CrossRef]
113. Cheung, C.K.H.; Not, C. Impacts of extreme weather events on microplastic distribution in coastal environments. *Sci. Total Environ.* **2023**, *904*, 166723. [CrossRef] [PubMed]
114. Chen, C.-F.; Ju, Y.-R.; Lim, Y.C.; Chen, C.-W.; Dong, C.-D. Seasonal variation of diversity, weathering, and inventory of microplastics in coast and harbor sediments. *Sci. Total Environ.* **2021**, *781*, 146610. [CrossRef] [PubMed]
115. Fischer, E.K.; Paglialonga, L.; Czech, E.; Tamminga, M. Microplastic pollution in lakes and lake shoreline sediments—A case study on Lake Bolsena and Lake Chiusi (central Italy). *Environ. Pollut.* **2016**, *213*, 648–657. [CrossRef] [PubMed]
116. Rodrigues, D.; Antunes, J.; Pais, J.; Pequeno, J.; Caetano, P.S.; Rocha, F.; Sobral, P.; Costa, M.H. Distribution patterns of microplastics in subtidal sediments from the Sado river estuary and the Arrábida marine park, Portugal. *Front. Environ. Sci.* **2022**, *10*, 998513. [CrossRef]
117. Liu, T.; Zhao, Y.; Zhu, M.; Liang, J.; Zheng, S.; Sun, X. Seasonal variation of micro-and meso-plastics in the seawater of Jiaozhou Bay, the Yellow Sea. *Mar. Pollut. Bull.* **2020**, *152*, 110922. [CrossRef] [PubMed]
118. Boni, W.; Arbuckle-Keil, G.; Fahrenfeld, N. Inter-storm variation in microplastic concentration and polymer type at stormwater outfalls and a bioretention basin. *Sci. Total Environ.* **2022**, *809*, 151104. [CrossRef] [PubMed]
119. Rimondi, V.; Monnanni, A.; De Beni, E.; Bicocchi, G.; Chelazzi, D.; Cincinelli, A.; Fratini, S.; Martellini, T.; Morelli, G.; Venturi, S. Occurrence and quantification of natural and microplastic items in urban streams: The case of Mugnone Creek (Florence, Italy). *Toxics* **2022**, *10*, 159. [CrossRef]
120. Uogintė, I.; Pleskytė, S.; Pauraitė, J.; Lujanienė, G. Seasonal variation and complex analysis of microplastic distribution in different WWTP treatment stages in Lithuania. *Environ. Monit. Assess.* **2022**, *194*, 829. [CrossRef]
121. Moses, S.R.; Löder, M.G.; Herrmann, F.; Laforsch, C. Seasonal variations of microplastic pollution in the German River Weser. *Sci. Total Environ.* **2023**, *902*, 166463. [CrossRef]
122. Moore, C.J. Synthetic polymers in the marine environment: A rapidly increasing, long-term threat. *Environ. Res.* **2008**, *108*, 131–139. [CrossRef]
123. Andrady, A.L. Microplastics in the marine environment. *Mar. Pollut. Bull.* **2011**, *62*, 1596–1605. [CrossRef]
124. Zhao, X.; Wang, J.; Yee Leung, K.M.; Wu, F. Color: An important but overlooked factor for plastic photoaging and microplastic formation. *Environ. Sci. Technol.* **2022**, *56*, 9161–9163. [CrossRef]
125. Jeppesen, E.; Meerhoff, M.; Holmgren, K.; González-Bergonzoni, I.; Teixeira-de Mello, F.; Declerck, S.A.; De Meester, L.; Søndergaard, M.; Lauridsen, T.L.; Bjerring, R. Impacts of climate warming on lake fish community structure and potential effects on ecosystem function. *Hydrobiologia* **2010**, *646*, 73–90. [CrossRef]
126. Chaturvedi, S.; Yadav, B.P.; Siddiqui, N.A.; Chaturvedi, S.K. Mathematical modelling and analysis of plastic waste pollution and its impact on the ocean surface. *J. Ocean Eng. Sci.* **2020**, *5*, 136–163. [CrossRef]
127. Kim, S.-K.; Kim, J.-S.; Kim, S.-Y.; Song, N.-S.; La, H.S.; Yang, E.J. Arctic Ocean sediments as important current and future sinks for marine microplastics missing in the global microplastic budget. *Sci. Adv.* **2023**, *9*, eadd2348. [CrossRef] [PubMed]
128. Zhang, Y.; Gao, T.; Kang, S.; Allen, S.; Luo, X.; Allen, D. Microplastics in glaciers of the Tibetan Plateau: Evidence for the long-range transport of microplastics. *Sci. Total Environ.* **2021**, *758*, 143634. [CrossRef]
129. Chen, G.; Li, Y.; Wang, J. Occurrence and ecological impact of microplastics in aquaculture ecosystems. *Chemosphere* **2021**, *274*, 129989. [CrossRef] [PubMed]
130. Ballent, A.; Corcoran, P.L.; Madden, O.; Helm, P.A.; Longstaffe, F.J. Sources and sinks of microplastics in Canadian Lake Ontario nearshore, tributary and beach sediments. *Mar. Pollut. Bull.* **2016**, *110*, 383–395. [CrossRef]
131. Lattin, G.L.; Moore, C.J.; Zellers, A.F.; Moore, S.L.; Weisberg, S.B. A comparison of neustonic plastic and zooplankton at different depths near the southern California shore. *Mar. Pollut. Bull.* **2004**, *49*, 291–294. [CrossRef]
132. Schwarz, A.E.; Ligthart, T.N.; Boukris, E.; Van Harmelen, T. Sources, transport, and accumulation of different types of plastic litter in aquatic environments: A review study. *Mar. Pollut. Bull.* **2019**, *143*, 92–100. [CrossRef]
133. Liu, Y.; Hao, R.; Shi, X.; Zhang, S.; Sun, B.; Zhao, S.; Huotari, J. Application of a microplastic trap to the determination of the factors controlling the lakebed deposition of microplastics. *Sci. Total Environ.* **2022**, *843*, 156883. [CrossRef] [PubMed]
134. Collignon, A.; Hecq, J.-H.; Glagani, F.; Voisin, P.; Collard, F.; Goffart, A. Neustonic microplastic and zooplankton in the North Western Mediterranean Sea. *Mar. Pollut. Bull.* **2012**, *64*, 861–864. [CrossRef] [PubMed]
135. Yuan, W.; Xu, E.G.; Li, L.; Zhou, A.; Peijnenburg, W.J.; Grossart, H.-P.; Liu, W.; Yang, Y. Tracing and trapping micro-and nanoplastics: Untapped mitigation potential of aquatic plants? *Water Res.* **2023**, *242*, 120249. [CrossRef] [PubMed]
136. Cesarini, G.; Scalici, M. Riparian vegetation as a trap for plastic litter. *Environ. Pollut.* **2022**, *292*, 118410. [CrossRef] [PubMed]
137. Helcoski, R.; Yonkos, L.T.; Sanchez, A.; Baldwin, A.H. Wetland soil microplastics are negatively related to vegetation cover and stem density. *Environ. Pollut.* **2020**, *256*, 113391. [CrossRef] [PubMed]

138. Sundbæk, K.B.; Koch, I.D.W.; Villaro, C.G.; Rasmussen, N.S.; Holdt, S.L.; Hartmann, N.B. Sorption of fluorescent polystyrene microplastic particles to edible seaweed Fucus vesiculosus. *J. Appl. Phycol.* **2018**, *30*, 2923–2927. [CrossRef]
139. Li, C.; Gao, Y.; He, S.; Chi, H.-Y.; Li, Z.-C.; Zhou, X.-X.; Yan, B. Quantification of nanoplastic uptake in cucumber plants by pyrolysis gas chromatography/mass spectrometry. *Environ. Sci. Technol. Lett.* **2021**, *8*, 633–638. [CrossRef]
140. Sun, X.-D.; Yuan, X.-Z.; Jia, Y.; Feng, L.-J.; Zhu, F.-P.; Dong, S.-S.; Liu, J.; Kong, X.; Tian, H.; Duan, J.-L. Differentially charged nanoplastics demonstrate distinct accumulation in Arabidopsis thaliana. *Nat. Nanotechnol.* **2020**, *15*, 755–760. [CrossRef]
141. Zhou, C.-Q.; Lu, C.-H.; Mai, L.; Bao, L.-J.; Liu, L.-Y.; Zeng, E.Y. Response of rice (*Oryza sativa* L.) roots to nanoplastic treatment at seedling stage. *J. Hazard. Mater.* **2021**, *401*, 123412. [CrossRef]
142. Gan, Q.; Cui, J.; Jin, B. Environmental microplastics: Classification, sources, fates, and effects on plants. *Chemosphere* **2023**, *313*, 137559. [CrossRef]
143. da Costa, J.P.; Chamkha, M.; Ksibi, M.; Sayadi, S. Effects of microplastics' physical and chemical properties on aquatic organisms: State-of-the-art and future research trends. *TrAC Trends Anal. Chem.* **2023**, *166*, 117192.
144. Azevedo-Santos, V.M.; Goncalves, G.R.; Manoel, P.S.; Andrade, M.C.; Lima, F.P.; Pelicice, F.M. Plastic ingestion by fish: A global assessment. *Environ. Pollut.* **2019**, *255 Pt 1*, 112994. [CrossRef] [PubMed]
145. Hossain, M.A.; Olden, J.D. Global meta-analysis reveals diverse effects of microplastics on freshwater and marine fishes. *Fish Fish.* **2022**, *23*, 1439–1454. [CrossRef]
146. Miranda, D.d.A.; de Carvalho-Souza, G.F. Are we eating plastic-ingesting fish? *Mar. Pollut. Bull.* **2016**, *103*, 109–114. [CrossRef] [PubMed]
147. Lusher, A.L.; Mchugh, M.; Thompson, R.C. Occurrence of microplastics in the gastrointestinal tract of pelagic and demersal fish from the English Channel. *Mar. Pollut. Bull.* **2013**, *67*, 94–99. [CrossRef]
148. Vroom, R.J.; Koelmans, A.A.; Besseling, E.; Halsband, C. Aging of microplastics promotes their ingestion by marine zooplankton. *Environ. Pollut.* **2017**, *231*, 987–996. [CrossRef] [PubMed]
149. Porter, A.; Godbold, J.A.; Lewis, C.N.; Savage, G.; Solan, M.; Galloway, T.S. Microplastic burden in marine benthic invertebrates depends on species traits and feeding ecology within biogeographical provinces. *Nat. Commun.* **2023**, *14*, 8023. [CrossRef]
150. Abbasi, S.; Soltani, N.; Keshavarzi, B.; Moore, F.; Turner, A.; Hassanaghaei, M. Microplastics in different tissues of fish and prawn from the Musa Estuary, Persian Gulf. *Chemosphere* **2018**, *205*, 80–87. [CrossRef]
151. Zhao, J.; Lan, R.; Wang, Z.; Su, W.; Song, D.; Xue, R.; Liu, Z.; Liu, X.; Dai, Y.; Yue, T. Microplastic fragmentation by rotifers in aquatic ecosystems contributes to global nanoplastic pollution. *Nat. Nanotechnol.* **2024**, *19*, 406–414. [CrossRef]
152. Donlan, R.M. Biofilms: Microbial life on surfaces. *Emerg. Infect. Dis.* **2002**, *8*, 881. [CrossRef]
153. Chen, C.-S.; Shiu, R.-F.; Hsieh, Y.-Y.; Xu, C.; Vazquez, C.I.; Cui, Y.; Hsu, I.C.; Quigg, A.; Santschi, P.H.; Chin, W.-C. Stickiness of extracellular polymeric substances on different surfaces via magnetic tweezers. *Sci. Total Environ.* **2021**, *757*, 143766. [CrossRef]
154. Wang, J.; Guo, X.; Xue, J. Biofilm-developed microplastics as vectors of pollutants in aquatic environments. *Environ. Sci. Technol.* **2021**, *55*, 12780–12790. [CrossRef]
155. He, S.; Jia, M.; Xiang, Y.; Song, B.; Xiong, W.; Cao, J.; Peng, H.; Yang, Y.; Wang, W.; Yang, Z. Biofilm on microplastics in aqueous environment: Physicochemical properties and environmental implications. *J. Hazard. Mater.* **2022**, *424*, 127286. [CrossRef]
156. Tu, C.; Chen, T.; Zhou, Q.; Liu, Y.; Wei, J.; Waniek, J.J.; Luo, Y. Biofilm formation and its influences on the properties of microplastics as affected by exposure time and depth in the seawater. *Sci. Total Environ.* **2020**, *734*, 139237. [CrossRef] [PubMed]
157. Leiser, R.; Wu, G.-M.; Neu, T.R.; Wendt-Potthoff, K. Biofouling, metal sorption and aggregation are related to sinking of microplastics in a stratified reservoir. *Water Res.* **2020**, *176*, 115748. [CrossRef] [PubMed]
158. Guan, J.; Qi, K.; Wang, J.; Wang, W.; Wang, Z.; Lu, N.; Qu, J. Microplastics as an emerging anthropogenic vector of trace metals in freshwater: Significance of biofilms and comparison with natural substrates. *Water Res.* **2020**, *184*, 116205. [CrossRef]
159. Meng, D.; Li, Y. Assessing the Settling Velocity of Biofilm-Encrusted Microplastics: Accounting for Biofilms as an Equivalent to Surface Roughness. *Environ. Sci. Technol.* **2024**, *58*, 1329–1337. [CrossRef]
160. Lobelle, D.; Cunliffe, M. Early microbial biofilm formation on marine plastic debris. *Mar. Pollut. Bull.* **2011**, *62*, 197–200. [CrossRef] [PubMed]
161. Nauendorf, A.; Krause, S.; Bigalke, N.K.; Gorb, E.V.; Gorb, S.N.; Haeckel, M.; Wahl, M.; Treude, T. Microbial colonization and degradation of polyethylene and biodegradable plastic bags in temperate fine-grained organic-rich marine sediments. *Mar. Pollut. Bull.* **2016**, *103*, 168–178. [CrossRef]
162. Cho, Y.; Shim, W.J.; Ha, S.Y.; Han, G.M.; Jang, M.; Hong, S.H. Microplastic emission characteristics of stormwater runoff in an urban area: Intra-event variability and influencing factors. *Sci. Total Environ.* **2023**, *866*, 161318. [CrossRef]
163. Singh, N.; Tiwari, E.; Khandelwal, N.; Darbha, G.K. Understanding the stability of nanoplastics in aqueous environments: Effect of ionic strength, temperature, dissolved organic matter, clay, and heavy metals. *Environ. Sci. Nano* **2019**, *6*, 2968–2976. [CrossRef]
164. Andersen, T.J.; Rominikan, S.; Olsen, I.S.; Skinnebach, K.H.; Fruergaard, M. Flocculation of PVC microplastic and fine-grained cohesive sediment at environmentally realistic concentrations. *Biol. Bull.* **2021**, *240*, 42–51. [CrossRef]
165. Li, Y.; Wang, X.; Fu, W.; Xia, X.; Liu, C.; Min, J.; Zhang, W.; Crittenden, J.C. Interactions between nano/micro plastics and suspended sediment in water: Implications on aggregation and settling. *Water Res.* **2019**, *161*, 486–495. [CrossRef] [PubMed]
166. Long, M.; Paul-Pont, I.; Hegaret, H.; Moriceau, B.; Lambert, C.; Huvet, A.; Soudant, P. Interactions between polystyrene microplastics and marine phytoplankton lead to species-specific hetero-aggregation. *Environ. Pollut.* **2017**, *228*, 454–463. [CrossRef]

167. Oriekhova, O.; Stoll, S. Heteroaggregation of nanoplastic particles in the presence of inorganic colloids and natural organic matter. *Environ. Sci. Nano* **2018**, *5*, 792–799. [CrossRef]
168. Kowalski, N.; Reichardt, A.M.; Waniek, J.J. Sinking rates of microplastics and potential implications of their alteration by physical, biological, and chemical factors. *Mar. Pollut. Bull.* **2016**, *109*, 310–319. [CrossRef] [PubMed]
169. Besseling, E.; Quik, J.T.; Sun, M.; Koelmans, A.A. Fate of nano-and microplastic in freshwater systems: A modeling study. *Environ. Pollut.* **2017**, *220*, 540–548. [CrossRef]
170. Xie, M.; Alsina, M.A.; Yuen, J.; Packman, A.I.; Gaillard, J.-F. Effects of resuspension on the mobility and chemical speciation of zinc in contaminated sediments. *J. Hazard. Mater.* **2019**, *364*, 300–308. [CrossRef]
171. Corcoran, P.L.; Belontz, S.L.; Ryan, K.; Walzak, M.J. Factors controlling the distribution of microplastic particles in benthic sediment of the Thames River, Canada. *Environ. Sci. Technol.* **2019**, *54*, 818–825. [CrossRef]
172. Dong, Z.; Qiu, Y.; Zhang, W.; Yang, Z.; Wei, L. Size-dependent transport and retention of micron-sized plastic spheres in natural sand saturated with seawater. *Water Res.* **2018**, *143*, 518–526. [CrossRef]
173. Zhang, S.; Shen, C.; Zhang, F.; Wei, K.; Shan, S.; Zhao, Y.; Man, Y.B.; Wong, M.H.; Zhang, J. Microplastics removal mechanisms in constructed wetlands and their impacts on nutrient (nitrogen, phosphorus and carbon) removal: A critical review. *Sci. Total Environ.* **2024**, *918*, 170654. [CrossRef] [PubMed]
174. Wang, Q.; Hernández-Crespo, C.; Du, B.; Van Hulle, S.W.; Rousseau, D.P. Fate and removal of microplastics in unplanted lab-scale vertical flow constructed wetlands. *Sci. Total Environ.* **2021**, *778*, 146152. [CrossRef] [PubMed]
175. Chen, Y.; Li, T.; Hu, H.; Ao, H.; Xiong, X.; Shi, H.; Wu, C. Transport and fate of microplastics in constructed wetlands: A microcosm study. *J. Hazard. Mater.* **2021**, *415*, 125615. [CrossRef] [PubMed]
176. Zhong, L.; Wu, T.; Sun, H.-J.; Ding, J.; Pang, J.-W.; Zhang, L.; Ren, N.-Q.; Yang, S.-S. Recent advances towards micro (nano) plastics research in wetland ecosystems: A systematic review on sources, removal, and ecological impacts. *J. Hazard. Mater.* **2023**, *452*, 131341. [CrossRef] [PubMed]
177. Di, M.; Wang, J. Microplastics in surface waters and sediments of the Three Gorges Reservoir, China. *Sci. Total Environ.* **2018**, *616*, 1620–1627. [CrossRef] [PubMed]
178. Watkins, L.; McGrattan, S.; Sullivan, P.J.; Walter, M.T. The effect of dams on river transport of microplastic pollution. *Sci. Total Environ.* **2019**, *664*, 834–840. [CrossRef] [PubMed]
179. Liu, Y.; Zhang, J.; Cai, C.; He, Y.; Chen, L.; Xiong, X.; Huang, H.; Tao, S.; Liu, W. Occurrence and characteristics of microplastics in the Haihe River: An investigation of a seagoing river flowing through a megacity in northern China. *Environ. Pollut.* **2020**, *262*, 114261. [CrossRef] [PubMed]
180. Kan, A.T.; Tomson, M.B. Ground water transport of hydrophobic organic compounds in the presence of dissolved organic matter. *Environ. Toxicol. Chem. Int. J.* **1990**, *9*, 253–263. [CrossRef]
181. Packman, A.I.; Salehin, M.; Zaramella, M. Hyporheic exchange with gravel beds: Basic hydrodynamic interactions and bedform-induced advective flows. *J. Hydraul. Eng.* **2004**, *130*, 647–656. [CrossRef]
182. Elliott, A.H.; Brooks, N.H. Transfer of nonsorbing solutes to a streambed with bed forms: Theory. *Water Resour. Res.* **1997**, *33*, 123–136. [CrossRef]
183. Ren, S.; Xia, Y.; Jin, X.; Sun, D.; Luo, D.; Wei, W.; Yang, Q.; Ding, J.; Lv, M.; Chen, L. Influence of microplastics on the availability of antibiotics in soils. *Sci. Total Environ.* **2024**, *924*, 171514. [CrossRef] [PubMed]

MDPI AG
Grosspeteranlage 5
4052 Basel
Switzerland
Tel.: +41 61 683 77 34

Water Editorial Office
E-mail: water@mdpi.com
www.mdpi.com/journal/water

MDPI

www.ingramcontent.com/pod-product-compliance
Lightning Source LLC
Chambersburg PA
CBHW061548210926
51850CB00009B/2036

* 9 7 8 3 7 2 5 8 5 4 0 1 1 *